Understanding Large Temporal Networks and Spatial Networks

WILEY SERIES IN COMPUTATIONAL AND QUANTITATIVE SOCIAL SCIENCE

Embracing a spectrum from theoretical foundations to real world applications, the Wiley Series in Computational and Quantitative Social Science (CQSS) publishes titles ranging from high level student texts, explanation and dissemination of technology and good practice, through to interesting and important research that is immediately relevant to social/scientific development or practice.

Other Titles in the Series

Vladimir Batagelj, Patrick Doreian, Anuška Ferligoj, Nataša Kejžar – Understanding Large Temporal Networks and Spatial Networks: Exploration, Pattern Searching, Visualization and Network Evolution

Gianluca Manzo (ed.) – Analytical Sociology: Actions and Networks

Rense Corten – Computational Approaches to Studying the Co-evolution of Networks and Behavior in Social Dilemmas

Danny Dorling – The Visualisation of Spatial Social Structure

Understanding Large Temporal Networks and Spatial Networks

Exploration, Pattern Searching, Visualization and Network Evolution

Vladimir Batagelj

Department of Mathematics, Faculty of Mathematics and Physics
University of Ljubljana, Slovenia

Patrick Doreian

Faculty of Social Sciences, University of Ljubljana, Slovenia and
Department of Sociology, University of Pittsburgh, USA

Anuška Ferligoj

Faculty of Social Sciences, University of Ljubljana, Slovenia

Nataša Kejžar

Faculty of Medicine, Institute for Biostatistics and Medical Informatics
University of Ljubljana, Slovenia

WILEY

This edition first published 2014
© 2014 John Wiley & Sons, Ltd

Registered office
John Wiley & Sons Ltd, The Atrium, Southern Gate, Chichester, West Sussex, PO19 8SQ, United Kingdom

For details of our global editorial offices, for customer services and for information about how to apply for permission to reuse the copyright material in this book please see our website at www.wiley.com.

Library of Congress Cataloging-in-Publication Data applied for

A catalogue record for this book is available from the British Library.

ISBN: 978-0-470-71452-2

Set in 10/12pt TimesLTStd-Roman by Thomson Digital, Noida, India.
Printed and bound in Singapore by Markono Print Media Pte Ltd.

1 2014

To Norm Hummon for his foundational work and inspiration for our research efforts.

Contents

Preface

Writing this book was a wild and exciting ride involving three of our favorite activities: thinking, computing, and scribbling. We did a lot of all three and, in doing so, had a lot of fun. Learning new ideas and new techniques, moving into new substantive domains, becoming utterly entranced once we entered these areas, and constructing new programs was exhilarating. It was challenging also as we encountered serious problems and difficulties in the form of seeming roadblocks. Yet solving such problems was more than half the fun. Temporary despair was followed by long-term joy, its own reward, one paving the way for new adventures.

This book concerns the understanding of large temporal networks and spatial networks. One of the networks we studied was a citation network which reinforced our understanding of science being a cumulative venture where new contributions build upon work already completed and available. The first debt of gratitude to acknowledge is the work of all of the people whose contributions we cite. Without such foundations there would be nothing upon which we could build. Our second thank you goes to Norm Hummon, to whom we dedicate this book, whose work on citation networks, along with a shared commitment regarding the inherent nature of empirical research, was an inspiration for us. We continue to mourn his untimely death.

The third round of thanks goes to all who were responsible for collecting the data we use in this book. While we had a lot of data processing to get these data into the form we needed for our work, had there been no data upon which we could work, there would have been nothing for us to do. Assembling data is time-consuming work, work that is often underappreciated when the glory goes to the publications resting on these data. We appreciate greatly the efforts going into the construction of the data proving so useful for us.

There are specific people whose contributions were of immense value. They include Andrej Mrvar whose programming and maintenance of Pajek was invaluable. We have lost track of the number of times our requests for new algorithms were handled quickly and well. While he does not work at the speed of light, it often seems he gets very close to this. Our description of collecting, cleaning, and preparing the WoS data understated the difficulties involved. For this immense task we thank Monika Cerinšek and Jernej Bodlaj for their efforts. Finally, Tine Jerman and Sara Atanasova helped us assemble the bibliography and went through the prose with a fine-tooth comb checking for errors.

Esther Sales read innumerable drafts with grace and good humor. She corrected prose, and pointed out ambiguities and lapses in our arguments. David Barnard read through a draft of the Supreme Court chapter as did Joe Labianca. Together, they prodded us to be clearer and more incisive in our arguments. All three helped us improve the manuscript greatly and we appreciate their contributions to our efforts.

While it may be impertinent to quote from a Johnny Nash song, popularized by Marvin Gaye, about there being 'more questions than answers,' we think we have more of the former than the latter. Of course, this is the nature of research, a kind of perpetual motion machine, but we end with so many more questions to explore, enough to fill several lifetimes. Adapting the title of another Johnny Nash song to claim that 'we can see clearer now' qualifies the sentiment of the first. Finally, without meaning any disrespect for a great U2 song, we can change 'I still haven't found what I am looking for' to 'we really have found what we were looking for' (especially during our exploratory mode) and are very eager to find so much more of the same.

Anyone wishing to join the adventure or wanting to check our results can visit the following URL, which contains supplementary materials and data sets:
http://pajek.imfm.si/doku.php?id=book

1

Temporal and spatial networks

Our primary concern is *understanding* both large temporal and large spatial networks in ways going beyond simple general descriptions of their structures. For the former, doing this amounts to discerning the structure(s) of such networks as they develop over time, and grasping the social forces driving these changes. For the latter, it involves understanding spatial social patterns and the processes by which they were generated. For both network types, these two broad tasks – delineating structures and understanding their formation – go hand in hand: doing one without the other leaves our understanding of these networks incomplete. However, in order to understand the impact of social forces, it is necessary to know the structure(s) of networks. We focus, initially, on outlining foundational network concepts in Chapter 2. A detailed presentation of methods for analyzing citation networks is included in Chapter 3. In the remaining chapters, we study how temporal networks change and social phenomena are distributed over spatial networks. We provide substantively based interpretations of the results we obtain. As is usually the case, for us, creating these understandings was an iterative process where empirical results led to substantive understandings which, in turn, triggered further analyses. We report results of these analytic sequences but without reporting the iterations.

1.1 Modern social network analysis

Freeman (2004) argued that four features define *modern* social network analysis (SNA). In a slightly expanded form they are:

1. SNA is founded on a 'structural intuition' regarding social ties linking social actors. This motivates the study of the social networks formed by these social ties when they form *coherent* wholes.

2. 'It is grounded in *systematic* empirical data (emphasis added).' Implicitly, network data must be meaningful for studying social networks: not all social network data sets are useful.

Understanding Large Temporal Networks and Spatial Networks: Exploration, Pattern Searching, Visualization and Network Evolution, First Edition. Vladimir Batagelj, Patrick Doreian, Anuška Ferligoj and Nataša Kejžar.

3. 'It draws heavily on graphical imagery' to represent these social networks and their salient features in useful ways. Visualization of these features is useful both for displaying results and for suggesting further avenues of inquiry.

4. 'It relies on the use of mathematical and/or computational models.' This dual reliance has grown even stronger since 2004.

We add the following three items:

1. Fully *understanding* social networks in time and across space requires a concern with substance.

2. When studying the operation of social *processes* creating, sustaining, and dissolving social networks, Doreian and Stokman (1997), the relevant network data *must* be temporal. Intuitively, a temporal network has units and relational ties distributed through time.

3. Given that most social networks are conditioned by the contexts within which they exist, ignoring these contexts imposes major constraints on understanding network phenomena. One contextual feature is the geographic space within which these networks are located. Spatial networks have units and relational ties distributed across geographical space.

While substance can never be ignored safely, we note that many networks have been studied without considering time. Other networks were studied while ignoring space. Quite often, neither time nor space had relevance for analyzing network data. This has changed dramatically in recent years with considerable attention being devoted to both space and time when studying social networks. Consistent with this new emphasis, the networks we consider here involve time or involve space and, occasionally, both. In the main, we focus on temporal networks.

Building upon the above seven items, our study of temporal networks and spatial networks is informed by four working assumptions:

1. Social networks form through the operation of *social* processes. These processes have direct relevance for studying networks, implying that substantive ideas really matter. In turn, the contexts within which social networks are generated are crucial for understanding network creation and the consequences they have for the people, groups, organizations, states, and nations located in them.

2. As Freeman noted, computation has been crucial. However, *practical* and *sound* computational methods are required for detecting useful structural patterns in networks. Developing these methods is necessary, even mandatory. Ideally, computational methods are informed by substantive concerns. However, we have no objection to developing methods for their own sake. Even so, the use of methods developed in this fashion requires some justification in terms of both substance and relevance, at least as far as understanding social network processes is concerned. Methods are more useful when coupled to the substantive issues for which analyses are performed.

3. Temporal network data have to be *meaningful* in terms of both social substance and social contexts. This implies that temporal network data need to be selected carefully in order to be relevant substantively. The same arguments hold for spatial networks.

4. Coupling substance, context, methods, and data is most effective when these items are combined into a single coherent framework.

1.2   Network sizes

We define the terms small, large, and huge for network sizes in Section 2.3. The networks discussed by Freeman (2004) are small. Indeed, for many decades, social network analysis dealt primarily with small or very small networks. This was driven by traditional ways of collecting data and by the technical constraints on the collection and analysis of social network data. However, since the 1990s, large networks (having from thousands to many millions of units) have become abundant, for which information technology (IT) has been particularly important in assembling these data. This development has serious implications both for visualizing networks and for implementing computational models. Many of the earlier traditional computational methods useful for studying small networks are now completely impractical for analyzing large networks. As a result, developing new practical computation methods has become essential. We focus on some newly designed computationally feasible methods for handling these networks and present the results of using them while being attentive to substantive concerns.

1.3   Substantive concerns

At face value, the only feature common to the networks studied here is being large. While the book title emphasizes this, its most important word is 'understanding.' The datasets we use were selected to cover different substantive domains, to have different sizes, and to be characterized by different structures. We have no single 'cookie cutter' method suitable for all temporal networks because of these differences. However, two methods used repeatedly in our analyses are line islands (used in Chapters 4–6) and clustering symbolic data (used in Chapters 5 and 8). We focus briefly on substantive concerns here and more fully in Chapters 4–9 where different combinations of methods, based on the foundations laid out in Chapter 2, and developed in Chapter 3, are used. Methods employed only in single chapters are presented therein.

1.3.1   Citation networks

Chapters 4–6 consider three distinct citation networks. While citation networks may seem to be the 'same' in their general structure, with later units (documents) citing earlier units, these three networks differ greatly with regard to their sizes and network structures in addition to substance.

1.3.1.1   Scientific citation networks

Hummon et al. (1990) noted three features of science: 1) 'Science is a cumulative venture where each new discovery or development depends on some prior work;' 2) 'The products are generated at the research fronts of specialty fields' where the products are scientific productions; and 3) 'The resulting written record, in the form of citation networks, left as research fronts move on, contains valuable information for understanding the processes of science.' The small citation network they studied concerned the centrality literature between the initial task-oriented group experiments of Bavelas (1948) through Freeman's (Freeman, 1979) systematic statement of three distinct operationalizations of centrality. This network, while complete, was small with only 119 scientific productions. Since then, centrality became accepted widely as one of the most important concepts in social network analysis. Certainly, it is one of the most frequently used ideas: in short, this literature has exploded. We were curious about the extent to which this literature grew and the structure of the resulting citation network.

Although, as documented by Freeman (2004), SNA has been in existence for over a century, physicists have developed a recent interest in networks within the rubric of what has become known as 'network science.' See, for example, Watts and Strogatz (1998), Newman (2001), Barabási (2003), and Newman et al. (2006). This interest was triggered, in part, by the availability of large networks obtained readily via electronic methods. Despite some notable exceptions, physicists have tended to ignore most of the prior social network literature while claiming the creation of a 'new' field resulting from their endeavors. In the memorable phrase of Bonacich (2004), this was (seen as) 'the invasion of the physicists.' We were curious as to the structure of the citation network of the SNA citation network following this 'invasion.' Some results of our analyses are laid out in Chapter 4 together with a narrative concerning the ways citation network structures are developed and some of the institutional forces involved in this process.

We study two scientific citation networks. One is restricted to the centrality literature while the other, more broadly, is the SNA literature. In the main, the former is located within the latter. However, as centrality was initially a very narrowly focused concept, there is interest value in looking at it especially as the concept has been applied in many substantive domains. Of additional interest is that we have learned (see Chapter 4) that the concept has been formulated and found valuable in areas remote from SNA. Indeed, tracking how a technical concept has been applied in different substantive areas adds to the value of looking at the citation network for centrality. One implication of the results reported in Chapter 4 is that the centrality literature in no longer located fully within the SNA literature.

We were interested also in the nature of the linkages between the traditional SNA literature, as seen by social network analysts, and the network science literature. To the extent that different fields merged in pursuit of studying social networks, it is reasonable to expect flows of ideas between them. A rival expectation is that the fields partially diverged while cleaving to their own conceptual frameworks, network interests, and methods.

As described below, both the patent citation network and the Supreme Court citation network were 'cleaned' explicitly (for patents) or implicitly (for the Supreme Court). This cannot be done fully with academic citation networks. Once publications are in the literature, these works remain. Of course, never-cited works can be removed as a part of a data analytic process removing them. Even so, apparent dead-end lines of work remain, a topic we do not pursue here.[1]

1.3.1.2   Patent citation networks

Patents are legal devices attempting to confer some protection of intellectual property rights for the inventors of new technological items. They have been seen as particularly interesting

---

[1] It remains an open question whether approaches simply dying out (for example, 'functional theories' in sociology), or because of scientific revolutions in their fields (Kuhn 1970), can be tracked in citation networks.

regarding their role in the study of innovation and technological change. Economists have long sought to link technological innovation to economic change and development. Examining patents and their role in triggering technological change has been a part of this effort. However, our concern differed: we sought an understanding of the temporal patterns of patents citing earlier patents. This includes when and how inventions protected by earlier patents become useful for later inventions and patent applications. More precisely, we sought an understanding of the *time scales* regarding how earlier inventions became useful for later inventions. We learned that there are at least four distinctive patterns to these time scales. Also of interest is the influence patterns between broad technological areas (they are significant), how they vary over time, and how specific technologies decline temporally, to be replaced by others in specific eras. We document some of these changes in Chapter 5.

We learned that inventors, or their proxies, applying for patents enter complex technical *and* legal arenas. The technological arena alone is intellectually complex as is documented in Chapter 5. As a result, applications for patents to protect inventions trigger a stringent review process performed by experts in the United States Patent and Trademark Office (USPTO), at least for patents issued in the USA.[2] The result – in addition to the granting of patents for inventions or deciding specific inventions cannot be patented – is a complete citation network that is 'efficient' in the sense of citations being made *only* to *all relevant* earlier patents. The institutionalized review process of the USPTO is the first phase for creating a cleaner citation network.

In addition, when existing patents are revoked, they and all citations to them, as well as citations from them, are removed from the patent citation database by the USPTO. As a result, only genuinely useful patents remain in the available data. This is particularly important because, in essence, the boundary problem (Laumann et al., 1979) – one posing problems for many social network studies – is solved completely: these patent data are the cleanest citation data we examine. Intuitively, the boundary problem for networks is simple to state and has two basic components. One is the exclusion of relevant data points (units) and all of the (real) network ties involving them. The other is the inclusion of data points that are not part of the network being studied but are included in the network along with their network ties. Both types of errors have great potential for distorting networks and analyses of them.[3]

However, more than technological issues are involved when considering patents. Many economic actors seek to capitalize on the patents they have by creating production processes and services, new physical products and new substances for economic gain. Conflict over them leads to legal issues involving both apparent and real patent infringements when other economic actors produce similar or identical products for sale. The parties involved in these disputes either attempt to protect inventions covered by patents they hold or challenge the legitimacy of patents already granted to others.

Conflicts over inventions and patents often lead to court cases. Some reach the US Supreme Court, where the Justices on this court weigh in on technological matters and the patentability of inventions. We sought an understanding of patent citation phenomena in technological contexts defined by the USPTO. The primary tools used were line islands (see Section 2.9) and clustering symbolic data (described in Section 3.10). The results of

---

[2] No doubt, this is true elsewhere, but our data are for US patents.

[3] There is also the separate measurement problem, even if the boundary problem has been solved, of erroneously excluding network ties, erroneously including them or recording ties inaccurately. These problems are solved also for the patent data.

using these methods for the patent data are presented in Chapter 5. The involvement of the Supreme Court in these technological matters added an unexpected (for us) connection between Chapters 5 and 6. The technique of identifying line islands was important for considering the role of the Supreme Court in evaluating rival claims over patents.

1.3.1.3   A US Supreme Court citation network

Fowler and Jeon (2008) compiled an extensive citation database for US Supreme Court (SC) decisions citing earlier SC decisions based on the content of their majority opinions and the citations they contain. At face value, this data set is complete also in including all decisions in a specified and very long period (1789–2002).[4] These citation data differ from the patent citation data in at least one important respect. There are no constraints on Supreme Court Justices writing opinions regarding which earlier decisions they cite as precedent nor on which earlier decisions they choose to ignore as precedent. This is a luxury unavailable in patent applications where, as noted above, relevant, and only relevant, citations to earlier patents are permitted. This adds an intriguing wrinkle for understanding this temporal network.

Fowler and Jeon's primary focus was the evolution of precedent as a legal concept: earlier decisions inform (and therefore constrain) later decisions. For analyzing precedent, the frequency of decisions being cited, especially by other salient decisions, takes centre stage. We take their results as a given.

Our interest takes a complementary form: we focused on subsets of Supreme Court decisions forming coherent *parts* within the overall citation network. In terms of methods, we did not focus on computing measures for single decisions but examined sets of decisions instead. We approached this by considering the extent to which earlier decisions are *co-cited* by later decisions. The rationale behind this interest was driven by a key intuition: decisions cited *together* have import by having common substantive or legal principles (or both) holding them together. To qualify for this additional closer scrutiny, earlier decisions have to be co-cited *frequently* (by pairs of subsequent decisions).[5] The primary method used in studying these coherent parts of the Supreme Court citation network was identifying line islands, a procedure described in Section 2.9.

By definition, never cited earlier decisions – of which there are very many – can never be co-cited. Similarly, decisions citing no earlier decisions cannot contribute much of interest regarding precedent nor for considering co-citating. Decisions neither citing other decisions nor receiving citations are easily removed. In effect, doing this helps 'clean' the US Supreme Court citation network to achieve implicitly an effect similar to the results due to the process enforced by USPTO's review process. Removing these isolated decisions, having no historical relevance, prunes the citation network. While this serves our purposes very well, this may affect the fitted distributions of measures computed for single decisions.

---

[4] On looking closer, and examining many decisions, we learned that some decisions were omitted from the Fowler and Jeon dataset. While we have inserted the missed decisions that we located, there is no guarantee that the list of decisions we studied is complete. Indeed, all large datasets contain errors as Fowler and Jeon note for the dataset they created. However, it is very close to being complete for *relevant* decisions. Never cited decisions are the most likely to be overlooked. These omissions are unimportant because they have no relevance in this citation network. A potentially much more serious data recording problem was unearthed, and this is examined in Section 6.6.

[5] It is possible to examine pairs of decisions in terms of frequently *co-citing* earlier decisions, a line of inquiry not pursued here.

The Supreme Court network can be looked at as a stand-alone entity to be studied by itself. However, the Supreme Court is only one of three 'top' branches for governing the US. The other branches are the President and the US Congress (made up of the House of Representatives and the Senate). In principle, the three branches were created as independent entities designed to constrain each other through the much discussed 'checks and balances.' As we show in Chapter 6, there were periods when the three branches acted in concert while being in sharp conflict at other times. Both of these contextual conditions have relevance for understanding the actions of the Supreme Court. The same applies more generally with regard to historical contexts for all social networks.

At face value, the task of the Supreme Court is simple: its members use the Constitution as the foundation for establishing the legitimacy (or not) of laws and the appropriateness (or not) of decisions rendered by lower courts. Unfortunately for this naive view, the Constitution is profoundly ambiguous, a built-in feature reflecting the greatly divergent positions of the rival parties and political interests represented by those involved in drafting this document. These deep conflicts have never been resolved. Indeed, it seems impossible to resolve them, and the vague (ambiguous) language of this constitutional document papered over these differences in order to get enough signatures to it. The deep political conflicts did not end with the signing of the Constitution, and the subsequent ratification process in the separate states was deeply conflictual within them. As a result, the whole judicial system, and the Supreme Court in particular, as authorized by the Constitution, was 'political' from its inception and has remained so. This alone implies caution in taking the Constitution as an 'objective' document free of biases and contradictions and viewing the resulting citation network as a simple record of the processes leading to its decisions. It cannot be studied as a network detached from broader social contexts.

Further, this citation network covers by far the longest time interval (more than two centuries) of the citation networks we consider. In this long period, the USA experienced great economic, political, and social change. In short, the context within which the Supreme Court operated changed dramatically over time. This suggests that the resulting citation network cannot be studied solely as if it were simply just another citation network. The changing contexts within which decisions were made matters greatly for understanding it. Some results using line islands in the Supreme Court citation network are presented in Chapter 6. With these identified islands, close attention was paid to the changing history of the USA and the Court in understanding both the citation network's structure and the more important actions of the Court. This included attention to 'accidents' (sudden deaths of Justices and Presidents, plus unanticipated electoral outcomes) and changes in the composition of the Court over time.

1.3.2   Other types of large networks

The US patent network is acyclic (a concept defined in Section 3.1). Both the Supreme Court network and the scientific citation network lack this feature which creates technical problems for analyzing them. Fortunately, such networks can be transformed easily to an acyclic form by using the methods described in Chapter 3. In order to obtain completely different types of temporal networks, we focused on different substantive issues where the techniques useful for citation networks were not relevant. These other networks are not acyclic networks and *cannot* be studied as such. Including them was important because 'large temporal networks' can take many forms, including some we do not consider here.

1.3.2.1   The movement of football players across the globe

For the first alternative type of network, we focused on the movement of football (soccer) players as they traveled across the globe to play football. Football players, almost from the formal inception of the game, following its codification, moved between clubs and countries. In doing so, they created club-to-club networks because clubs had to agree on the movements of players between them. We conceive of football players' careers fundamentally as *movements between clubs*.[6] These movements define and create basic social networks linking clubs and, secondarily, countries. While this is our primary interest, we use these movements also to examine some common presumptions about 'the beautiful game.' Alas, no single systematic nor reliable dataset comprising these movements exists. So, we constructed one, albeit with a geographically restricted focus. The details of this database are described fully in Appendix A.4, along with the many difficulties encountered while constructing it. These data are unique. We recognize that this claim can be made for any dataset but this one was created by combining information from over a thousand different sources.

Recognizing that the game is played in over 200 countries, we restricted attention to just one country because one of us (Doreian) has had a long-term interest in football played in England. Our data were defined by *all* of the football players who appeared in the English Premier League (EPL) during its first 15 seasons (1992/3 through 2006/7).[7] This temporal restriction was, primarily, a practical issue – although we did track these players through to the end of 2012. To set the broad background for the analyses that we pursue, we describe football in England in Chapter 7 as a local institutionalized representation of the so-called world game. Additional reasons for focusing on football in England, beyond familiarity, stem from its unique history, one having major impacts on player movements within and to England. This is part of the context for these player-induced networks.

English officials administering the game were, and continue to be, acutely aware of the game having been invented in their land. Of course, the game quickly spread to many other places but this diffusion was ignored largely by these administrators for close to a century. As a result of the assumed historical primacy of their legacy and a presumption of English 'superiority' regarding 'their' game, these officials assumed that they had little to learn from developments in football at other places on the globe. Indeed, they attempted, with great success, to keep (most, but not all) 'foreigners' out of the game as well as 'foreign' conceptions of strategy, tactics, and styles for playing football.

This fundamental restrictive control of the game in England was shattered by a series of court cases. One was indigenous to England (resulting in the Eastman decision as described in Chapter 7). Far more importantly, the European Union (EU) ruled on labor practices for all its members in the Bosman Decision, also described in Chapter 7. The so-called 'transfer and control' system by which football clubs controlled their players did not come close

---

[6] Other conceptions of careers focus on player and club performances on the field in terms of goals scored, defensive plays made, appearances for national teams, club victories, and trophies won. This conception of careers in terms of success has little interest for us beyond operationalizing temporal sequences of club success (overall ranks) to characterize player careers as sequences of moves between the clubs, in nationally and internationally stratified systems, for which they play. This is described more fully in Chapter 8.

[7] For this study period, the EPL is an accurate label. Much more recently, Swansea City and Cardiff City, both located in Wales, were promoted to this top league. The term Premiership is now used frequently rather than EPL as a label. An alternative label is BPL, presumably for 'British Premier League,' although this usage ignores the Scottish Premiership.

to conforming to EU rules, especially with regard to people moving between its members to find employment. These court decisions changed forever the relations between football players and the clubs for which they played. Suddenly, players had greater (but not complete) freedom to move between clubs (and between countries). Some of this changes are described in Chapter 7. We were curious about the resulting network *patterns* of player movements and what this implied for the organization of football.

Much has been written about player movements *to* England, the nature of football in England, and the impact of TV money flooding into the game since the late 1980s. These include arguments about the impact of this flow of foreign football players on football in the place where it was invented. Some concern the widely held belief about the EPL being the 'best' league on the world and how beneficial this has been locally. Other arguments claim that these flows were disastrous for *English* football, especially at the national level. Clearly, skepticism is merited regarding these arguments. Kuper and Szymanski (2009) exemplify this skepticism, a stance which prompted some of our analyses. As conventional (assumed) wisdom often rests on ignoring relevant information, our interest was piqued by these rival claims.

We examine some claims about modern football using the network of players moves that we constructed, together with some ancillary data. Our results and conclusions are reported in Chapters 7 and 8. Some of the hypotheses were confirmed and some failed while others turned out to be untestable.

1.3.2.2 A large US spatial network

As an example of a large spatial network, we examined the network defined for all US counties in the Continental USA.[8] We had two motivations for considering these data. One was substantive while the other was methodological. The Continental USA has 3111 counties. Pairs of counties are linked through sharing a common border. This adjacency in geographical space defines an unambiguous spatial relation linking counties. The Continental USA is divided also into 48 states each made up of counties. Each state has its 'own' history. In these histories, events and outcomes are described frequently as being unique to the 'proud' history of each state. Yet, on the ground, the boundaries between many pairs of states are evident only by signs marking them.[9] Certainly, social processes operate across the boundaries between these large aggregates. Attempts to understand these broad social processes need to move beyond state boundaries.[10]

There have been two broad approaches to characterizing the spatial distribution of the large social, economic, and political diversity within the USA. One attempts to map broad contiguous areas of the landscape within which greater homogeneity is thought to exist. Two examples of doing this are Garreau (1981) who defined and delineated Nine Nations covering the USA, Canada, Mexico, and the Caribbean Islands, and Woodard (2011) who argued for there being eleven such nations. Their general argument has appeal, with both authors assembling considerable qualitative evidence in support of their theses. While there

---

[8] Hawaii and Alaska were excluded, for obvious reasons.

[9] Rivers are one of the exceptions when they form clear boundaries between states. Occasionally lakes do this.

[10] The same argument can be made with regard to counties. However, as we claim in Chapter 9, counties represent a reasonable compromise between large heterogeneous areas like states and very small potentially more homogenous local areas for which systematic data do not exist.

are some commonalities to the two sets of nations they defined, there are also considerable differences. This alone merits a closer examination of their detailed delineation of nations within North America.

A second broad approach is exemplified by Chinni and Gimpel (2010) who eschewed geography during their detailed data analysis. After assembling statistical data for counties, Chinni and Gimpel clustered them using these constructed variables. They then plotted these clusters of counties in geographical space to describe a 'patchwork' nation with very different patches distributed across the nation and within states.

It seemed reasonable to seek a middle ground between focusing solely in large contiguous regions and focusing solely on the attributes of the units (counties) located in geographical space. The general problem is one of clustering units based on measured variables while being attentive to relations among the units. Although it was not proposed initially for dealing with spatially distributed data, one method for doing this – clustering with relational constraints – was proposed by Ferligoj and Batagelj (1982, 1983). It clusters units based on a set of measured variables, consistent with the approach of Chinni and Gimpel (2010), while constraining cluster memberships according a relation linking the units being clustered. The obvious relation in the US context is the spatial adjacency of counties. However, the method, as initially formulated, is impractical for any large network, especially for one as large as this spatial network. The technical concern motivating our analysis was establishing a practical computational method for networks of this size while remaining faithful to the core conception of clustering with relational constraints. The newly developed algorithms and the results of applying them are described in Chapter 9.

1.4 Computational methods

We develop extensive formal foundations for the methods we use in Chapter 2. It serves as a preliminary introduction to graph theoretical representations of networks. We then extend this systematically to deal with temporal networks, as defined in Section 2.2 and detailed in Chapter 3. Our focus on large networks is driven, primarily, by the intriguing computational difficulties of handling them efficiently. The notion of a 'large network' is defined in Section 2.3 to include networks with many millions of units. In terms of computation, the central workhorse for the empirical results we present is Pajek (Batagelj and Mrvar, 1998; de Nooy et al., 2012). Indeed, Pajek was designed *explicitly* for analyzing large networks efficiently.

We do not claim that Pajek is the only useful software for this purpose: it was simply the one we chose for our computational efforts when analyzing large networks. It served our purposes well. Doreian (2006) noted that Pajek is not a 'one button' set of routines. Instead, the results obtained from most of the analyses we present were completed by *combining sets of commands*. This design feature of Pajek facilitates great flexibility. However, it also requires users to understand the program's logical structure. Given this, we include Pajek commands wherever they are appropriate so that readers can do the analyses leading to our results for themselves if they wish – either on the data we used or with data of their own.

Figure 1.1 shows the primary (initial) dialogue box for Pajek. There are two distinct listings of objects. The general concept of a network is that it is composed of vertices (representing units) and lines representing relational ties between units. These terms are defined fully in Chapter 2. On the left (reading from the top), is a column listing objects: Networks (described by vertices and lines); Partitions (assigning values to units to split them into clusters); Vectors (assigning numerical values to units); Permutations (to rearrange the order of units as they are

Figure 1.1   The Pajek main dialogue box.

stored in files); Cluster; and Hierarchy. The icons for each of these objects are used (reading from the left) for reading objects saved as Pajek files, saving objects as Pajek files, examining the contents of objects, and getting summary information about these objects.

Across the top of the main dialogue box, the listed objects are (reading from the left) File, Network, Networks (for handling more than one network), Operations (using different combinations of networks, partitions, vectors, permutations, and hierarchies), Partition, Partitions (for using multiple partitions), Vector, Vectors (for using more than one vector), Permutation, Permutations (for utilizing multiple permutations), Cluster and Hierarchy. Clicking on these icons produces drop-down menus with more detailed data analytic options. These icons on the top row are followed by Options, Draw (for visualizing networks), Macro, Info, and Tools (for exporting selected information to other programs, including R and SPSS, for supplementary analyses) which are concerned also with mobilizing procedures. Clicking the Macro icon presents a list of prepared and stored sets of commands presented in Pajek. Users can define and save their own macros for combinations of commands they use often enough to merit the construction of macros. Clicking on each of these opens a dialogue box for working with, and using, objects, pairs of objects, or triples of objects. When we present commands for using Pajek, we use primarily the items across the top of this dialogue box followed by the relevant options.

In Section 2.5, we distinguish statistical summaries of network features and summaries formed through network analytic methods. While they differ in the analyses performed, these methods are most effective when coupled. In the main, for the former, we used R.[11] Where necessary, we provide the R code used for some of our analyses, as Pajek permits easy transitions to analyses using network outputs within R.

---

[11] See http://www.r-project.org/.

Given our focus on large citation networks, Chapter 3 builds on Chapter 2 to lay out formal tools designed for examining such temporal networks. In general, methods used in multiple chapters are described in these two chapters. As noted above, methods specific to single substantive chapters are presented in those chapters, especially for the patent citation network (Chapter 5), the EPL football player movement network (Chapter 8), and the US spatial network (Chapter 10).

We are fully aware of a wider literature discussing the topology of large temporal networks, the importance of which we do not deny. See, for example, all of the contributions brought together in Newman et al. (2006) and the many scientific productions built on these foundations. Our goal here is *not* to provide a broad comparison with all of the methods used in other literatures.[12] Instead, in the spirit of 'letting many flowers bloom' we lay out another complete framework for studying large temporal networks. If the results of using this approach do have value, then comparisons between different approaches will have considerable merit. It is simply too early to impose a single approach for studying large networks on the study of all such networks. Of course, we are not proposing the methods introduced here as the only appropriate ones for studying large networks.

1.5   Data for large temporal networks

The datasets we use fall into two categories. The first contains data defined by the substantive interests outlined in Section 1.3. These data are used for the analysis and results presented in Chapters 4–9. The other (secondary) category[13] has data used for *illustrating* concepts and methods introduced in Chapters 2 and 3. We know that the term 'interesting' (when it is not used as cover for not expressing an opinion one way or another) is in the eye of the beholder. The distinction between main (primary) and secondary datasets is not intended as an evaluative statement about their relative merits even though we do insist that the data considered here need to be relevant for specific substantive concerns. The secondary data sets have different substantive interests and technical issues in mind.

1.5.1   The main datasets

We describe briefly these main datasets, each driven by substantive interests, and present their dimensions here. Appendix A contains detailed descriptions of them, including how they were obtained and the data processing for getting them into the form we use. Their initial[14] dimensions are provided in Table 1.1. Some[15] of these data are freely available at Pajek datasets (see http://vlado.fmf.uni-lj.si/pub/networks/data/sport/football.htm).

---

[12] We have noticed in submitted manuscripts involving blockmodeling (see Doreian et al. (2005)), reviewers often *demand* a *full coverage* of the community detection literature (created mainly by physicists) ideas even when community detection ideas are tangential. While there is, at face value, some commonality between these approaches, the differences are quite marked and rather subtle. Such broad summaries often are distractions – and, when complied with, can affect citation networks.

[13] With a few exceptions, we maintain this distinction to have our substantively relevant results remain within single chapters.

[14] For some analyses, not all of these data were used. For other analyses various subsets were used and the results combined. (See Section 2.4 for a description of the 'divide and conquer' strategy that we employ for simplifying large networks.)

[15] The exception is the football data because we intend to explore them further before making them available publicly.

Table 1.1 Dimensions of the datasets used in Chapters 4–10.

Substantive network	Number of units	Number of lines	Pajek data-set name
Patent	> 3.2 million	> 32 million	`patent.net`
Supreme Court	30,288	216,758	`allcitesV4.net`
Centrality	995,783	1,856,102	`Cite.net`
SNA	193,376	324,616	`SN5cite.net`
Football	2355	40,246	not available
Spatial USA	3111	7101	`UScounties.zip`

The patent citation network (for patents citing earlier granted patents) features patents issued in the USA. The time period is relatively short, covering 1976–2006, a mere 30 years. However, this network is the largest dataset we consider for the substantive chapters, having more than 3.2 million patents linked by over 32 million citation links. The US Supreme Court citation network, in contrast, is much smaller with more than 30,000 units and over 216,000 citation links. However, it covers more than 200 years; this is by far the longest time span of all of the networks we study.

There are intrinsic differences beyond their sizes of these two networks. We noted in Section 1.3.1 the strict constraints on patent citations, in contrast to the freedom that SC Justices have in citing prior decisions. There are many SC decisions that neither made nor received citations. One practical consequence is that the relevant citation network has fewer units than the number listed in Table 1.1. However, the long time span and the depth (defined in Chapter 3) of the SC citation network created technical problems requiring attention before the general methods for acyclic networks presented in Chapter 3 could be used. The patent citation network was acyclic as received. This was not the case for the SC data: some decisions handed down by the same Court in a short period of time do cite each other, a phenomenon present also in the scientific citation data for publications appearing in the same year. Solutions for handling this problem are described in Chapter 3 and mobilized in the analyses of both the centrality and the broader SNA literature, in addition to the Supreme Court network. In analysis of centrality and SNA literature we used also some other bibliometric networks.

The football data that we constructed have a far more complex structure, featuring football players, football clubs, and countries. It was defined by the 3749 football players playing in any of the first 15 seasons of the EPL. These players had 148 nationalities (dual citizenship is precluded for defining the nationality of players). Even though the player network is defined by these players, our primary interest centered on the clubs for which they played. More specifically, the network ties for this network are the links between these clubs as created by players moving between them. The number of clubs involved in their migrations to and from the EPL was 2355. These clubs are located in 152 countries. The total number of player moves between clubs was 40,246. We also used ancillary data (described in Appendix A.4) on clubs and player presence by nationality in other top European leagues for additional analyses.

Our example of a large spatial network features all of the counties of the contiguous USA and was motivated by trying to reconcile two very different approaches to mapping social diversity in geographic space. The substantive problem has intrinsic interest, and the network we study is one of the larger substantively interesting networks we have located.

Table 1.2 Dimensions of the illustrative datasets.

Substantive network	Number of units	Number of lines	Pajek data set name
EAT	23,219	325,589	`eatSR.net`
NBER	174	11755	`NBERwt.zip`
KEDS	325	78,667	`BalkanDays.net`
e-companies	219	631	`krebs.paj`

1.5.2   Secondary datasets

We report our results for the networks described in Section 1.5.1 extensively in the relevant chapters. To avoid repetition of results, we used data from other sources to illustrate our methods in Chapters 2 and 3. These data, the dimensions of which are listed in Table 1.2, came from the following sources.

1.5.2.1   The Edinburgh Associative Thesaurus (EAT)

The primary goal of the EAT project[16] was to understand how words in the English language are coupled. This was done by examining empirical 'associations' between words. The approach taken to obtain these word associations was straightforward. Subjects were shown a word and then asked to provide the first word coming to their minds. The procedure presented batches of words to each subject. The presented words were regarded as stimuli and the words offered by subjects as responses. The established links between stimuli and responses were provided by subjects. There were no imposed rules dictating the nature (appropriateness) of the responses. The pairings of stimuli and responses were simply empirical associations. For each pair of words, they were aggregated across subjects as a way of quantifying these associations. For example, some frequent couplings included 'husband' in response to 'wife' and 'cheddar' in response to 'cheese'.

The resulting Edinburgh Thesaurus association norms were started from a nucleus set of words. Further associations were collected by expanding from the nucleus: initial words were used to obtain further responses, together with additional words. The EAT website reports this cycle was repeated about three times. By then, the number of different responses became so large they could not be reused as stimuli in a systematic fashion. The EAT data collection stopped after 8400 stimulus words were used.[17] The result was a total of 23,219 words in the Thesaurus network linked by 325,589 associations. The database has two files: one is a SR (stimulus–response) file, with the other being a RS (response–stimulus) file. These data are used in Section 2.5.

1.5.2.2   The NBER-United Nations Trade Data, 1962–2000

This network was used for illustrative purposes in Section 2.6. The network ties are trade exchanges (exports and imports) between nations. The data we used came from 1999: there are

---

[16] See http://www.eat.rl.ac.uk/ for a description of this project.

[17] Each stimulus word was presented to 100 different subjects. Their website reports that the subjects were mostly undergraduates from many British universities whose ages ranged from 17 to 22 with a modal age of 19. The sex distribution was about 64 per cent male and 36 per cent female. The data were collected between June 1968 and May 1971. Any bias in the distribution of associations due to using university students as subjects has no relevance for our *illustrative* purposes regarding methods.

174 vertices and 11755 trade flows linking nations. The weight of the arcs are trade values in $US1000. The source for these data is http://cid.econ.ucdavis.edu/data/undata/undata.html. The complete dataset is available as the zipped Pajek project file listed in Table 1.2.

1.5.2.3   The Kansas Event Data (KEDS)

The data in this resource are the results of a 20-year project, originally based in the Department of Political Science at the University of Kansas. This project and its data were known as the Kansas Event Data System (KEDS), a label we use here. It was moved to the Department of Political Science at Pennsylvania State University in January 2010 (http://eventdata.psu.edu/). The project uses automated coding of English-language news reports from a variety of news resources to generate political event data focusing on the Middle East, the Balkans, and West Africa. These data were designed primarily for use in statistical early warning models to predict political change in these regions with attention given to suggestions and policies for mediating conflicts. The units for this network are nations and organizations. The relations include ties between nations in the form of actions by one nation directed towards another nation, as described by verbs. These actions include visits, seeking information, issuing warnings, and expelling persons. Data for the Balkans for KEDS are used in Section 2.2. The full dataset is available also at the KEDS website.

1.5.2.4   Krebs Internet industry partnerships

Valdis Krebs collected in 2002 (http://www.orgnet.com/netindustry.html) a network of Internet industry partnerships. Two companies are linked with a line if they have announced a joint venture, strategic alliance, or other partnership during the period 1998–2001. The companies are classified into three classes: 1 – content, 2 – infrastructure, 3 – commerce.

1.5.2.5   Data archives

There are variety of sources containing many datasets, both large and small, but with a primary focus on large datasets. One is SNAP, the Stanford Large Network Dataset Collection maintained by Jure Leskovec. It is documented at http://snap.stanford.edu/data/. The topics covered include on line social networks, communication networks, citation networks, and collaboration networks. There are also graphs of the internet and physical road systems. Signed networks are included in this archive. KONECT, the Koblenz Network Collection, contains large network datasets assembled at the Institute of Web Science and Technologies at the University of Koblenz-Landau. As stated on its website (http://konect.uni-koblenz.de/): 'KONECT contains over a hundred network datasets of various types, including directed, undirected, bipartite, weighted, unweighted, signed and rating networks. The networks of KONECT are collected from many diverse areas such as social networks, hyperlink networks, authorship networks, physical networks, interaction networks and communication networks.'

These archives of datasets are used in Section 2.3 when describing the distribution of network sizes in terms of the number of units and relational ties. Networks are sparse when they have roughly the same number of units and relations ties. More specifically, the numbers of these ties are not orders of magnitude larger than the number of units. Networks being sparse is crucial for developing efficient methods for analyzing large networks.

1.6 Induction and deduction

Throughout this book we have been both inductive and deductive in our approach, with a strong bias in favor of being inductive when examining large networks. However, our uses of both induction and deduction were based on substance and driven by curiosity. For the three citation networks, induction reigned. When examining the social network citation networks, we wanted to learn the citation structure following two salient events. One was the formalization of centrality (and network centralization) in the Freeman (1977, 1979) papers. As we have noted, this triggered an explosion of work both extending and using these ideas. The other event was the recent interest of physicists in studying social networks. We stated rival expectations regarding the possible convergence or divergence of the traditional SNA and network science literatures. But, lacking anything beyond broad statements about this question in the literature, we had no foundation for a specific hypothesis.

This was just as well. One of our expectations was that the physicist conception regarding network science supplanted the SNA conception, especially regarding centrality. This was born out but, as noted in Chapter 4, this was not the end of the story. The full disciplinary sequence, for the centrality citation data ending in early 2013, was SNA → network science → neuroscience. We learned that the general concept of 'centrality' has multiple sources. There are parts of the broader centrality literature having *nothing* to do with traditional SNA concerns. Worries about 'the invasion of the physicists' may be a somewhat parochial conception within the older SNA community. This issue is explored more extensively in Chapter 4.

Our exploration of the patent citation network was inductive also. Given the four broad technological domains for 'utility patents' defined by USPTO and described in Section 5.1, we were curious about the flow of ideas between these broad technological areas as reflected by citations between patent applications and how they changed over time. Further, as technologies change over time, specific inventions are likely to have a limited shelf life. One crucial feature related to this is the lag between patents being granted and their ideas being picked up and used fruitfully for later inventions. Our interest centered on the distribution of these lags and their temporal dynamics.

Again, we were inductive in our approach to the Supreme Court citation network. However, there was an implicit hypothesis – about the line islands we identified having coherence – which underlay our analysis. Alternatively put, we gambled on this hypothesis being correct. If the gamble was lost, then this approach would be seen as severely flawed. Fortunately, thus far, *every* line island we have examined has a singular coherence even though the specific nature of their coherence differs by island. Establishing the presence of coherence of decisions being co-cited frequently was a purely inductive, but not surprising, outcome. This coherence among a set of frequently co-cited decisions comes either from the constitutional principles underlying these sets of decisions, the substantive domains of the decisions, or both. Induction of a different sort followed the identification of coherent patches in this citation network. Having identified them, we sought to understand both the decisions and the citations between them in their historical, social, and political contexts. Beyond the line islands considered in Chapter 6, two we considered were technologically driven. One concerned railraids when rail was an emerging technology with great commercial and social implications. Another featured maritime law, first defined over centuries for travel on seas and oceans, and then adapted as internal waterways – rivers and lakes – were used in the USA, especially for commerce.

For the football network, our approach was completely deductive. Based on our reading of the literature regarding football player moves, we formulated an explicit set of hypotheses. We knew that players move in the hope of advancing their careers, while clubs recruited players with the intent of achieving greater success (or avoiding failure) on the field. Coupling the decisions of players and the decisions of clubs is a highly uncertain processes for reasons outlined in Chapter 7. We state 21 hypotheses in Chapter 7 and test them. The results of these tests are reported in Chapters 7 and 8. While some hypotheses were obvious, others were counter to conventional thinking about player movements. Many hypotheses passed muster, some failed miserably, and others, while sounding plausible, turned out to be untestable in the sense of there being both supporting and refuting evidence about them. Not surprisingly, regarding the failed hypotheses and the untestable hypotheses, conventional wisdom about football in England does tend to be supported by selective attention to the evidence.

In essence, we returned to induction for our analysis of the large US spatial network. Indeed, we state no hypotheses. Our intent was to combine two broad– seemingly incompatible– approaches to mapping spatial diversity. The resulting compromise led to results sitting between these two broad approaches. Of course, this does not have surprise value because we were more attentive to *both* network geography (adjacency in space) and also appropriate statistical data. As a result, our results provide the foundations for a deeper characterization of the spatial distribution of diversity in the USA.

The final chapter provides a partial summary of the results provided in Chapters 4–9, together with commentary on the utility of the methods used throughout this book. Also proposed in the final chapter are some suggestions for further work. Despite all that is accomplished here, one salutary implication is that much more needs to be done. Pursuing these issues has immense appeal.

2

Foundations of methods for large networks

Many new methods are defined and mobilized in this book to study the large networks outlined in Chapter 1. Their presentation is divided into two chapters, even though there is no rigid divide between them. We present basic network analytic ideas and methods in this chapter: they serve as foundations for the methods we use in Chapters 4–10. Methods used in these chapters are presented fully in Chapter 3. Some of the foundational methods presented in this chapter can be used for large networks. As Chapters 4–6 are explicitly citation networks, the break between this chapter and the next is marked by a sustained consideration of these networks in Chapter 3. Some methods specific to single chapters are provided there. The Chapters 2 and 3 are mostly based on lectures given by Vladimir Batagelj at the ECPR summer school, Ljubljana, 2006–2013.

Nearly all of the network analyses we present were performed in Pajek (Batagelj and Mrvar 1998; de Nooy et al. 2012), a program suite[1] designed to handle large networks very efficiently. Given our extensive use of Pajek, we include Pajek commands, where appropriate, both here and in other chapters. As R was also used extensively, we provide some R commands.[2]

2.1 Networks

A *network* is based on two sets: a set of *vertices* (nodes), representing the selected *units* or *actors* and a set of *lines* (links) representing *ties* between units. They determine a *graph* (defined formally below). If a line is *directed*, it is an *arc*. An *undirected* line is an *edge*.

---

[1] Pajek is expanded and updated regularly.
[2] Some secondary analyses were completed by using Stata. We do not include those commands; they can be found in the Stata manuals.

Understanding Large Temporal Networks and Spatial Networks: Exploration, Pattern Searching, Visualization and Network Evolution, First Edition. Vladimir Batagelj, Patrick Doreian, Anuška Ferligoj and Nataša Kejžar.
© 2014 John Wiley & Sons, Ltd. Published 2014 by John Wiley & Sons, Ltd.

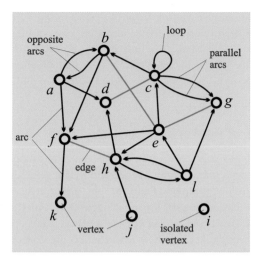

Figure 2.1   Some basic terms for graphs.

A graph describes the structure of a network. When additional data about vertices and lines are obtained via measurement, they determine their *properties* (attributes). Properties can also be computed from the graph or other available properties. For example, names/labels, vertex/line types, vertex/line values, and temporal information for the graph are properties. A network is the graph and its properties. Informally, a **Network = Graph + Data**. We make extensive use of both items in the networks studied in this book.

Formally, a *network* $\mathcal{N} = (\mathcal{V}, \mathcal{L}, \mathcal{P}, \mathcal{W})$ consists of:

- a *graph* $\mathcal{G} = (\mathcal{V}, \mathcal{L})$ where $\mathcal{V}$ is the set of vertices, $\mathcal{A}$ is the set of arcs, $\mathcal{E}$ is the set of edges, with $\mathcal{L} = \mathcal{E} \cup \mathcal{A}$ the set of lines, $\mathcal{E} \cap \mathcal{A} = \emptyset$. We denote the number of vertices by $n = |\mathcal{V}|$ and the number of lines by $m = |\mathcal{L}|$.

- $\mathcal{P}$ is a set of *vertex value functions*, called properties, mapping vertices to a domain A_p: $p : \mathcal{V} \to A_p$.

- $\mathcal{W}$ is a set of *line value functions*, called weights, mapping lines to a domain B_w: $w : \mathcal{L} \to B_w$.

The sets A_p and B_w are the sets in which properties, p, and weights, w, take their values. A graph in which $\mathcal{A} = \emptyset$ is called an *undirected* graph. If $\mathcal{E} = \emptyset$ it is a *directed* graph.

Some basic definitions follow. They are illustrated in Figure 2.1 where vertices are labeled *a* through *l*.

An arc leading from its *initial* vertex, *u*, to its *terminal* vertex, *v*, is denoted by (u, v). An example in Figure 2.1 is (a, f). Similarly, the edge linking its *end-vertices*, *u* and *v*, is denoted by $(u : v)$ (or equivalently by $(v : u)$). In Figure 2.1, $(f : h)$ is an edge. Both the initial and terminal vertices of arcs are their end-vertices. An end-vertex of an edge is both its initial and terminal vertex. A line having both end-vertices the same is called a *loop*. There is a loop on vertex *c* in Figure 2.1. A vertex that is not an end-vertex of any line is called an *isolated* vertex (or an *isolate*, a more frequently used term). Vertex *i* is an isolate in Figure 2.1.

In a network, arcs (or edges) leading from the same initial vertex to the same terminal vertex can exist. We call them *parallel* lines. The arcs (c, g) in Figure 2.1 are parallel arcs. To distinguish parallel lines we usually use indices: $(c, g)_q$ – the q-th arc from c to g, or label the lines: $l(c, g)$ – the arc from c to g with label l. A pair of *opposite* arcs links a pair of vertices in opposite directions. See the arcs between vertices a and b (and between h and l) in Figure 2.1.

2.1.1   Descriptions of networks

The graph in the left upper panel of Figure 2.2 repeats the graph in Figure 2.1. The left lower panel gives its formal mathematical description. The right side of this figure has the transcription of the graph description into Pajek's input format. It has descriptions of the sets of vertices, arcs, and edges. Describing the set of vertices starts with the keyword `*Vertices` followed by the number of vertices, n. The following (text) lines contain the vertex number (index) followed by the vertex label. The vertex numbers are consecutive integers from 1 to n. The description of the set of arcs starts with the keyword `*Arcs`. The following lines, one for each arc, have pairs of integers: the first is the index of an arc's initial vertex, the second is the arc's terminal vertex index. Similarly, the description of the set of edges starts with the keyword `*Edges`. In each of the following lines, with each edge on a separate line, there is a pair of integers: the indices of its end-vertices. The descriptions of lines can also contain a third number (integer or real) representing the weight of the line. All Pajek files have the extension 'net', e.g. `example.net`.

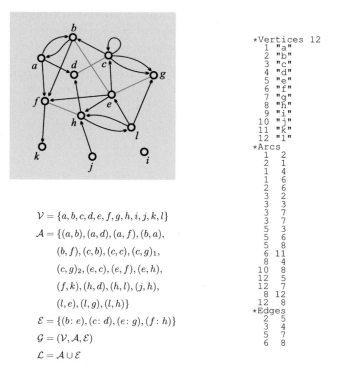

$$\mathcal{V} = \{a, b, c, d, e, f, g, h, i, j, k, l\}$$
$$\mathcal{A} = \{(a, b), (a, d), (a, f), (b, a),$$
$$(b, f), (c, b), (c, c), (c, g)_1,$$
$$(c, g)_2, (e, c), (e, f), (e, h),$$
$$(f, k), (h, d), (h, l), (j, h),$$
$$(l, e), (l, g), (l, h)\}$$
$$\mathcal{E} = \{(b : e), (c : d), (e : g), (f : h)\}$$
$$\mathcal{G} = (\mathcal{V}, \mathcal{A}, \mathcal{E})$$
$$\mathcal{L} = \mathcal{A} \cup \mathcal{E}$$

```
*Vertices 12
   1 "a"
   2 "b"
   3 "c"
   4 "d"
   5 "e"
   6 "f"
   7 "g"
   8 "h"
   9 "i"
  10 "j"
  11 "k"
  12 "l"
*Arcs
   1   2
   2   1
   1   4
   1   6
   2   6
   3   2
   3   3
   3   7
   3   7
   5   3
   5   6
   5   8
   6  11
   8   4
  10   8
  12   5
  12   7
   8  12
  12   8
*Edges
   2   5
   3   4
   5   7
   6   8
```

Figure 2.2   Graph descriptions: a picture, its mathematical statement, and Pajek's input file.

v	deg	indeg	outdeg
a	4	1	3
b	5	3	3
c	6	3	4
d	3	3	1
e	6	3	5
f	5	4	2
g	4	4	1
h	6	4	3
i	0	0	0
j	1	0	1
k	1	1	0
l	4	1	3

```
*vertices 12
1
3
3
3
3
4
4
4
0
0
1
1
```

Figure 2.3   Degree distributions for the graph in Figure 2.2.

2.1.2   Degrees

Examples of numerical properties of vertices are degree measures. The *degree* of vertex v, $\deg(v)$, counts the number of lines with v as an end-vertex. The *indegree* of vertex v, $\mathrm{indeg}(v)$, counts the number of lines with v as a terminal vertex. The *outdegree* of vertex v, $\mathrm{outdeg}(v)$, counts the number of lines with v as an initial vertex. Three degree tables for our example graph from Figure 2.2 are presented[3] on the left of Figure 2.3. On the right is a Pajek file listing indegrees.

Other types of degrees can be defined when needed. For example, restricting the indegree to arcs, $\mathrm{indeg}_A(v)$, counts the number of arcs with v as a terminal vertex.

Since each initial vertex is counted once on each arc and twice on each edge,

$$\sum_{v \in \mathcal{V}} \mathrm{indeg}(v) = \sum_{v \in \mathcal{V}} \mathrm{outdeg}(v) = |\mathcal{A}| + 2|\mathcal{E}|,$$

and similarly

$$\sum_{v \in \mathcal{V}} \deg(v) = 2|\mathcal{L}| - |\mathcal{E}_0|,$$

where $\mathcal{E}_0$ is the set of undirected loops.

In a graph $\mathcal{G} = (\mathcal{V}, \mathcal{L})$, a vertex v is called an *initial* vertex of $\mathcal{G}$ iff[4] $\mathrm{indeg}(v) = 0$. It is a *terminal* vertex of $\mathcal{G}$ iff $\mathrm{outdeg}(v) = 0$.

2.1.3   Descriptions of properties

Vertex properties are represented in Pajek with one of three data objects depending on their measurement scales. Numerical properties are represented with *vectors*, ordinal or nominal properties by *partitions* and linear orderings by *permutations*. A *label* can also be attached to each vertex in the network description file.

---

[3] For this network, $n = 12$, $m = 23$, $\mathrm{indeg}(e) = 3$, $\mathrm{outdeg}(e) = 5$, and $\deg(e) = 6$.
[4] We use the standard convention of using 'iff' for 'if and only if.'

All three data objects have the same Pajek file structure:

```
*vertices n
   v₁
...
   vᵢ
...
   vₙ
```

A **VEC**tor file has *numeric* data for *all* vertices where $v_i \in \mathbb{R}$: the property has value v_i on vertex i;

A **CLU**stering file has a partition of vertices representing *nominal* or *ordinal* data about vertices where $v_i \in \mathbb{N}$: vertex i belongs to the cluster v_i;

A **PER**mutation file has an ordering of vertices where $v_i \in \mathbb{N}$: vertex i is at the v_i-th position.

Vector files have the extension 'vec' (e.g. example.vec), clustering files have the extension 'clu' (e.g. example.clu), and permutation files have the extension 'per' (e.g. example.per). The Pajek file on the right of Figure 2.3 is a vector file. It could also be a clustering file if the analyst wanted to cluster vertices according to their degrees.

2.1.4 Visualizations of properties

Any numerical property can be displayed in a picture of a network. A vertex can be shown by its *size* (width and height) and by its *coordinates* (*x*, *y*, *z*). A nominal property can be shown as a *color* or a *shape* or by its *label* (content, size, and color).

We can assign numerical values to links in Pajek. A link can be displayed as a *value*, its *thickness*, or by a *gray level*. Nominal vales can be assigned as a *label*, a *color* (and also as a *line pattern*,[5] e.g. solid, dotted, straight, etc.).

2.2 Types of networks

In addition to ordinary (directed, undirected, mixed) networks some extended types of networks are also useful. We use them extensively in this book:

- *temporal networks* (or dynamic networks) which change over time;

- *multirelational networks* having different relations (represented by distinct sets of lines) over the same set of vertices;

- *two-mode networks* with lines between two disjoint sets of vertices (of different types); and

---

[5] See the Pajek manual (Batagelj and Mrvar 1996–2013), Section 5.3.

- specialized networks, for example representation of genealogies as *p-graphs*, *Petri's nets*, etc.

The network (input) file formats must provide the means for representing all these types of networks for subsequent analyses.

2.2.1 Temporal networks

In a temporal network, the presence and activity of vertices and lines can change through time. Pajek supports two types of descriptions of temporal networks based on *presence* and on *events*. Here, we describe only an approach to capturing the presence of vertices and lines.

A *temporal network*

$$\mathcal{N}_T = (\mathcal{V}, \mathcal{L}, \mathcal{P}, \mathcal{W}, \mathcal{T})$$

is obtained by attaching the *time*, $\mathcal{T}$, to an ordinary network, where $\mathcal{T}$ is a set of *time points*, $t \in \mathcal{T}$.

Given a temporal network where vertices $v \in \mathcal{V}$ and lines $l \in \mathcal{L}$ are not necessarily present or active in all time points, a way of handling this is necessary. Let $T(v)$, $T \in \mathcal{P}$, be the activity set of time points for vertex v, and $T(l)$, $T \in \mathcal{W}$, the activity set of time points for line l. The following *consistency condition* is imposed: If a line $l(u, v)$ is active in time point t then its end-vertices u and v should be also active in time t. Formally we express this by

$$T(l(u, v)) \subseteq T(u) \cap T(v).$$

2.2.1.1 Some examples of temporal networks

Citation networks (WoS)
    $\mathcal{V} = \{ \text{works} \}$
    $\mathcal{L} = \{(u, v) : u \text{ cites } v\}$
    $\mathcal{T} = \{\text{dates (years)}\}$
    $T(v) = [\text{publication date (year) of } v, *]$
    $T(u, v) = [\text{publication date (year) of } u, *, *]$

Project collaboration networks (EU Research Projects/CORDIS)
    $\mathcal{V} = \{ \text{institutions} \}$
    $\mathcal{L} = \{(u, v) : u \text{ and } v \text{ work on a joint project}\}$
    $\mathcal{T} = \{\text{dates}\}$
    $T(v) = \mathcal{T}$
    $T(u, v) = \{[a, b] : \text{exists a project } P \text{ such that } u \text{ and } v \text{ are partners on } P; a \text{ is the start and } b \text{ is the finish date of } P\}$

Genealogies (GEDCOMs)
    $\mathcal{V} = \{ \text{people} \}$
    $\mathcal{L} = \mathcal{L}_1 \cup \mathcal{L}_2$
    $\mathcal{L}_1 = \{(u, v) : u \text{ is a parent } v\}$
    $\mathcal{L}_2 = \{(u, v) : u \text{ and } v \text{ are married}\}$

$\mathcal{T} = \{\text{days}\}$

name function $v \in \mathcal{P}$: $v(v) = $ name of person v, $v \in \mathcal{V}$

gender function $g \in \mathcal{P}$: $g : \mathcal{V} \rightarrow \{M, F\}$

$T(v) = [\text{birth}(v), \text{death}(v)]$

for $(u, v) \in \mathcal{L}_1$: $T(u, v) = [\text{birth}(v), \min(\text{death}(u), \text{death}(v))]$

for $(u, v) \in \mathcal{L}_2$: $T(u, v) = [\text{marriage}(u, v), \min(\text{death}(u), \text{death}(v), \text{divorce}(u, v))]$ [6]

We denote a network consisting of lines and vertices active in time, $t \in \mathcal{T}$, by $\mathcal{N}(t)$, and call it the (network) *time slice* or *footprint* in t. Let $\mathcal{T}' \subset \mathcal{T}$ (for example, a time interval). The notion of a time slice is extended to $\mathcal{T}'$ by

$$\mathcal{N}(\mathcal{T}') = \bigcup_{t \in \mathcal{T}'} \mathcal{N}(t).$$

The time $\mathcal{T}$ is usually either a subset of integers, $\mathcal{T} \subseteq \mathbb{Z}$, or a subset of reals, $\mathcal{T} \subseteq \mathbb{R}$. $T(v)$ and $T(l)$ are usually described as a sequence of intervals.

In Pajek, the time set, T, is discrete and consists of a subset of natural numbers $T \subset \mathbb{N}$. Its interpretation is left to the user. In Pajek's input format the data about the times when an element is present (active) are given in the continuation of the line describing the element inside the brackets [and]. Time periods are separated by commas, ,. Continuous time periods between a starting time, s, to the ending time, e, can be written as s-e. Pajek uses the symbol, *, for 'infinity' or, in most practical situations, the last time point for a dataset.

Consider a simple example with only three vertices:

```
*Vertices 3
1 "a" [5-10,12-14]
2 "b" [1-3,7]
3 "c" [4-*]
*Edges
1 2 1 [7]
1 3 1 [6-8]
```

In this example, the vertex a is present in time points 5, 6, 7, 8, 9, 10, 12, 13, and 14. Vertex b is present at time points 1, 2, 3, and 7. The only time both of these vertices are present is the seventh time point. Vertex c is present from the fourth time point and thereafter. An edge exists between a and b at time point 7 only. The edge $(1:3)$ is present in time points 6, 7, 8.

To get time slices in Pajek the relevant command is:

Network/Temporal Network/Generate in time

The generating in time operation creates a sequence of temporal networks for subsequent study. These networks can range from small to large in size.[7] Specific methods for temporal networks are considered more fully in Chapter 3.

At the ECPR Summer School, Ljubljana, 2007 we obtained from Roberto Franzosi the data on temporal network describing the violent actions among political actors in Italy in the period January 1919 to December 1922. He collected the data from Italian newspapers published in that period.

---

[6] Other (basic) kinship relations $\mathcal{L}_i$ can be included.

[7] A detailed specification of network sizes is presented in Section 2.3.

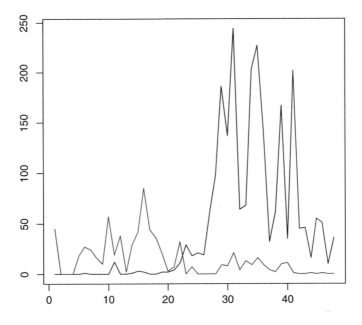

Figure 2.4   Police (blue) and Fascists (black) activity.

In Figure 2.4 we present the diagram representing the activity of Police (blue) and Fascists (black) obtained from time slices (by months). We see that after November 1920 Fascists took over the role of Police in suppressing the protests.

Figure 2.5 displays the time slice for December 1922.

2.2.2 Multirelational networks

A *multirelational network* is denoted by

$$\mathcal{N}_T = (\mathcal{V}, (\mathcal{L}_1, \mathcal{L}_2, \dots, \mathcal{L}_k), \mathcal{P}, \mathcal{W}).$$

It contains different relations (sets of lines), $\mathcal{L}_i$, $1 \leq i \leq k$, over the same set of vertices. Also, the weights from $\mathcal{W}$ are defined for different relations or their unions.

Examples of multirelational networks include: transportation systems in a city (stations/stops and lines[8]); networks of words in statements (e.g. WordNet with words, semantic relations: synonymy, antonymy, hyponymy, meronymy, etc.); the KEDS network (described briefly in Chapter 1) where actors are states and organizations and relations between them: visiting, asking for information, warning, expelling person(s), etc.

An often used way for obtaining networks is by *computer-assisted text analysis* (CaTA). Popping (2000) distinguishes three main approaches to CaTA: *thematic* TA, *semantic* TA, and *network* TA. The *terms* considered in TA are collected into a *dictionary* (which can

---

[8] On the London Underground, all successive pairs stations between Baker Street and Liverpool Street have edges corresponding to the Circle, the Metropolitan, and the Hammersmith and City lines.

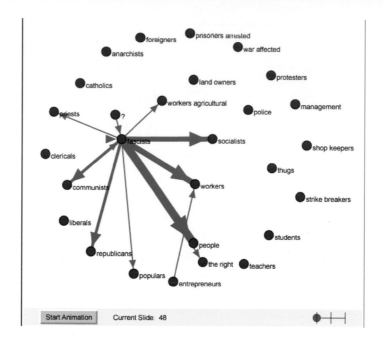

Figure 2.5   Time slice December 1922.

be fixed in advance, or built dynamically). The main two problems with terms are the *equivalence* of different words (synonyms) representing the same term and the *ambiguity* with the same word (homonym) representing different terms. These problems imply a need for the *coding* – transformation of raw text data into formal *descriptions* – to be done often manually or semiautomatically. The *unit*s of TA are usually clauses, statements, paragraphs, news, messages, etc.

Hitherto, both thematic and semantic TAs mainly used statistical methods for analyzing coded data. In thematic TA, the units are organized in a rectangular matrix, *Text units* × *Concepts*, which is essentially a two-mode network.[9] In semantic TA the units (usually clauses) are encoded according to the S-V-O (*Subject-Verb-Object*) model or its improvements.[10] A generic encoding form is:

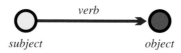

The encoded text can be directly considered as a multirelational network with *Subjects* ∪ *Objects* as vertices and *Verbs* as relations.

---

[9] Examples of the available software include: M.M. Miller: VBPro, H. Klein: Text Analysis/ TextQuest.
[10] Examples include: Roberto Franzosi; *KEDS* and *Tabari*; RDF triples in semantic web, SPARQL.

In Pajek, a line can be assigned to a relation in two ways:

- By adding to a keyword for descriptions of lines (*arcs, *edges, *arcslist, *edgeslist, *matrix) the number of a relation followed by its name:

  ```
  *arcslist :3 "sent a letter to"
  ```

  All lines controlled by this keyword belong to the specified relation.

- Any line controlled by *arcs or *edges can be assigned to a selected relation by starting its description by the number of this relation (followed by a colon, :,). For example:

  ```
  3: 47 14 5
  ```

  states: a line with end-vertices 47 and 14 and weight 5 belongs to relation 3.

2.2.2.1 An example of a multirelational temporal network

The KEDS project, described in Chapter 1, includes political event data focused on the Middle East, the Balkans, and West Africa. Figure 2.6 shows how KEDS data can be transformed into Pajek's temporal multirelational network for an example concerning the Balkans. At the bottom right side of Figure 2.6 in each line the corresponding text encoding on KEDS file is presented. It has 325 vertices representing different countries, their institutions and

```
% Recoded by WEISmonths, Sun Nov 28 21:57:00 2004
% from http://www.ku.edu/~keds/data.dir/balk.html
*vertices 325
1 "AFG" [1-*]
2 "AFR" [1-*]
3 "ALB" [1-*]
4 "ALBMED" [1-*]
5 "ALG" [1-*]
...
318 "YUGGOV" [1-*]
319 "YUGMAC" [1-*]
320 "YUGMED" [1-*]
321 "YUGMTN" [1-*]
322 "YUGSER" [1-*]
323 "ZAI" [1-*]
324 "ZAM" [1-*]
325 "ZIM" [1-*]
*arcs :0 "*** ABANDONED"
*arcs :10 "YIELD"
*arcs :11 "SURRENDER"
*arcs :12 "RETREAT"
...
*arcs :223 "MIL ENGAGEMENT"
*arcs :224 "RIOT"
*arcs :225 "ASSASSINATE TORTURE"
*arcs
224: 314 153 1 [4]      890402 YUG    KSV     224 (RIOT)           RIOT-TORN
212: 314 83 1 [4]       890404 YUG    ETHALB  212 (ARREST PERSON) ETHALB JAILED
224: 3 83 1 [4]         890407 ALB    ETHALB  224 (RIOT)           RIOTS
123: 83 153 1 [4]       890408 ETHALB KSV     123 (INVESTIGATE)    PROBING
...
42: 105 63 1 [175]      030731 GER    CYP     042 (ENDORSE)        GAVE SUPPORT
212: 295 35 1 [175]     030731 UNWCT  BOSSER  212 (ARREST PERSON) SENTENCED
43: 306 87 1 [175]      030731 VAT    EUR     043 (RALLY)          RALLIED
13: 295 35 1 [175]      030731 UNWCT  BOSSER  013 (RETRACT)        CLEARED
121: 295 22 1 [175]     030731 UNWCT  BAL     121 (CRITICIZE)      CHARGES
122: 246 295 1 [175]    030731 SER    UNWCT   122 (DENIGRATE)      TESTIFIED
121: 35 295 1 [175]     030731 BOSSER UNWCT   121 (CRITICIZE)      ACCUSED
```

Figure 2.6 A fragment of the Balkan KEDS network.

international organizations. The first six and last eight vertices are listed in Figure 2.6. Each vertex is assumed to be active all the time [1-*]. The vertices are followed by a list of *arcs keywords linking the relation number to the corresponding label. The events data follow the simple *arcs keyword and have the form

$r: u\ v\ 1\ [t]$

In words: the arc (u, v) with weight 1 belonging to relation r is active in time point t.

2.2.3   Two-mode networks

Examples of such networks include: persons belonging to societies with years of membership as weights, consumers buying goods with quantity or value as weights, Supreme Court justices 'voting' on decisions with agreeing or dissenting as signed weights, authors publishing in journals with publication counts as weights.

Formally, in a *two-mode* network $\mathcal{N} = ((\mathcal{U}, \mathcal{V}), \mathcal{L}, \mathcal{P}, \mathcal{W})$, the set of vertices consists of two disjoint sets of vertices $\mathcal{U}$ and $\mathcal{V}$, and all the lines from $\mathcal{L}$ have one end-vertex in $\mathcal{U}$ and the other in $\mathcal{V}$. Often also a *weight* $w : \mathcal{L} \rightarrow \mathbb{R} \in \mathcal{W}$ is given; if not, we assume $w(u, v) = 1$ for all $(u, v) \in \mathcal{L}$. A two-mode network can be described also by a rectangular matrix $\mathbf{A} = [a_{uv}]_{\mathcal{U} \times \mathcal{V}}$.

$$a_{uv} = \begin{cases} w_{uv} & (u, v) \in \mathcal{L} \\ 0 & \text{otherwise} \end{cases}$$

In Pajek, a two-mode network is announced by *vertices n $n_{\mathcal{U}}$ with the keyword *vertices followed by two numbers: the total number of vertices[11] $n = |\mathcal{U}| + |\mathcal{V}|$ and the number of vertices in the first set $n_{\mathcal{U}} = |\mathcal{U}|$. It is followed by the list of all vertices from the first set, followed by the list of all vertices from the second set.

An example of a two-mode network is shown in Figure 2.7. The data came from the *Deep South* study of Davis and Gardner (1941). The figure is a redrawn version of Figure 8.3 from Doreian et al. (2005). Part of the Pajek file for this network is in Figure 2.8. There were $n_{\mathcal{U}} = 18$ women (shown as ellipses) who attended $n_{\mathcal{V}} = 14$ events (shown as boxes); $n = 32$.

Three numbers following the vertex label are its x, y, z coordinates.[12] The tags ellipse and box determine the shapes of the vertices and are valid until changed by another tag. Figure 2.8 shows a partition of the women into two subsets and the events into three subsets according to which women tended to attend which events.

2.3   Large networks

The size of a network/graph is expressed by two numbers: the number of vertices $n = |\mathcal{V}|$ and the number of lines $m = |\mathcal{L}|$. In a simple undirected graph (with neither parallel edges

---

[11] The use of || is the conventional shorthand for indicating 'the size of' a set.

[12] The coordinates belong to the interval [0, 1]. Most of the figures we present are two-dimensional and use only x and y as coordinates. For such networks, using $z = 0.5$ for all vertices is a convention as shown in Figure 2.8.

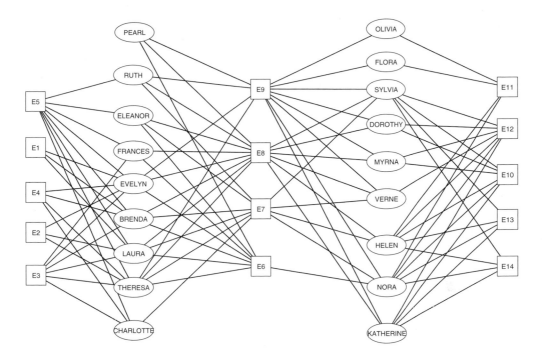

Figure 2.7   A two-mode network (the *Deep South* Network).

```
*Vertices 32 18
    1 "EVELYN"        0.2254    0.4723    0.5000 ellipse
    2 "LAURA"         0.2254    0.6524    0.5000
  ....
   17 "OLIVIA"        0.7104    0.0705    0.5000
   18 "FLORA"         0.7104    0.1574    0.5000
   19 "E1"            0.0400    0.3778    0.5000 box
   20 "E2"            0.0400    0.5970    0.5000
   21 "E3"            0.0400    0.7078    0.5000
  ....
   31 "E13"           0.9400    0.5668    0.5000
   32 "E14"           0.9400    0.6877    0.5000
*Edges
    1   19 1
    1   20 1
  ....
   18   27 1
   18   29 1
```

Figure 2.8   A fragment of the Pajek file for Figure 2.7.

nor loops) $m \leq \frac{1}{2}n(n-1)$; and in a simple directed graph (with no parallel arcs) $m \leq n^2$. For a simple two-mode network, $n = |\mathcal{U}| + |\mathcal{V}|$ and $m \leq |\mathcal{U}||\mathcal{V}|$.

2.3.1 Small and middle sized networks

Small networks (having some tens of vertices) can be represented easily by pictures and analyzed by many algorithms included in programs such as Pajek, UCInet, and NetMiner. Also, *middle-size* networks (having some hundreds of vertices) can still be represented by pictures although drawing many of them for visual clarity is not straightforward. However, some analytical procedures are no longer practical for middle size networks.[13]

Until 1990 most networks were small: their data were collected by researchers using surveys, observations, archival records, etc. The advances in IT in the 1990s allowed the creation of networks from data already available in computer(s). *Large* networks became a reality. Large networks are too big to be displayed in all their details: special algorithms are needed for their analysis. Pajek is a program developed for this purpose.

2.3.2 Large networks

Large networks have from several thousands to many millions of vertices. The upper bound for 'large' is the maximum size of a network that can be stored in a computer's memory. Any network larger than this is a *huge* network. Of course, the notion of what is large for a network is technology dependent. The 32-bit computers had the upper limit of 4 GiB to the memory size. The new 64-bit architecture increased this limit: 64-bit operating systems can support up to a theoretical 16 EiB[14] of RAM. Windows 8 supports up to 512 GiB and Windows Server 2012 up to 4 TiB of RAM. A Dell Precision T5600 workstation can have up to 128 GiB RAM. The real problem is physical support/realization of large computer memories.

Listings of networks, for example those in SNAP and KONECT (discussed in Chapter 1), can be ordered by their average degrees $\bar{d}$ where

$$\bar{d} = \frac{1}{n} \sum_{v \in V} \deg(v) = \frac{2m}{n}.$$

Historically, most networks studied by social network analysts were small with low average degree, a continuing focus today. Most real-life large networks are *sparse* – the number of vertices and lines are of the same order. This property is known also as Dunbar's number (Dunbar 1992). The basic idea is simple. Each actor (represented by a vertex) has to spend a certain amount of 'energy' (e.g. time) to maintain each link to selected other actors. The energy available of a specific relation is constrained by the limited energy at an actor's disposal. This implies that the number of links for actors will be limited. Additionally, remembering networks is constrained. In human networks, Dunbar's number is thought to be between 100 and 150.

Figure 2.9 shows the distribution of $\bar{d}$ for the SNAPS and KONECT collections of networks. For the huge majority of these networks, $\bar{d}$ is below 50. All these networks are sparse. Network sparsity is fundamental to the design of fast algorithms for large networks.

2.3.3 Complexity of algorithms

Informally, when some algorithms take too long they are impractical. This is especially true for large networks. As some new algorithms build on old algorithms created for

---

[13] In the drop-down menus of Pajek, these operations are marked by an asterisk. A similar convention is used in UCInet.

[14] kibi Ki = $2^{10} \approx 10^3$ = K kilo, mebi Mi = $2^{20} \approx 10^6$ = M mega, gibi Gi = $2^{30} \approx 10^9$ = G giga, tebi Ti = $2^{40} \approx 10^{12}$ = T tera, pebi Pi = $2^{50} \approx 10^{15}$ = P peta, exbi Ei = $2^{60} \approx 10^{18}$ = E exa.

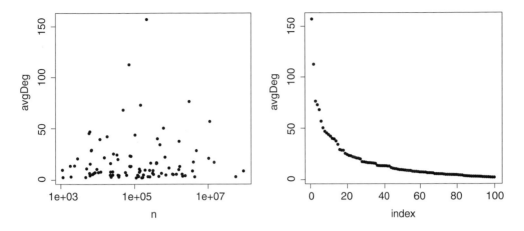

Figure 2.9 Average degrees of the SNAP and KONECT networks.

small networks it is necessary to consider practicality within the rubric of algorithmic complexity.

The time complexities of some typical algorithms are listed in Table 2.1. They were computed for imaginary algorithms assuming that they were executed on a computer capable of performing 10^9 basic algorithmic steps per second (an optimistic assumption for current PCs). Impractical times are marked in maroon and totally impractical running times are marked in red. Note the running times in the bottom row of Table 2.1 for which matrix multiplication using the formal definition is an example. Using such an algorithm would rule out the matrix multiplication used throughout this book. For interactive analysis of large graphs even quadratic algorithms, $O(n^2)$, are too slow. The analysis of large networks has to be based on *subquadratic* algorithms with a complexity lower than $O(n^2)$. Some of these algorithms are integral to Pajek. The design criteria for Pajek are outlined in the next section.

Tools designed to handle large networks can be used also for small networks in the seamless presentation of methods within Pajek: there was no point in dividing methods by their complexity. However, there are some methods implemented in Pajek in order to include

Table 2.1 Algorithmic complexities.

	$T(n)$	1000	10000	100000	1000000	10000000
LinAlg	$O(n)$	0.00 s	0.015 s	0.17 s	2.22 s	22.2 s
LinLogAlg	$O(n \log n)$	0.00 s	0.06 s	0.98 s	14.4 s	2.8 m
LinSqrtAlg	$O(n\sqrt{n})$	0.01 s	0.32 s	10.0 s	5.27 m	2.78 h
QuadAlg	$O(n^2)$	0.07 s	7.50 s	12.5 m	20.8 h	86.8 d
CubAlg	$O(n^3)$	0.10 s	1.67 m	1.16 d	3.17 y	3.17 ky

Some examples of algorithms in these categories are: LinAlg, finding elements of maximal value in an unordered list of *n* numbers; LinLogAlg, fast algorithm for sorting a list of *n* numbers; QuadAlg, bubble sort; CubAlg, using the formal definition to compute the product of two square matrices of order *n*.

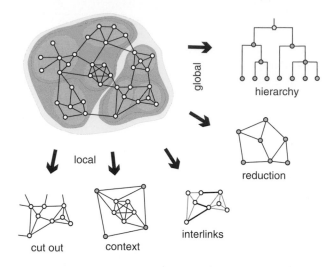

Figure 2.10   Decomposition and parts of network.

some traditional problems. These are marked by asterisks when they cannot be performed for large networks, for example generalized blockmodeling (Doreian et al. 2005).

2.4   Strategies for analyzing large networks

When analyzing a *large* network, we cannot display *all* its details in a *single* pass. Some other approach is required. There are two broad strategies for examining large networks. One is using statistical methods to generate summary descriptions of networks. The second is identifying smaller *interesting* subnetworks for which more sophisticated methods can be mobilized. While we employ both strategies throughout this book, more attention is given to the second. Dealing with large structures in this fashion is a *divide and conquer* strategy (known well to the Romans). A large structure is divided into smaller parts. Further, if an obtained part is not small enough, it can be divided into subparts. This process can be continued until we get subparts small enough for using more sophisticated methods. The descriptions of these smaller networks are useful in their own right. More importantly, they can be combined to obtain fuller understandings of the structure(s) of networks.

Figure 2.10 illustrates some of the options under the divide and conquer strategy. A network is shown in the upper left quadrant from which parts can be extracted. One option is to 'cut out' a part of a network, while noting that there are ties to the rest of the network. Another is to extract a part along with its immediate context. Yet another is extracting *parts* of networks to examine closely the links between them. The vertices of Figure 2.10 have been drawn in areas marked by different colors. The finest grained partition has vertices in the mustard yellow area. When the vertices are merged[15] into a single vertex, a reduction of the network is obtained. Using the areas marked in gray provides a different reduction. The result of a decomposition process at a selected level can be described with a partition of vertices or lines and the complete decomposition as a hierarchy.

---

[15] We call this 'shrinking' the network, and define the operation in Section 2.6.2.

The above concerns led to formulating the main goals in the design of Pajek:

- to support abstraction by (recursive) *decomposition* of a large network into several smaller networks that can be treated further using more sophisticated methods;
- to provide the user with some powerful *visualization* tools; and
- to implement a selection of efficient *subquadratic* algorithms for analysis of large networks.

As illustrated in Figure 2.10, Pajek can be used to *find* clusters (components, 'important' vertices, their neighborhoods, cores, etc.) in a network, *extract* vertices belonging to clusters and *show* them separately with parts of the context (a local view), *shrink* vertices from the same cluster and show relations among the shrunken clusters (a global view).

Within Pajek, the broad strategy of divide and conquer creates a very flexible framework for analyzing networks within which many different types of tools can be mobilized and their results combined. See also Borgatti, Everett and Johnson (2013) regarding this when using UCInet.

2.5   Statistical network measures

There are two broad types of summary network statistics: global and local. *Global* computed properties (e.g. numbers of vertices, edges/arcs, components; maximum core number, ...) are obtained by using Pajek commands or can be seen using the Info option. Applying *repetitive* commands such measures can be stored in vectors. *Local* properties (e.g. degrees, cores, indices (betweenness, hub scores, authority scores, ...) are computed by using Pajek's commands and stored as vectors or partitions. Information about their distributions is also obtained from the Info option.

To illustrate these ideas, we use an example from the EAT database described in Chapter 1. It is a network of word association as collected from subjects. The weights on the arcs are counts of pair wise word associations. These data were stored in a Pajek network file, eatSR.net. In Pajek, after reading eatSR.net, general information for the network is obtained by using the network Info button. The network has 23,219 vertices and 325,593 arcs (of which 564 are loops). The number of lines with value = 1 is 227,459, with the remaining 98,134 arcs having greater values.

Many networks have labeled vertices. In such networks, the *extreme elements* can have interesting value in terms of a selected property or a weight. Such a property is the degrees of vertices. The ten vertices with the largest degree in eatSR.net in ranked order are

Rank	Vertex	Value	Id
1	12720	1108	ME
2	12459	1074	MAN
3	8878	878	GOOD
4	18122	875	SEX
5	13793	803	NO
6	13181	799	MONEY
7	23136	732	YES
8	15080	723	PEOPLE
9	13948	720	NOTHING
10	22973	716	WORK

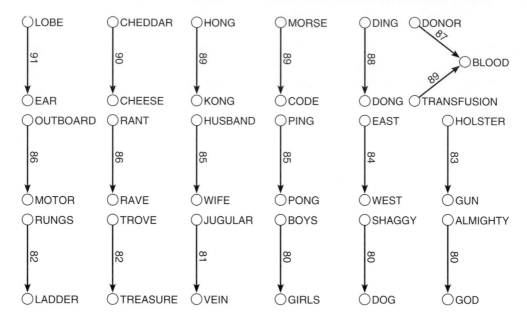

Figure 2.11   The most frequent associations in EAT.

The Pajek commands, equally useful for small networks, for doing this are:

```
Network/Create Vector/Centrality/Degree/All
Vector/Info  +10
```

Obtaining the weight distribution on arcs is done by selecting the option:

```
Network/Info/Line Values
```

When this was done for the EAT data, the values of arc weights ranged from 1 to 91. Specific arcs in any part of this range can be selected and examined. There are few lines having values above 80. To examine only those, all arcs with values below 80 were removed from the network along with the resulting isolated vertices. The Pajek commands for doing this are:

```
Network/Create New Network/Transform/Remove/Lines with Value/
  Lower than [80][Yes]
Network/Create Partition/Degree/All
Operations/Network+Partition/Extract Subnetwork [1-*]
```

The obtained arcs are displayed in Figure 2.11. None of the top ten most frequent words appear in any of the 19 arcs with the highest weights. (This changes when different thresholds are used.) In eatSR.net, the arcs lead from the Stimulus (a word given to a subject) to the Response (a word provided by a subject). Labeled vertices allow us also to *inspect* a selected vertex and explore its position/role in the network by examining its value in partitions, vectors, values of lines, etc. Such information is obtained by using the following command:

```
Network/Info/Vertex Label -> Vertex Number
```

2.5.1   Using Pajek and R together

Several *structural properties* of the vertices and lines in a network can be computed. Examining them more closely includes inspecting their distributions. In Pajek, statistical analyses are 'delegated' to statistical programs linked with Pajek[16] as tools by using its menu: `Tools`. The degrees of vertices determined in Pajek can be submitted to R with:

```
Network/Info/General
Network/Create Vector/Centrality/Degree/All
Tools/R/Send to R/Current Vector
```

Their distributions[17] can be plotted in R by using:

```
summary(v1)
t <- table(v1)
x<-as.numeric(names(t))
plot(x,t,log='xy',main='degree distribution',
  xlab='deg',ylab='freq')
```

The resulting picture can be saved with `File/Save` as in a selected format. Available formats include *.pdf or *.ps for LaTeX (and a Windows metafile format for inclusion in Word). Also, *.eps can be brought into Word.

The all-degree distribution for network `eatSR.net` is presented in Figure 2.12. The distribution is bi-modal indicating the presence of two 'generators'. Since the network is directed (having only arcs as lines), $\deg(v) = \mathrm{indeg}(v) + \mathrm{outdeg}(v)$, it is the combination of the input and ouput degree distributions (see Figure 2.13). The out-degree distribution is 'scale-free'-like, except for vertices with $\deg(v) = 1$. While having such distributions has visual appeal, estimated curves have greater utility.

2.5.2   Fitting distributions

When the shape of the empirical distribution is close to a well-known theoretical distribution, or it follows from a theory of some process, determining the parameters of the theoretical distribution best fitting the empirical data is useful.

To illustrate this, we examine the data about the year of publication of papers from the 2008 version of network on centrality literature[18] contained in the file `Year.clu`. The corresponding distribution of *papers by years* has a shape similar to the lognormal distribution.

The following code in R constructs the empirical distribution t and using nonlinear (weighted) least-squares procedure `nls` fits it with the *lognormal distribution*.

```
> setwd("C:/work/WoS/Central")
> years <- read.table(file="Year.clu",header=FALSE,skip=2)$V1
> t <- table(years)
> year <- as.integer(names(t))
```

---

[16] Pajek supports interaction with SPSS, as well as R and other external programs.

[17] Some care is required when doing this. For example, vertices of degree 0 are problematic if taking logarithms of degrees.

[18] This network is described in Appendix A.1.1 and used in Chapter 4.

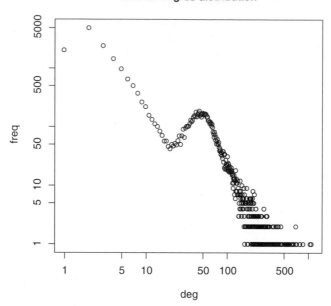

Figure 2.12   The all-degree distribution for the EAT data.

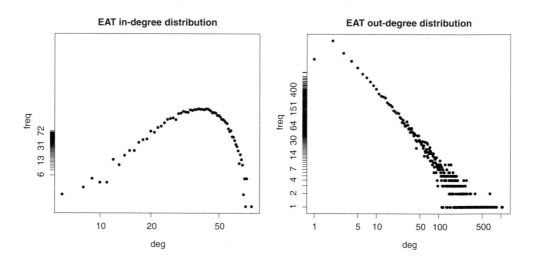

Figure 2.13   The indegree and outdegree distributions for the EAT data.

```
> freq <- as.vector(t[1950<=year & year<=2009])
> y <- 1950:2009
> plot(y,freq)
> model <- nls(freq~c*dlnorm(2010-y,a,b),start=list(c=350000,a=2,b=0.7))
> model
Nonlinear regression model
```

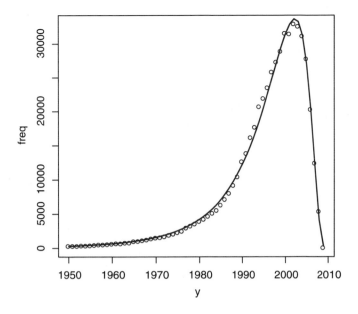

Figure 2.14   The distribution of papers by years for the 2008 centrality network.

```
model:   freq ~ c * dlnorm(2010 - y, a, b)
 data:   parent.frame()
        c          a           b
5.427e+05 2.491e+00 6.624e-01
 residual sum-of-squares: 20474181

Number of iterations to convergence: 7
Achieved convergence tolerance: 3.978e-06
> lines(y,predict(model,list(x=2010-y)),col='red')
```

Figure 2.14 shows that this distribution can be well approximated by the lognormal distribution. It can be well approximated also by the *generalized reciprocal power exponential curve* $c * (x + d)^{\frac{a}{b+x}}$.

2.6   Subnetworks

There are several ways of extracting subnetworks from a network. Some are presented in the following sections along with illustrations of their uses.

2.6.1   Clusters, clusterings, partitions, hierarchies

Some new concepts beyond those listed already are needed for describing decompositions of networks. A non-empty subset $C \subseteq \mathcal{V}$ is called a *cluster*. A non-empty set of clusters $\mathbf{C} = \{C_i\}$ forms a *clustering*. A clustering $\mathbf{C} = \{C_i\}$ is a *partition* iff

$$\cup \mathbf{C} = \bigcup_i C_i = \mathcal{V} \quad \text{and for all } i \text{ and } j \quad i \neq j \Rightarrow C_i \cap C_j = \emptyset.$$

In words, every unit is in a cluster and no unit belongs to more than one cluster.

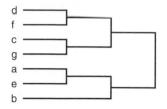

Figure 2.15   The dendrogram of a hierarchy.

A clustering $\mathbf{C} = \{C_i\}$ is a *hierarchy* iff for all i and j

$$C_i \cap C_j \in \{\emptyset, C_i, C_j\}.$$

In words, for any pair of clusters, C_i and C_j, either they are disjoint or one is contained in the other. A hierarchy $\mathbf{C} = \{C_i\}$ is *complete*, iff $\cup\mathbf{C} = \mathcal{V}$; and is *basic* if for all $v \in \cup\mathbf{C}$ also $\{v\} \in \mathbf{C}$. A simple example of these ideas is the following:

Vertex set: $\mathcal{V} = \{a, b, c, d, e, f, g\}$
Cluster / group / class: $C_2 = \{c, g\}$
Partition / clustering: $\mathbf{C} = \{\{a, b, e\}, \{c, g\}, \{d, f\}\}$
Hierarchy: $\mathbf{H} = \{\{a\}, \{b\}, \{c\}, \{d\}, \{e\}, \{f\}, \{g\}, \{a, e\}, \{c, g\}, \{d, f\}, \{a, b, e\},$
$\{c, d, f, g\}, \{a, b, c, d, e, f, g\}\}$

The hierarchy can be represented by the corresponding (clustering) tree or *dendrogram*. See Figure 2.15 for a dendrogram of the hierarchy $\mathbf{H}$.

2.6.2   Contractions of clusters

A *contraction* of a cluster C is called a graph, labeled $\mathcal{G}/C$, in which all vertices of the cluster C are replaced by a single new vertex, say c. More precisely:

$\mathcal{G}/C = (\mathcal{V}', \mathcal{L}')$, where $\mathcal{V}' = (\mathcal{V} \setminus C) \cup \{c\}$ and $\mathcal{L}'$ consists of lines from $\mathcal{L}$ that have both end-vertices in $\mathcal{V} \setminus C$. Also, it contains a 'star' with center c and: arcs (v, c), if $\exists p \in \mathcal{L}, u \in C : p(v, u)$; or arc (c, v), if $\exists p \in \mathcal{L}, u \in C : p(u, v)$. There is a loop (c, c) in c if $\exists p \in \mathcal{L}, u, v \in C : p(u, v)$.

For a network over a graph $\mathcal{G}$ it is necessary to specify how the values (weights) are determined in the shrunken part of the network and the values of the properties of vertex c. Usually, this is the sum or maximum/minimum of the original values.

Figure 2.16 shows a toy example of shrinking a network. At the top of the figure, there are two 'parts' where the vertices are marked by ellipses and boxes. The parts can be identified in a variety of ways. If ties are weighted, these weights could be used to identify the parts (clusters). Another option could be having more internal linkages than external linkages. Alternatively, they could be two sets of vertices distinguished by some vertex property, for example Democrat and Republican politicians in the US Congress for communication ties between them. The idea is to shrink each part to a single vertex to create a two-vertex shrunken network. To do this, a partition is required and the shrunken network is obtained in Pajek with:

```
Operations/Network + Partition/Shrink Network
```

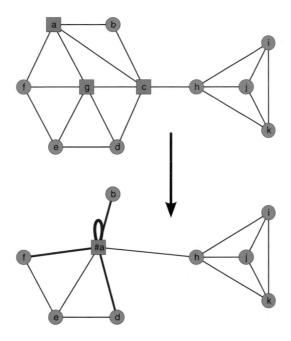

Figure 2.16   Shrinking a network.

Two options exist when shrinking a network: 1) choosing the minimum number of ties between shrunken vertices, and 2) choosing which cluster is not shrunk. In this toy example, we contracted cluster 2 (boxes). If all clusters are to be contracted then some 'non-existent' cluster number is chosen in (2). The shrunken network is shown at the bottom of Figure 2.16. The contracted vertex is marked as the representative vertex for identifying its cluster.[19] In Figure 2.16, #a identifies the contracted cluster. When the shrunken network is saved, the vertex names can be edited to be more meaningful. This is done in the following example.

2.6.2.1   Contracted clusters – international trade

The NBER-UN Trade Data described in Chapter 1 can be contracted usefully. These data are for 1999.[20] The project file also contains a partition by continents (with the number of nations in each continent in parentheses): 1 – Africa (47), 2 – North America (5), 3 – South America (32), 4 – Asia (44), 5 – Europe (38), 7 – Oceania (8). The results we report come from Hidalgo et al. (2007) and take the form of matrix arrays (for a very dense network). The first panel reports raw results in $US with the second panel having these numbers in a more digestible form with trillions of $US. The diagonal elements report trade within continents. The off-diagonal elements report trade between continents.

---

[19] The shrinking operation creates loops on the vertices if there are ties within clusters. These loops can be retained or removed.

[20] This network for 1999 was extracted where the groups of countries (World, Asia NES, China FTZ, China HK SAR, Eur. EFTA NS, etc.) were removed. These data are available as a Pajek project file: WorldTrade1999.paj

	Asia	Europe	Africa	SAmerica	Oceania	NAmerica
1 Asia	643584718	270159754	25876345	25805314	39845497	221904362
2 Europe	342269459	1742643337	63078518	47256144	11059853	211315613
3 Africa	24966099	59284150	7530980	2811868	1151109	12214387
4 South America	49920062	63102534	3574453	46908302	1370294	167982856
5 Oceania	34347059	20088174	727978	830356	8590894	17636492
6 North America	447312810	255313217	19209351	182867144	8932059	347719491

in 1000,000,000 USD

	Asia	Europe	Africa	SAmerica	Oceania	NAmerica
1 Asia	643.6	270.2	25.9	25.8	39.8	222.0
2 Europe	342.3	1742.6	63.1	47.3	11.1	211.3
3 Africa	25.0	59.3	7.5	2.8	1.2	12.2
4 South America	50.0	63.1	3.6	46.9	1.4	168.0
5 Oceania	34.3	20.1	0.7	0.8	8.6	17.6
6 North America	447.3	255.3	19.2	182.9	8.9	347.7

Figure 2.17 displays the original trade matrix in the top panel where countries within continents have been grouped together. A network picture of the contracted network is in the lower panel where continents are the vertex names. Given the sizeable levels of trade within continents, the loops created by shrinking the trade network were retained. The latter image is far more informative regarding trade between continents. Actual trade volumes were used in this shrinkage. There are alternative data manipulations that can be used prior to shrinking. For example, the matrix can be row normalized to focus on exports or column normalized to focus on imports. Additionally, if attention is focused only on trade between continents then trade within continents could be removed before normalizing the trade matrix.

2.6.3   Subgraphs

A subgraph of a given graph is obtained by deleting some lines or some vertices from it. If a vertex is deleted, all its incident lines are deleted also. Formally, a *subgraph* $H = (V', L')$ of a given graph $G = (V, L)$ is a graph whose set of lines is a subset of set of lines of G, $L' \subseteq L$, its vertex set is a subset of set of vertices of G, $V' \subseteq V$, and it contains all end-vertices of L'. An example is shown in Figure 2.18.

A subgraph can be *induced* by a given subset of vertices or lines.

In the subgraph $H = (V', L')$ induced by subset of vertices V', the subset of lines L' contains exactly the lines from L that have both end-vertices in V'.

In the subgraph $H = (V', L')$ induced by subset of lines L', the subset of vertices V' consists exactly of all end-vertices of the lines from L.

A subgraph is a *spanning* subgraph iff $V' = V$.

Usually, an induced subgraph is obtained by extracting selected cluster(s) from a given network using the Pajek command:

```
Operations/Network+Partition/Extract Subnetwork
```

Recall, **network = graph + data**. To obtain a *subnetwork*, it is necessary to extract also the corresponding subpartitions, subvectors, etc. This is done in Pajek with:

```
Partitions/Extract Subpartition
Operations/Vector+Partition/Extract Subvector
```

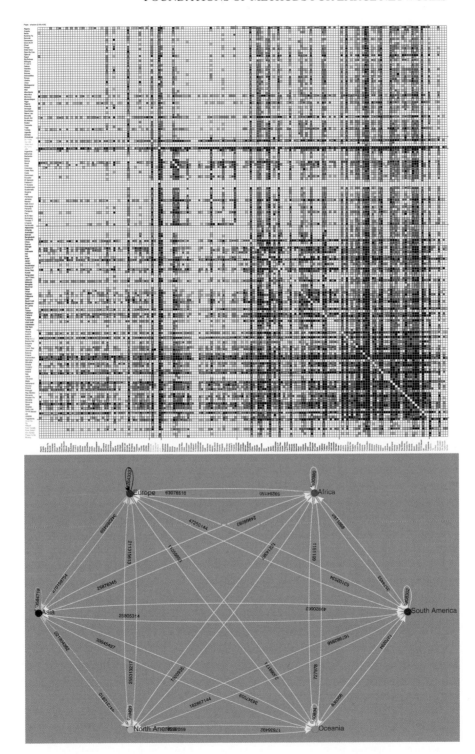

Figure 2.17   Contracted clusters using continents – World Trade 1999.

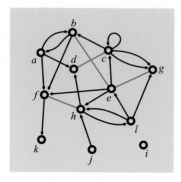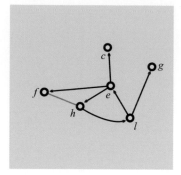

Figure 2.18   A graph and one of its subgraphs.

The graph on the left is the same as the graph shown in Figure 2.2.

Returning to the NBER-UN Trade Data, suppose that interest is focused only on Africa. The African countries can be extracted from the trade network in Pajek with:[21]

```
Operations/Network+Partition/Extract Subnetwork [1]
```

The result is the induced subgraph, the *cut out* of Africa, shown in Figure 2.19.

Trade between pairs of continents (e.g. Africa and South America) can be examined in the same way. Given that two regions have been defined, the links between them are called *interlinks*. To obtain them, two steps are used: 1) extract the subgraphs for these two regions, and 2) remove the lines inside each region. This is done in Pajek with:[22]

```
Operations/Network+Partition/Extract Subnetwork    [1,3]
Operations/Network+Partition/Transform/Remove lines/
   Inside clusters   [1,3]
```

The result is presented in Figure 2.20.

2.6.4 Cuts

One standard approach for finding interesting groups inside a network is based on properties and/or weights reflecting (or coupled closely to) substantive concerns, problems, or questions. These items can be *measured* while collecting the network data or *computed* from the structure of a network.

For example, one popular model of network structure is based on the assumption that the important parts of a network are denser compared to the rest of the network. There are different measures of local density including degrees, coreness, the number of short rings (triangles, quadrangles), and the (corrected) clustering coefficient. (Measures not already defined are presented in the following sections.) Commands for obtaining them in Pajek include:

---

[21] As shown in the commands, these countries belong to Cluster 1. Cluster numbers are listed on page 39.
[22] South America is Cluster 3.

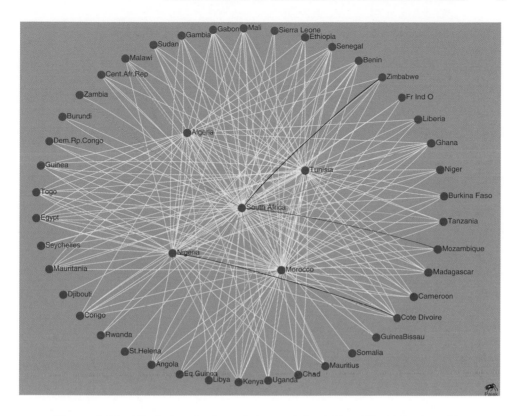

Figure 2.19   The induced subgraph (cut out) for Africa: World Trade 1999.

```
Network/Create Vector/Centrality/Degree
Network/Create Partition/k-Core
Network/Create New Network/with Ring Counts stored as Line Values
Network/Create Vector/Clustering Coefficients
```

The *vertex-cut* of a network $\mathcal{N} = (\mathcal{V}, \mathcal{L}, p)$, for a property $p : \mathcal{V} \to \mathbb{R}$, at a selected level (threshold), t, is a subnetwork $\mathcal{N}(t) = (\mathcal{V}', \mathcal{L}(\mathcal{V}'), p)$, determined by the set

$$\mathcal{V}' = \{v \in \mathcal{V} : p(v) \geq t\}$$

and $\mathcal{L}(\mathcal{V}')$ is the set of lines from $\mathcal{L}$ that have both end-vertices in $\mathcal{V}'$. In words, those vertices whose values fall below the threshold ($p(v) < t$) are removed along with *all* lines incident to them.

The *line-cut* of a network $\mathcal{N} = (\mathcal{V}, \mathcal{L}, w)$, for a weight $w : \mathcal{L} \to \mathbb{R}$, at a selected level, t, is a subnetwork $\mathcal{N}(t) = (\mathcal{V}(\mathcal{L}'), \mathcal{L}', w)$, determined by the set

$$\mathcal{L}' = \{e \in \mathcal{L} : w(e) \geq t\}$$

and $\mathcal{V}(\mathcal{L}')$ is the set of all end-vertices of the lines from $\mathcal{L}'$. Lines whose values are below the threshold ($w(e) < t$) are removed along with any created isolates.

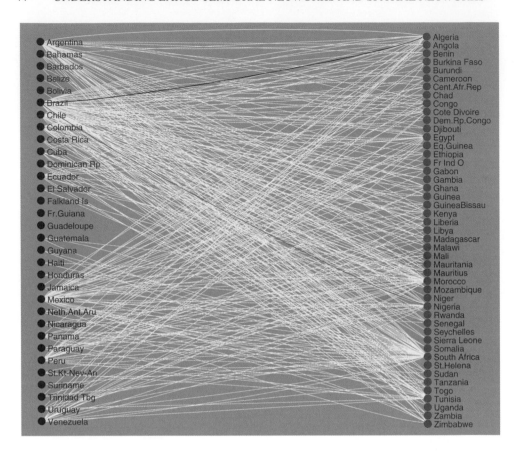

Figure 2.20   Interlinks for Africa and South America: World Trade 1999.

The values of the thresholds, t, can be determined by inspecting the distributions of vertex and line values. Obtaining a vertex-cut in a network using a vertex property, p, stored in a vector is done in Pajek with:

```
Vector/Info   [+10] [#10]
Vector/Make Partition/by Intervals/Selected Thresholds   [t]
Operations/Network + Partition/Extract Subnetwork   [2]
```

Obtaining a line-cut from a network with weights, w, is done with the following Pajek commands:

```
Network/Info/Line values    [#10]
Network/Create New Network/Transform/Remove/Lines with Value/
   lower than   [t]
Network/Create Partition/Degree/All
Operations/Network + Partition/Extract Subnetwork   [1-*]
```

We illustrate these ideas with the network shown in Figure 2.21. The vertices have different weights, as shown by their sizes. Vertices a, b, c, f are above a threshold, set

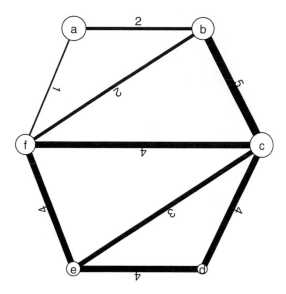

Figure 2.21 A network with values on vertices and lines.

The size of the vertices show their values. The line weights are shown by line thickness.

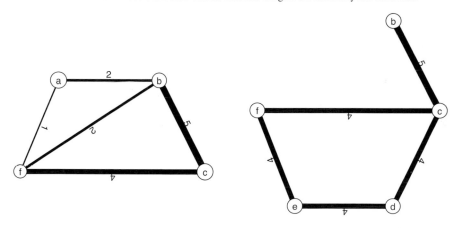

Figure 2.22 Vertex and line cuts of Figure 2.21.

The vertex cut subnetwork is on the left and the line cut subnetwork is on the right.

at $t = 3$ for this example. The weights on these edges are shown by their values and the thickness of the lines. The threshold for a line cut was set at $t = 4$.

The results are shown in Figure 2.22, where the vertex cut with $t = 3$ is on the left and the line cut with $t = 4$ is on the right. The vertices of Figure 2.22 are drawn without having their sizes represented by weights. Vertices d and e were dropped from the graph produced by the vertex cut while only a was dropped from the network produced by the line cut.[23]

---

[23] The last two lines of the Pajek commands for producing the line-cut graph are needed to remove isolates, in this case a.

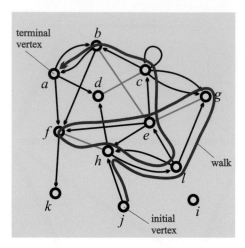

Figure 2.23   A walk (marked in red) of length 11 from j to a.

2.7 Connectivity properties of networks

As networks can be 'connected' in many different ways, terms are needed to specify precisely the meaning(s) of this term. Some of these concepts are presented in this section.

2.7.1 Walks

A *walk*, s, from its *initial* vertex, u, to its *terminal* vertex, v, is a finite sequence of vertices and lines

$$s = (v_0, e_1, v_1, e_2, v_2, \ldots, v_{k-1}, e_k, v_k)$$

such that for all $i = 1, 2, \ldots, k$ where $e_i(v_{i-1}, v_i)$ is the line e_i leads from vertex v_{i-1} to vertex v_i. The *length* $|s| = k$ of the walk s is equal to the number of lines it contains.

Often, when the individual lines are not important or are uniquely determined by their end-vertices, the walk is described by only the vertices through which it passes. For example, the walk marked in red in Figure 2.23, $s = (j, h, l, g, e, f, h, l, e, c, b, a)$ has length $|s| = 11$.

A walk is *closed* iff its initial and terminal vertices coincide. If the direction of the lines in the walk is irrelevant it is called a *semiwalk* or a *chain*.

There are some special types of walks: a *trail* is a walk in which all lines are different; a *path* is a walk with all vertices different; and a *cycle* is a closed walk in which all internal vertices are different. As noted earlier, a graph is *acyclic* if it contains no cycles.

A shortest path[24] from u to v is called also a *geodesic* from u to v. Its length is denoted by $d(u, v)$. If there is no walk from u to v then $d(u, v) = \infty$. For example, for the graph in Figure 2.23, $d(j, a) = |(j, h, d, c, b, a)| = 5$ and $d(a, j) = \infty$.

---

[24] There can be more than one shortest path between a pair of vertices.

The measure $d(u, v)$ is not symmetric: in general, $d(u, v) \neq d(v, u)$. Therefore, this measure is not a distance. Defining $\hat{d}(u, v)$ by

$$\hat{d}(u, v) = \max(d(u, v), d(v, u))$$

creates a distance: $\hat{d}(u, v)$ has the required properties: $\hat{d}(v, v) = 0$, $\hat{d}(u, v) = 0 \Rightarrow u = v$, $\hat{d}(u, v) = \hat{d}(v, u)$ and $\hat{d}(u, v) \leq \hat{d}(u, t) + \hat{d}(t, v)$. In an undirected graph $\hat{d} = d$.

The *diameter* of a graph is equal to the distance between the most distant pair of vertices: $D = \max_{u, v \in \mathcal{V}} d(u, v)$.

2.7.2 Equivalence relations and partitions

A relation, R, on $\mathcal{V}$ is an *equivalence* relation iff it is *reflexive* ($\forall v \in \mathcal{V} : vRv$), *symmetric* ($\forall u, v \in \mathcal{V} : uRv \Rightarrow vRu$), and *transitive* ($\forall u, v, z \in \mathcal{V} : uRz \wedge zRv \Rightarrow uRv$). Each equivalence relation determines a partition into *equivalence classes* $[v] = \{u : vRu\}$.

Each partition **C** determines an equivalence relation, R, defined as $uRv \Leftrightarrow \exists C \in \mathbf{C} : u \in C \wedge v \in C$.

The *k-neighbors* of v is the set of vertices at 'distance' k from v, $N^k(v) = \{u \in v : d(v, u) = k\}$. The set of all sets of k-neighbors of v, $k = 0, 1, \dots$ is a partition of $\mathcal{V}$.

Using k-neighbors we can define the *k-neighborhood* of v as $N^{(k)}(v) = \{u \in v : d(v, u) \leq k\} = \bigcup_{i=0}^{k} N^i(v)$. Obtaining k-neighbors of a vertex in Pajek uses the commands:

```
Network/Create Partition/k-Neighbors
Operations/Network + Partition/Extract Subnetwork [0-k]
```

As an example, in Figure 2.24 the Motorola's neighborhood in the Krebs Internet industries network is presented. The thickness of edges are square roots of their values.

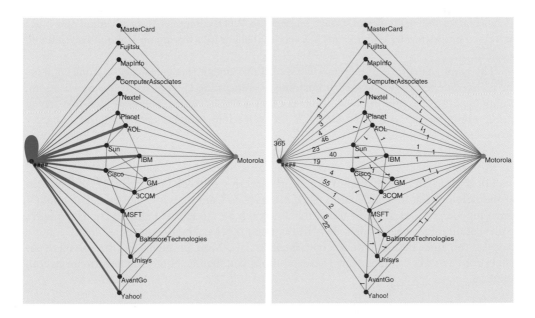

Figure 2.24   Motorola's neighborhood in the Krebs Internet industries network.

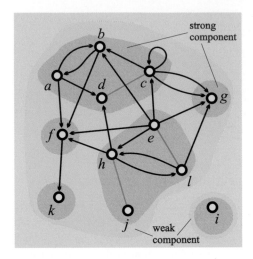

Figure 2.25   Strong and weak connectivity.

2.7.3   Connectivity

A vertex, *u*, is *reachable* from a vertex, *v*, iff there exists a walk whose initial vertex is *v* and its terminal vertex is *u*. A vertex *v* is *weakly connected* with vertex *u* iff there exists a semiwalk with *v* and *u* as its end-vertices. The vertex *v* is *strongly connected* with vertex *u* iff they are mutually reachable.

Weak and strong connectivity both define equivalence relations. The corresponding equivalence classes induce weak and strong *components*. Weak components are essentially the basic parts of a graph. Strong components are maximal subgraphs having only mutually reachable vertices.

The graph in Figure 2.25 contains two weak components (the green regions): one contains the isolated vertex *i* with the other containing the remaining 12 vertices. There are six strong components (in the pink regions). Four contain a single vertex. There are two non-trivial strong components: $\{a, b, c, d\}$ and $\{e, h, j, l\}$. Obtaining components in Pajek is done by using:

```
Network/Create Partition/Components/Weak
Network/Create Partition/Components/Strong
```

For a matrix representation of a network in terms of weak components,[25] the vertices of the network can be reordered to place all vertices from the same cluster (weak component) together. An example of doing this is shown on the left of Figure 2.26. This matrix representation has weak components as diagonal blocks. The off-diagonal blocks contain only 0s (no lines between weak components exist). Most problems can be solved separately for each weak component. These solutions can then be combined to form an overall solution.

---

[25] This strategy can be used for any partition of a network, e.g. for clusters established in a blockmodel or as in Figure 2.16, bottom panel.

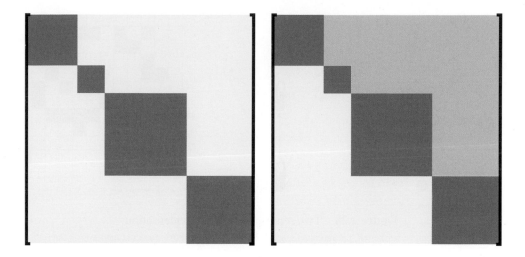

Figure 2.26   Weak and strong components in matrix representations.
The areas marked in yellow are null.

2.7.4   Condensation

If every strong component of a given directed graph is shrunk into a vertex, all loops are deleted, and parallel arcs identified, the contracted graph is called a *condensation*. This graph is acyclic (Harary et al. 1965). The picture in Figure 2.27 presents the condensation of the graph in Figure 2.25.

The condensation of a graph being acyclic is very important. It implies every directed graph has a composition of some equivalence (in this case, strong connectivity) and a hierarchy (acyclic graph).

The vertices of an acyclic graph can be ordered (assigned numbers from 1 to n, $i : \mathcal{V} \to \mathbb{N}$) in such a way that:

$$(u, v) \in \mathcal{A} \Rightarrow i(u) < i(v).$$

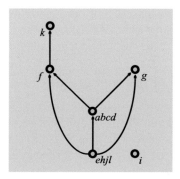

Figure 2.27   Condensation of the network in Figure 2.25.
The labels of the vertices are the vertex labels of the vertices going into the condensed vertices.

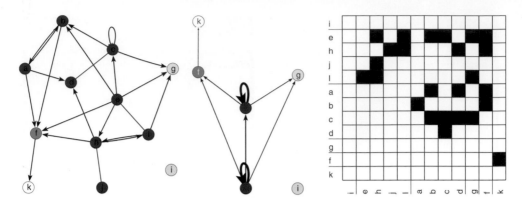

Figure 2.28   Two representations of condensation.

The original graph is one the left, the shrunken graph is in the middle, and the reordered matrix is on the right.

Since the condensation is an acyclic graph, such a numbering exists for the strong components of a graph. When the vertices of the original graph are reordered according to the numbers assigned to their strong component, the result is a matrix representation in which all blocks below the diagonal contain only 0s. This is shown on the right of Figure 2.26.

The commands in Pajek for doing this are:

```
Network/Create Partition/Components/Strong [1]
Operations/Network+Partition/Shrink Network   [1][0]
Network/Create New Network/Transform/Remove/Loops [yes]
Network/Acyclic Network/Depth Partition/Acyclic
Partition/Make Permutation
Permutation/Inverse Permutation
select partition [Strong Components]
Operations/Partition+Permutation/Functional Composition Partition*Permutation
Partition/Make Permutation
select [original network]
File/Network/Export Matrix to EPS/Using Permutation
```

Figure 2.28 presents the condensation of our (continuing) example graph. The strong components are color-coded on the right. This color-coding is used for the contracted graph shown in the middle of Figure 2.28. The matrix representation is on the right.

2.7.5   Bow-tie structure of the web graph

The *bow-tie structure* has its origins in the studies of web graphs. However, it can be used for any directed graph as shown in Figure 2.29. Let S be the *largest strong component* in network $\mathcal{N}$; $\mathcal{W}$ the weak component containing S; $\mathcal{I}$ the set of vertices not in S from which S can be reached; $\mathcal{O}$ the set of vertices not in S reachable from S; $\mathcal{T}$ (tubes) set of vertices not in S on paths from $\mathcal{I}$ to $\mathcal{O}$; $\mathcal{R} = \mathcal{W} \setminus (\mathcal{I} \cup S \cup \mathcal{O} \cup \mathcal{T})$ (tendrils); and $\mathcal{D} = \mathcal{V} \setminus \mathcal{W}$ the disconnected components. The partition

$$\{\mathcal{I}, S, \mathcal{O}, \mathcal{T}, \mathcal{R}, \mathcal{D}\}$$

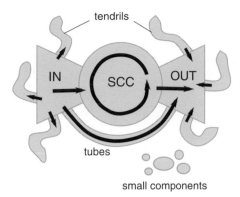

tendrils

IN   SCC   OUT

tubes

small components

Figure 2.29   Broder et al. (2000): Graph structure in the Web.

is called the *bow-tie* partition of $\mathcal{V}$. This partition is obtained in Pajek with the command:

```
Network/Create Partition/Bow-Tie
```

2.7.6   The internal structure of strong components

Condensations of graphs ignore the internal structure of strong components. However, it is often useful to examine the internal structure of these clusters.[26] One such internal structure features periodic pattern.

Let d be the largest common divisor of lengths of all closed walks in a strong component. The component is said to be *simple*, iff $d = 1$; otherwise it is *periodic* with a period d.

The set of vertices $\mathcal{V}$ of strongly connected directed graph $\mathcal{G} = (\mathcal{V}, R)$ can be partitioned into d clusters $\mathcal{V}_1, \mathcal{V}_2, \ldots, \mathcal{V}_d$, s.t. for every arc $(u, v) \in R$ holds $u \in \mathcal{V}_i \Rightarrow v \in \mathcal{V}_{(i \bmod d)+1}$ (Bonhoure et al. 1993). See Figure 2.30, where there is a clear internal structure to the strong component. The Pajek command for this is:

```
Network/Create Partition/Components/Strong-Periodic
```

2.7.7   Bi-connectivity and *k*-connectivity

While weak and strong components are important for delineating the structure of graphs, certain vertices and lines may have additional structural importance. To identify them, a more general conception of 'connectivity' is needed. Vertices u and v are *bi-connected* iff they are connected in both directions by two independent paths (having no common internal vertex). Bi-connectivity is an equivalence relation on the set of lines and determines a partition of them. The Pajek command for identifying them is:

```
Network/Create New Network/with Bi-Connected Components
```

A vertex is an *articulation* vertex iff its deletion increases the number of weak components in a graph. This is shown on the left of Figure 2.31. Suppose that G_1 and G_2 are weakly

---

[26] A more general point can be made: examining the internal structure of all identified clusters has great merit. However, we restrict our attention here to strong components.

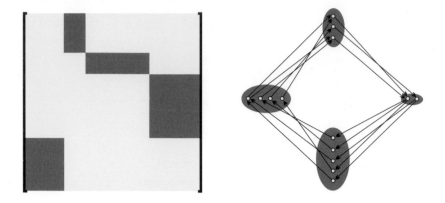

Figure 2.30   An example of periodic components.

Figure 2.31   An articulation vertex and a bridge.

connected subnetworks. Because all of the semipaths between vertices in G_1 and vertices in G_2 go through the vertex v, then its removal disconnects G_1 and G_2 so they become separate weak components. The vertex v serves as a *broker* between subnetworks G_1 and G_2.

A line is a *bridge* iff its deletion increases the number of weak components in a graph. On the right of Figure 2.31, if G_1 and G_2 are weakly connected where (u, v) is the only link between G_1 and G_2 then the removal of (u, v) disconnects G_1 and G_2.

In order to provide general measures of the connectivity of a graph, the concept of a complete graph is useful. It is a simple undirected graph in which every pair of distinct vertices is linked by an edge. The complete graph on n vertices is denoted by K_n. This is shown on the left in Figure 2.32 for $n = 7$. Note that K_1 is a trivial one-vertex graph. Also

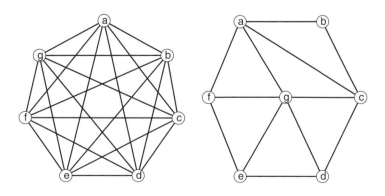

Figure 2.32 K_7 (right) and G_7 (left).

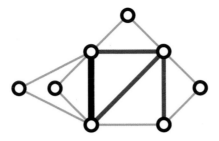

Figure 2.33   A network with triangular weights represented by line thickness.

in the figure (on the right) is a part of the example graph for shrinking a network from Figure 2.16, labeled G_7 for some comparisons.

The *vertex connectivity* $\kappa(G)$ of a graph, G, is equal to the smallest number of vertices that, if deleted, induce a disconnected graph or the trivial graph K_1. The vertex connectivity of K_n is $n - 1$ (so for K_7 it is 6) and for G_7 it is 2 (with the removal of a and c).

Line connectivity $\lambda(G)$ of graph G is equal to the smallest number of lines that, if deleted, induce a disconnected graph or the trivial graph K_1. The line connectivity of K_n is $n - 1$ (6 for K_7) and 2 for G_7 (with the removal of $(a : b)$ and $(b : c)$).

Whitney's inequality is: $\kappa(G) \leq \lambda(G) \leq \delta(G)$ where $\delta(G)$ is the minimum degree of G. With regard to Figure 2.32: in general, $\kappa(K_n) = \lambda(K_n) = \delta(K_n) = n - 1$ so for K_7 all values are 6. For G_7, $\kappa(G_7) = \lambda(G_7) = \delta(G_7) = 2$.

Graph G is *(vertex) k-connected*, if $\kappa(G) \geq k$ and is *line k-connected*, if $\lambda(G) \geq k$.

The Whitney / Menger theorem is: A graph G is vertex (line) k-connected iff every pair of vertices can be connected with k vertex (line) internally disjoint semiwalks.

2.8   Triangular and short cycle connectivities

There are additional useful conceptions of connectivity based on triangles and other small subgraphs (Batagelj and Zaveršnik 2007). A *triangle* in an undirected graph is a subgraph isomorphic[27] to K_3. Let G be a simple undirected graph. For $e \in \mathcal{E}$, the triangular weight $w(e)$ equals the number of different triangles in G to which e belongs. An example is shown in Figure 2.33. A subgraph $H = (\mathcal{V}', \mathcal{E}')$ of $G = (\mathcal{V}, \mathcal{E})$ is *triangular* if every vertex and every edge belongs to at least one triangle in H.

A sequence $(T_1, T_2, \ldots, T_s)$ of triangles of G *(vertex) triangularly connects* vertices $u, v \in \mathcal{V}$ iff $u \in T_1$ and $v \in T_s$ or $u \in T_s$ and $v \in T_1$ and $\mathcal{V}(T_{i-1}) \cap \mathcal{V}(T_i) \neq \emptyset, i = 2, \ldots s$. See the left panel of Figure 2.34. Such a sequence is called a *triangular chain*. Also, *edge triangularly connects* vertices $u, v \in \mathcal{V}$ iff a stronger version of the second condition holds $\mathcal{E}(T_{i-1}) \cap \mathcal{E}(T_i) \neq \emptyset, i = 2, \ldots s$. See the right panel of Figure 2.34. Vertex triangular connectivity is an equivalence on $\mathcal{V}$; and edge triangular connectivity is an equivalence on $\mathcal{E}$.

---

[27] Two graphs are isomorphic if one can be obtained from the other by relabeling the vertices. For two graphs, G_1 and G_2, there is a one-to-one correspondence between the vertices of G_1 and the vertices of G_2 preserving the adjacency and non-adjacency of all vertices.

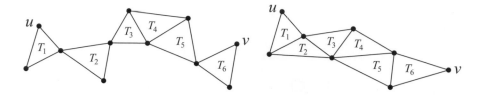

Figure 2.34   Examples of triangular connectivity.

Vertex triangular connectivity is on the left. Edge triangular connectivity is on the right.

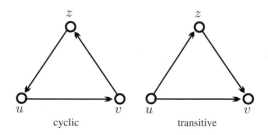

Figure 2.35   Triangular connectivity in directed graphs.

Triangular weights can be used to efficiently identify dense, clique-like parts of a graph. If an edge e belongs to a k-clique[28] in $\mathcal{G}$ then $w(e) \geq k - 2$.

The notion of triangular weights can be extended also to directed networks. If the graph $\mathcal{G}$ is mixed (having both arcs and edges), we can replace edges with pairs of opposite arcs. In the following, let $\mathcal{G} = (\mathcal{V}, \mathcal{A})$ be a simple directed graph without loops. There are only two different types of directed triangles: **cyc**lic and **tra**nsitive. These are shown in Figure 2.35. For each type, we get the corresponding triangular weights w_{cyc} and w_{tra}.

Short cycle connectivity is a generalization of ordinary connectivity – two vertices have to be connected by a sequence of short cycles, in which two consecutive cycles have at least one common vertex. If all consecutive cycles in the sequence share at least one edge, we talk about edge short cycle connectivity. Short cycle connectivity can be extended to directed graphs (by using cyclic and transitive connectivity). This is useful in hierarchical decompositions of networks as long cycles can be violations of the assumed hierarchical structure. This can be extended to the notion of *short (semi) cycle connectivity*. Instead, on short cycles we can base the connectivity definition also on other small subgraphs (Batagelj and Zaveršnik 2007).

2.9 Islands

Islands are particularly useful for examining networks by extracting coherent 'well connected' part of networks for further study. They are used extensively in Chapter 4 (for bibliographic networks), Chapter 5 (for the US patent citation network), and Chapter 6 (for the US Supreme Court citation network).

---

[28] These are defined formally in Section 2.10.

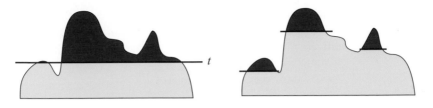

Figure 2.36   A visualization of the islands concept.

2.9.1   Defining islands

In a cut, $\mathcal{N}(t)$, obtained by the cuts approach outlined in Section 2.6.4, we examined its parts. Their number and sizes depend on a threshold, t. Usually, there are many small parts and some large parts. It is more useful to consider only parts of size at least k but not exceeding K where $k < K$. The parts of size smaller than k (and not containing the elements with the largest property values) are usually discarded as 'uninteresting'. Parts with sizes larger than K are often too large, and so are unwieldy for detailed analysis.[29] In general, the values of thresholds such as t, k, and K are determined by inspecting the distribution of vertex or line values, and the distribution of part sizes. Also relevant for specifying these thresholds is additional knowledge regarding the network, substantive concerns, and the goals of an analysis.

Islands represent another approach to identifying parts of networks. A simple visual image for thinking of islands stems from representing the values (given or computed) of vertices and representing them with heights. The same can be done for the values of lines. If the network is immersed in water, the level of the water acts as a threshold. Items above the surface are visible while items below it are not. Each potential level of the water is a threshold. The network forms a kind of landscape. When one is selected, there is a cut: items 'high' enough relative to the threshold form *islands*. Varying the water level yields different islands. There can be a single threshold for determining islands or a set of thresholds. This visual imagery is illustrated in Figure 2.36 where, on the left, there is a single threshold and, on the right, three thresholds. Very efficient algorithms have been developed and implemented in Pajek to determine the islands hierarchy and to list all the islands of selected sizes.

Islands are a very general and efficient approach for determining 'important' subnetworks in a given network. To do this, it is necessary to express the analytic goals of an analysis using relevant properties of vertices or weights of lines. Using a selected property allows analysts to determine the islands of an appropriate size (in the interval k to K). In large networks, it is possible to get many islands which are then inspected individually and interpreted in terms of their content. This is done extensively in Chapter 6 for islands in the Supreme Court citation network. An important property of the islands is that they identify locally important subnetworks at different levels. Therefore they can also be used to detect emerging groups or phenomena.

---

[29] They can be cut again at some other higher level, say t', where $t' > t$ but this can be cumbersome as a general strategy.

A set of vertices $C \subseteq \mathcal{V}$ is a *regular vertex island* in network $\mathcal{N} = (\mathcal{V}, \mathcal{L}, p)$, for a property $p : \mathcal{V} \rightarrow \mathbb{R}$ iff it induces a connected subgraph and the vertices from the island are 'higher' than the neighboring vertices

$$\max_{u \in N(C)} p(u) < \min_{v \in C} p(v)$$

where $N(C)$ is the neighborhood of the set C

$$N(C) = \{u \in \mathcal{V} \setminus C : \exists v \in C : (u : v) \in \mathcal{L}\}.$$

A set of vertices $C \subseteq \mathcal{V}$ is a *regular line island* in network $\mathcal{N} = (\mathcal{V}, \mathcal{L}, w)$, for a weight $w : \mathcal{L} \rightarrow \mathbb{R}$ iff it induces a connected subgraph and the lines inside the island are 'more strongly linked' among themselves than with the neighboring vertices. Formally, in $\mathcal{N}(C)$ there exists a spanning tree $\mathcal{T}$ over C such that

$$\max_{(u,v) \in \mathcal{L}, u \notin C, v \in C} w(u, v) < \min_{(u,v) \in \mathcal{T}} w(u, v).$$

An example of islands extracted from the EAT network is shown in Figure 2.37. The bottom left island deals with the word WORK and relates it to studying activities – an indication that the data were collected in a student population.

2.9.2   Some properties of islands

In his PhD thesis Matjaž Zaveršnik proved the following properties of islands:

- The sets of vertices of connected components of a vertex-cut or a line-cut, at a selected level t, are regular vertex islands or line islands, respectively.

- The set $\mathcal{H}_p(\mathcal{N})$ of all regular vertex islands of network $\mathcal{N}$ is a complete hierarchy:

  - two islands are disjoint or one of them is a subset of the other; and

  - each vertex belongs to at least one island.

- The set $\mathcal{H}_w(\mathcal{N})$ of all non-degenerate regular line islands of a network $\mathcal{N}$ is a hierarchy (but not necessarily a complete hierarchy):

  - two islands are disjoint or one of them is a subset of the other.

- Vertex and line islands are invariant for the strictly increasing transformations of the property p or weight w, respectively.

- Two linked vertices cannot belong to two disjoint regular vertex/line islands.

A *simple island* is an island with only one peak. Usually, a simple island deals with one substantive topic, while general islands combine related topics. The Pajek commands for obtaining islands are:

```
Operations/Network+Vector/Islands/Vertex Property
Network/Create Partition/Islands/Line Weights
```

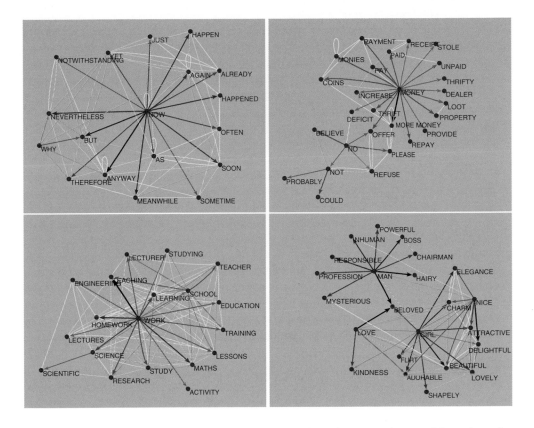

Figure 2.37   Islands from The Edinburgh Associative Thesaurus for transitive triangular weights.

2.10   Cores and generalized cores

As noted above, one principle for identifying important parts in a network is looking for its denser parts. Activity in a network often creates, among other things, additional lines, especially within denser patches. Many concepts have been formulated to formalize the seemingly imprecise notion of denser parts. The earliest such concept was a *clique*: a clique with k members is isomorphic to K_k, a maximal complete subgraph on k vertices. In practice, many empirical networks had a very large number of (overlapping) cliques, making interpretation very difficult. Indeed, there exist graphs with exponentially many cliques (Moon and Moser 1965). To grapple with this problem, the clique was generalized to a k-clique (Luce 1950), a concept which enjoyed some popularity.

Unfortunately, producing the list of all k-cliques is an NP-complete problem (Garey and Johnson 1979). In computational complexity theory, the complexity class known as NP-complete, where NP denotes non-deterministic polynomial time, is a class of problems. While any given solution to an NP-complete problem can be verified easily, there is no known efficient way of locating such solutions: the time required to solve the problem using any currently known algorithm increases very quickly as the size of the problem grows. Even for

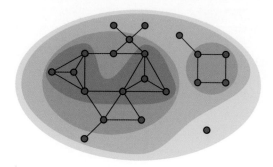

Figure 2.38   An example with cores of a graph.

moderately sized problems, solving them would take billions of years. In short, there is no efficient algorithm for identifying them. Even when all k-cliques can be identified in *small* enough graphs, most often the results are uninterpretable when many exist. In short, k-cliques are seldom, if ever, practical for dealing with real-life networks.

Other concepts, including k-clans, k-plexes, cores, and lambda sets, have been proposed for better grasping denser parts of networks (Wasserman and Faust 1994), usually without considering the computational aspects of identifying them. Most fell short in terms of practicality. Among them, *only* cores can be efficiently determined. They are considered next.

2.10.1   Cores

The notion of core was introduced by Seidman (1983). Let $G = (\mathcal{V}, \mathcal{L})$ be a graph. A subgraph $\mathcal{H} = (W, \mathcal{L}|W)$ induced[30] by the set W is a k-core or a *core of order k* iff for all $v \in W$: $\deg_{\mathcal{H}}(v) \geq k$, and $\mathcal{H}$ is a maximal subgraph with this property. For simplicity, we use core for k-core. The core of maximum order is the *main* core.

The *core number* of vertex v is the highest order of core containing this vertex. The degree, $\deg(v)$, can be replaced by in degree, out degree, (in degree + out degree), etc. for determining different types of cores.

Figure 2.38 provides some examples of cores with 0, 1, 2, and 3 cores. They are identified with different gray shades in the figure. The entire network is the 0-core and is marked with the lighest shade. The 1-core excludes the isolate. There is one 2-core (with two components) and the darkest shade is for the 3-core (left). Figure 2.38 illustrates the following properties of cores:

- The cores are nested: $i < j \implies \mathcal{H}_j \subseteq \mathcal{H}_i$.

- Cores are not necessarily connected subgraphs.

An efficient algorithm for determining the cores hierarchy is based on the following property (Batagelj et al. 1999):

---

[30] In $\mathcal{H} = (W, \mathcal{L}|W)$, the expression $\mathcal{L}|W$ means the edges of L are restricted to W.

If from a given graph $G = (V, L)$ we recursively delete all vertices, and edges incident with them, of degree less than k, the remaining graph is the k-core.

As cores can be determined very efficiently, they are one of the few concepts capable of providing meaningful decompositions of large networks. Different approaches to the analysis of large networks can be built on using cores. For example, we get the following bound on the chromatic number of a given graph G (the smallest number of colors needed to color the vertices of the graph with no two adjacent vertices sharing the same color):

$$\chi(G) \leq 1 + \mathrm{core}(G)$$

Cores can also be used to localize the search for interesting subnetworks in large networks: if they exist, both a k-component and a clique of order $k + 1$ are contained in a k-core.

2.10.2 Generalized cores

The notion of core can be generalized to networks (Batagelj and Zaveršnik 2011). Let $\mathcal{N} = (V, L, w)$ be a network, where $G = (V, L)$ is a graph and $w : L \rightarrow \mathbb{R}$ is a function assigning values to lines. A *vertex property function* on **N**, or a *p-function* for short, is a function $p(v, U)$, $v \in V$, $U \subseteq V$ with real values. It is useful to define $N_U(v) = N(v) \cap U$. All of the vertex degree measures defined earlier form vertex property functions. The central idea here is having a systematic way of working with vertex weights once they are defined. Additional examples of *p*-functions, include:
Sums of line weights:

$$p_S(v, U) = \sum_{u \in N_U(v)} w(v, u), \quad \text{where } w : L \rightarrow \mathbb{R}_0^+$$

The maximum weight among a set of lines:

$$p_M(v, U) = \max_{u \in N_U(v)} w(v, u), \quad \text{where } w : L \rightarrow \mathbb{R}$$

The number of k-cycles through a vertex:

$$p_k(v, U) = \text{number of cycles of length } k \text{ through vertex } v \text{ in } (U, L|U)$$

Relative density at vertices:

$$p_\gamma(v, U) = \frac{\deg(v, U)}{\max_{u \in N(v)} \deg(u)}, \quad \text{if } \deg(v) > 0; 0, \text{ otherwise}$$

Diversity across vertices:

$$p_\delta(v, U) = \max_{u \in N_U^+(v)} \deg(u) - \min_{u \in N_U^+(v)} \deg(u)$$

Average weight of a set of vertices:

$$p_a(v, U) = \frac{1}{|N_U(v)|} \sum_{u \in N_U(v)} w(v, u), \quad \text{if} N_U(v) \neq \emptyset; 0, \text{otherwise}$$

The subgraph $\mathcal{H} = (C, \mathcal{L}|C)$ induced by the set $C \subseteq \mathcal{V}$ is a *p-core at level* $t \in \mathbb{R}$ iff for all $v \in C : t \leq p(v, C)$ and C is a maximal such set.

This raises an obvious question: Can we use a similar algorithm as for ordinary cores also for determining generalized cores? Fortunately, the answer is yes for many interesting property functions, *p*. Two additional concepts are useful.

The function *p* is *monotone* iff it has the property

$$C_1 \subset C_2 \Rightarrow \forall v \in \mathcal{V} : (p(v, C_1) \leq p(v, C_2)).$$

The degrees and the functions p_S, p_M and p_k are monotone. The function p_a is not monotone. For a monotone function the *p*-core at level *t* can be determined, as in the ordinary case, by successively deleting vertices with value of *p* lower than *t*; and the cores on different levels are nested:

$$t_1 < t_2 \Rightarrow \mathcal{H}_{t_2} \subseteq \mathcal{H}_{t_1}$$

The *p*-function is *local* iff $p(v, U) = p(v, N_U(v))$.

All the listed property functions are local, except p_k; p_k is **not** local for $k \geq 4$. For a local *p*-function an $O(m \max(\Delta, \log n))$ algorithm for determining the *p*-core levels exists, assuming that $p(v, N_C(v))$ can be computed in $O(\deg_C(v))$. See Batagelj and Zaveršnik (2011).

The Pajek commands for obtaining cores and generalized cores are:

```
File/Network/Read  [Geom.net]
Net/Partitions/Core/All
Info/Partition
Operations/Extract from Network/Partition [13-*]
Draw/Draw-Partition
Layout/Energy/Kamada-Kawai
Options/Values of lines/Similarities
Layout/Energy/Kamada-Kawai
Operations/Extract from Network/Partition [21]
Draw
Layout/Energy/Kamada-Kawai
Options/Values of lines/Forget
Layout/Energy/Kamada-Kawai
[select Geom.net]
Net/Vector/PCore/Sum/All
Info/Vector
Vector/Make Partition/by Intervals/Selected Thresholds [45]
Info/Partition
Operations/Extract from Network/Partition [2]
Draw
Options/Values of lines/Similarities
Layout/Energy/Fruchterman-Reingold
```

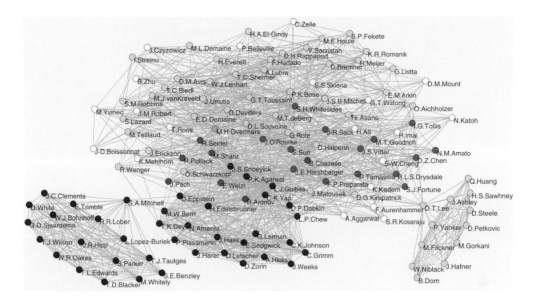

Figure 2.39   Cores of orders 10–21 in the Computational Geometry.

An example of detected cores in a network is shown[31] in Figures 2.39 and 2.40 for the collaboration network in the field of Computational Geometry. The degree core identifies authors who wrote papers with at least 10 others from the core; the p_S-core identifies authors who wrote at least 46 papers with others from the core.

2.11   Important vertices in networks

There has long been an interest in identifying 'salient' or 'significant' vertices in networks. For such vertices, we use the terms importance and/or distinctiveness. Measures of importance and distinctiveness for vertices in networks, for example degrees (Section 2.1.2) and core numbers (Section 2.10.1), are called *indices*.[32] One very important distinction between different *vertex indices* is determined by whether the network is directed or undirected. There are two main types of indices:

- For directed networks there are two subgroups of indices of *importance*; measures of *influence*, based on out going arcs; and measures of *support*, based on incoming arcs[33];

- For undirected networks: measures of *centrality*, based on all lines.

---

[31] Some of the above commands, starting with `draw`, pertain to drawing networks and strategies for doing this. A detailed account of visualization is beyond the scope of this book. See the Pajek manual (Batagelj and Mrvar 1996–2013), Section 4, for guidance.

[32] They play a crucial role in other areas using network ideas including chemistry (Puzyn et al. (2010); Todeschini and Consonni (2009)) where they are used to predict molecular activity on the basis of structural properties expressed by indices. This is used to filter out from a large number of possible molecules those potentially producing a desired effect.

[33] If the direction of all arcs is reversed, measures of influence become measures of support, and vice versa.

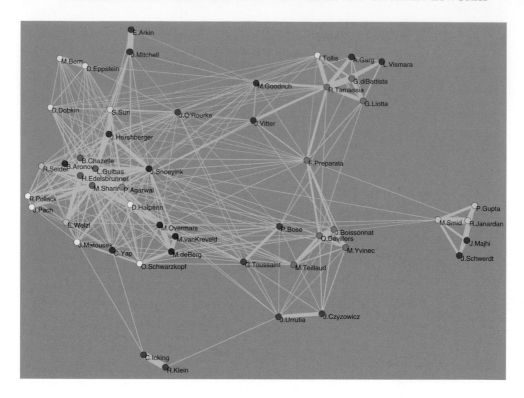

Figure 2.40 p_S-core at level 46 of the Computational Geometry network.

Many vertex indices are well-defined only for connected (sometimes only for strongly connected) networks. See also the extended discussion in Borgatti and Everett (2006).

The substantive meaning of an importance measure depends on the relation described by a network. In terms of degree as an index, for a directed advice network, the most 'important' person usually is one providing the most advice to others. The one receiving the most advice is also distinctive. In contrast, for a network defined by the relation 'does not want to work with', its most distinctive person is the one with whom others do not want to work.

Let $p : \mathcal{V} \to \mathbb{R}$ be an index in network $\mathcal{N} = (\mathcal{V}, \mathcal{L}, p)$. To compare an index, p, across different networks varying in size, they have to be comparable. Usually, this is achieved by some *normalization* of p. Let $\mathcal{N} \in \mathbf{N}(\mathcal{V})$, where $\mathbf{N}(\mathcal{V})$ is a selected family of networks over the same set of vertices $\mathcal{V}$,

$$p_{max} = \max_{\mathcal{N} \in \mathbf{N}(\mathcal{V})} \max_{v \in \mathcal{V}} p_{\mathcal{N}}(v) \quad \text{and} \quad p_{min} = \min_{\mathcal{N} \in \mathbf{N}(\mathcal{V})} \min_{v \in \mathcal{V}} p_{\mathcal{N}}(v)$$

then a generic normalized index is

$$p'(v) = \frac{p(v) - p_{min}}{p_{max} - p_{min}} \in [0, 1].$$

Some examples of normalization follow.

2.11.1   Degrees, closeness, betweenness and other indices

The simplest indices are the degrees of vertices (see Section 2.1.2). For simple undirected networks, $\deg_{min} = 0$ and $\deg_{max} = n - 1$, the corresponding normalized indices[34] are (using $'$ to denote the normalized indices):

$$\text{centrality} \qquad \deg'(v) = \frac{\deg(v)}{n - 1}$$

(for simple directed networks, the denominator is $\deg_{max} = n$)

$$\text{support} \qquad \text{indeg}'(v) = \frac{\text{indeg}(v)}{n}$$

$$\text{influence} \qquad \text{outdeg}'(v) = \frac{\text{outdeg}(v)}{n}$$

Degrees of a network can be considered also with respect to the reachability relation (transitive closure) or k-neighbors.

Many indices are based on the distance, $d(u, v)$, between vertices in a network $\mathcal{N} = (\mathcal{V}, \mathcal{L})$. Such measures are the *farness* $S(v) = \sum_{u \in \mathcal{V}} d(v, u)$ and the *diameter* of a network defined as $D = \max_{u, v \in \mathcal{V}} d(v, u)$. Sabidussi (1966) introduced a related measure $1/S(v)$. A vertex, v, is close to the other vertices if its total distance to them (farness) is small. The normalized form of this measure is

$$\text{closeness} \qquad cl(v) = \frac{n - 1}{\sum_{u \in \mathcal{V}} d(v, u)}$$

If some other vertices are not reachable from vertex v then $cl(v) = 0$. All vertices have non-zero closeness values iff the network is strongly connected.

Using the imagery of information flowing through a communication network, Freeman (1977) proposed an index of betweenness to operationalize an index of vertex importance. Vertices thought to exert control over the flow of information in a network, by virtue of being on certain paths, are seen as important. Assuming this flow uses only the geodesics (shortest paths) leads to a measure of *betweenness* (Anthonisse 1971; Freeman 1977):

$$b(v) = \frac{1}{(n - 1)(n - 2)} \sum_{\substack{u, t \in \mathcal{V} : g_{u,t} > 0 \\ u \neq v, t \neq v, u \neq t}} \frac{g_{u,t}(v)}{g_{u,t}}$$

where $g_{u,t}$ is the number of geodesics from u to t; and $g_{u,t}(v)$ is the number of those among them that pass through vertex v. Formulated initially for undirected networks, this concept has been extended to directed networks. Brandes (2001) developed a fast algorithm for computing betweenness.

Kleinberg (1998) proposed another approach for evaluating the importance of web pages in a network where web pages link to other such pages. To each vertex v of a network $\mathcal{N} = (\mathcal{V}, \mathcal{L})$ two values can be assigned to capture: 1) the 'quality' of its content (*authority*), x_v, and 2) the 'quality' of its references (*hub*), y_v. A good authority is selected by good hubs with good hubs pointing to good authorities.

$$x_v = \sum_{u : (u,v) \in \mathcal{L}} y_u \qquad \text{and} \qquad y_v = \sum_{u : (v,u) \in \mathcal{L}} x_u$$

---

[34] Comparisons using normalized indices are most useful for small networks of sizes less than Dunbar's number. For large networks, the normalized indices are very small and need to be multiplied by some large number. Even so, distinctive vertices can still be identified.

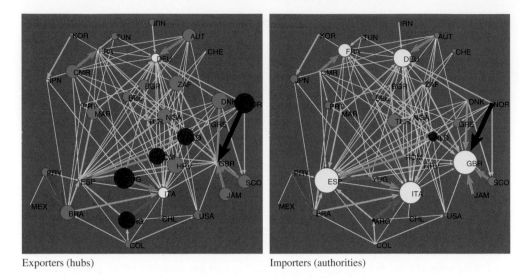

Exporters (hubs)   Importers (authorities)

Figure 2.41   Hubs and authorities for the 1998 World Cup football squads.

Let $\mathbf{W}$ be a matrix of network $\mathcal{N}$ and let $\mathbf{x}$ and $\mathbf{y}$ be authority and hub vectors. The two vectors are linked by $\mathbf{x} = \mathbf{W}^T\mathbf{y}$ and $\mathbf{y} = \mathbf{W}\mathbf{x}$. The normalized solution of these two equations can be determined by iteration. The limit vector $\mathbf{x}^*$ is the principal eigenvector of $\mathbf{W}^T\mathbf{W}$, and $\mathbf{y}^*$ is the principal eigenvector of $\mathbf{W}\mathbf{W}^T$. This approach is useful also for other directed networks and has been extended further to weighted networks and two-mode networks.

Chapters 7 and 8 examine the flow of football players to the English Premier League. More generally, moves of players between clubs and countries area are a feature of organized football, especially in recent times.[35] Football players playing professionally outside their country of origin is a global phenomenon. The 1998 World Cup finals in Paris were contested by 22 nations. Many of these international players played for clubs in other countries. While far removed from website networks, identifying hubs and authorities is useful for this football network. Hubs are countries exporting top players to clubs in other countries while authorities are the national teams. The hubs and authorities for the Krempel's data (see http://vlado.fmf.uni-lj.si/pub/networks/data/sport/football.htm) are shown in Figure 2.41. The largest hubs, marked in red in the left panel of Figure 2.41, are Yugoslavia, Norway, the Netherlands, Argentina, and Romania. The largest authorities, marked in yellow in the right panel, are Spain, England (labeled as GBR), Italy, Germany, and France. These five countries have the top football leagues on the planet.[36]

The Pajek commands for obtaining these indices are:

```
Network/Create New Network/Subnetwork with Paths/Info on Diameter
Network/Create Vector/Centrality/Closeness
```

---

[35] Hughes (2013) examined the Premier League and noted 'The lament is that homebred players players can scarcely get on the ball, scarcely get a game in their own league, now that the big clubs hire two thirds of their talents from overseas'.

[36] More data regarding this are presented in Chapter 7.

```
Network/Create Vector/Centrality/Betweeness
Network/Create Vector/Centrality/Hubs-Authorities
```

2.11.2   Clustering

In thinking about 'parts' of networks, we mentioned more dense areas of networks. Indeed, it is well known that in most real-world social networks, nodes tend to create tightly knit groups characterized by a relatively high density of ties. Holland et al. (1983) noted that the likelihood of ties forming in these dense parts tended to be larger than for the likelihood of ties forming between random pairs of individuals. Watts and Strogatz (1998) built on this insight to formalize their concepts of clusterability. The global variant provides information of the tendency for a network to be clustered (with denser parts linked to the rest of the network). In contrast, the local variant operationalizes the embeddedness of vertices by quantifying the extent to which the neighbors of a vertex form a clique.

Let $G = (V, E)$ be simple undirected graph. *Clustering* (local density) at vertex v is usually measured with clustering coefficient $C(v)$ defined as a quotient between the number of edges in subgraph $G^1(v) = G(N^1(v))$ induced by the neighbors of vertex v and the number of edges in the complete graph on these vertices:

$$C(v) = \begin{cases} \dfrac{2|\mathcal{E}(G^1(v))|}{\deg(v)(\deg(v) - 1)} & \deg(v) > 1 \\ 0 & \text{otherwise} \end{cases}$$

For a simple directed graph $G = (V, A)$ without loops we get:

$$C(v) = \begin{cases} \dfrac{|\mathcal{A}(G^1(v))|}{\deg(v)(\deg(v) - 1)} & \deg(v) > 1 \\ 0 & \text{otherwise} \end{cases}$$

The clustering coefficient attains its largest values mainly on vertices with small degrees – the probability that the subgraph $G^1(v)$ is complete is decreasing with the degree of v. Therefore, it is not very useful for the data analytic task of identifying large locally dense parts of a network. However, this important task can be achieved by using the *corrected clustering coefficient*

$$C_1(v) = \frac{\deg(v)}{\Delta} C(v)$$

where Δ is the maximum degree in graph G. This measure attains its largest value in vertices that belong to an isolated clique of size Δ.

The Pajek command for obtaining this index is:

```
Network/Create Vector/Clustering Coefficients
```

The network used in Figure 2.16 is shown again in Figure 2.42 to illustrate the computed values of selected indices. We use it also in Section 2.11.3. Corrected clustering coefficients,

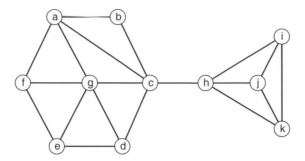

Figure 2.42   A simple network.

This network is from Figure 2.16.

Table 2.2   Values of selected indices for the simple network.

i	v	deg	cl	b	CC1	CC1'	CC2	CC2'	L
1	a	4	0.5263	0.1000	0.5000	0.40	0.3333	0.2667	50
2	b	2	0.4545	0.0000	1.0000	0.40	0.1250	0.0500	24
3	c	5	0.6667	0.5889	0.3000	0.30	0.2143	0.2143	66
4	d	3	0.5000	0.0611	0.6667	0.40	0.2000	0.1200	38
5	e	3	0.4000	0.0111	0.6667	0.40	0.2857	0.1714	34
6	f	3	0.4167	0.0167	0.6667	0.40	0.2222	0.1333	36
7	g	5	0.5556	0.1556	0.5000	0.50	0.6250	0.6250	66
8	h	4	0.5556	0.4667	0.5000	0.40	0.3000	0.2400	48
9	i	3	0.4000	0.0000	1.0000	0.60	0.7500	0.4500	32
10	j	3	0.4000	0.0000	1.0000	0.60	0.7500	0.4500	32
11	k	3	0.4000	0.0000	1.0000	0.60	0.7500	0.4500	32

as computed in Pajek, take two forms. The first, labeled CC1 uses 1-neighborhoods for the computation and the second, CC2 uses 2-neighborhoods. The results are shown in Table 2.2.

It seems that for such a small network, using 1-neighborhoods is more sensible – a reminder that all methods need to be chosen carefully taking into account the network studied and the questions asked.

2.11.3   Computing further indices through functions

Once indices for vertices have been obtained, they can be used to construct other indices. Pajek contains basic arithmetic operations and functions for working with vectors. These include:

```
Vector/Transform/
Operations/Vector/
Vectors/
```

This allows users to define additional indices and to compute them inside Pajek. One such example is found in Qi et al. (2013) who considered terrorist networks in order to identify vertices whose removal would disrupt the network. The new centrality measure, $L(v)$, called *Laplacian centrality*, is based on Laplacian energy: formally,

$$L(v) = \deg(v)(\deg(v) + 1) + 2 \sum_{u \in N(v)} \deg(u).$$

The commands in Pajek for doing this are:
select the network
```
Network/Create Vector/Centrality/Degree/All
Vector/Transform/Add Constant [1]
Operations/Network+Vector/Neighbours/Sum/All [No]
Vector/Transform/Multiply by [2]
```
select the degree vector as First
select the degree vector plus 1 as Second
```
Vectors/Multiply (First*Second)
```
select the 2*sum on neighbors as Second
```
Vectors/Add (First+Second)
File/Vector/Change Label [Laplace All centrality]
```

A Pajek macro `laplace.mcr` is available for doing this. The results of computing the *closeness*, *betweenness*, and *Laplace indices* for Figure 2.42 are also included in Table 2.2. It would seem, by visual inspection (of deg, *cl*, *b*, and *L* values), the removal of vertices, *c* and *h* would be the most disruptive. There are several items to note about these results. First, both degree and Laplacian degree identify vertex *c* but not *h* as the most distinctive vertices. Second, it is not clear how much is gained by using Laplacian degree, an impression reinforced by the product-moment-correlation between degree and Laplacian degree being 0.98. Indeed, Qi et al. (2013) computed a variety of degree measures for a variety of networks. The ranked distributions, overall, tended to agree strongly, with only minor differences. Third, in contrast, betweenness measures identify exactly *c* and *h*. Both are cut vertices (they disconnect the network) and, together, they form a bridge. These concepts provide another way of identifying such vertices.

Continuing the imagery of disrupting networks, suppose both *c* and *h* are truly elusive but *g* is not. Removing *g*, identified by the degree-based measures, would be disruptive: while not disconnecting the network, it would make communication within the left-hand denser patch more circuitous.

The basic message of this section is that it can be useful to construct indices from other indices. However, some caveats are worth noting. First, each index needs to be justified on both conceptual and empirical grounds. Second, doing this is more subtle than usually anticipated. Third, each index operationalizes different ideas, and the value of each index, relative to other indices, has to be justified in terms of superior performance. Fourth, it is worth exploring where indices point to different outcomes. While both the *degree-based measures* and betweenness measures can be seen as relating to identifying distinctive vertices, both may be useful empirically because they operationalize different aspects of network structure.

2.12   Transition to methods for large networks

The ideas, concepts, and tools outlined in this chapter provide a foundation for a sustained presentation of the methods for large networks presented in the next chapter. Section 2.3 presented a preliminary discussion of large networks and provided some results for analyzing a large network. Also, Section 2.5, devoted to statistical measures for networks, presented results for large networks. We noted that there was no rigid boundary between this chapter and the next in terms of technical concepts. Even so, every topic considered in Chapter 3, is designed for the analysis of large networks despite being useful also for smaller networks.

3

Methods for large networks

The first three temporal networks considered in this book are citation networks. Contemporary citation network analysis has quite old origins. Bernal (1953) and Asimov (1963) provided important foundations. Asimov's *The Genetic Code* provided a listing of key manuscripts, in one area of science, from which Garfield et al. (1964) extracted citation links between these documents.[1] This idea was the inspiration for a *citation network* as a concept and for creating citation networks. Asimov's history of DNA was one of a set of demonstration projects for the viability of this approach to studying the history of science. They showed that their analysis was able to achieve a high degree of coincidence between the citation relationships they established between publication events and a historian's account of these events. As they formulated their approach, the basic goal of citation network analysis is to identify the main scientific productions and the main branches or themes in the development of a field. Recent overviews and discussions of approaches to citation network analysis can be found in Lucio-Arias and Leydesdorff (2008) and Calero-Medina and Noyons (2008).

Figure 3.1 shows the small citation network taken from the DNA literature (see Garfield (1979) for details) as drawn by Hummon and Doreian (1989). The years of publication ranged from 1820 to 1962 for important publications regarding the development of DNA theory. Selected decades are marked to show approximate times of the publications.[2]

The arcs in citation networks can be interpreted in two, essentially equivalent, ways. The arcs can flow backward by marking the actual citations from one document to an earlier document. Alternatively, the arcs can represent the flow of knowledge forward in time. Citations mark the importance[3] of earlier knowledge contained in the cited document for developing the knowledge in the citing document. Figure 3.1 shows the latter, as was done by Hummon and Doreian (1989). (The figure of Garfield has the ties in the reverse

---

[1] They credited Gordon Allen for coupling the citations in reference lists and diagramming the citation links.

[2] When documents are few in some periods but concentrated in others, it is useful to use a 'distorted' timescale.

[3] While citations to earlier documents can be empty, self-serving self-cites, or negative cites, overwhelmingly, citations do acknowledge important earlier work.

Understanding Large Temporal Networks and Spatial Networks: Exploration, Pattern Searching, Visualization and Network Evolution, First Edition. Vladimir Batagelj, Patrick Doreian, Anuška Ferligoj and Nataša Kejžar.
© 2014 John Wiley & Sons, Ltd. Published 2014 by John Wiley & Sons, Ltd.

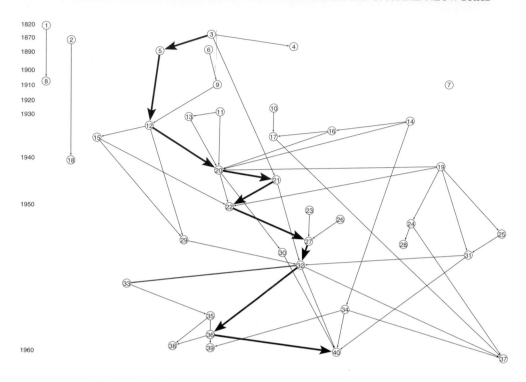

Figure 3.1   An example of a citation network: DNA literature.

direction.) Vertices can be differentially important, and some paths in a citation network may be distinctive. The idea of 'main paths' for the flow of knowledge was formulated to capture the main flows of knowledge in a citation network. The main path for the DNA network is shown with thicker arcs.

The citation network shown in Figure 3.1 is disconnected with two separate arcs on the left and an isolate on the right. Having them present is a reminder of several potential features of citation networks as representatives of a historical record for a field. First, knowledge can build from separate starting points through the combination of distinct ideas. Second, items thought to be important in written historical accounts may remain uncited. Third, ideas in manuscripts thought to be important at one point in time can lead nowhere and die out. Usually, when analyzing citation networks the disconnected fragments are ignored because cumulated knowledge cannot flow through them. Only the large weak components, such as the middle one in Figure 3.1, are considered.

In principle, citation networks have no cycles because later productions cite earlier productions. Many useful tools have been developed for analyzing acyclic networks. Section 3.1 presents these networks, and some methods for analyzing them follow in Section 3.2. The patent citation network of Chapter 5 is acyclic. The blue edge in Figure 3.1 shows a pair of papers (32 and 33) citing each other to form a 2-cycle. Both Chapter 4 on scientific citation networks and Chapter 6 on the Supreme Court have cycles of varying lengths. Their presence creates a problem for the methods described in Section 3.2. This problem can be handled by the methods described in Section 3.4.

Section 3.5 presents materials for the analysis of two-mode networks. Especially important are methods for multiplying two-mode matrices. We noted, when discussing algorithmic complexity in Chapter 2, that multiplying matrices by using the standard formula was impractical for large matrices. Sparse matrix methods for multiplying matrices are described within Section 3.5 together with other useful methods for handling two-mode networks. This is followed by a discussion of the pathfinder algorithm in Section 3.8. For Chapters 5 and 8, methods for clustering symbolic data play a central role and are laid out in Section 3.10.

3.1 Acyclic networks

A directed graph $\mathcal{G} = (\mathcal{V}, R)$, $R \subseteq \mathcal{V} \times \mathcal{V}$ is *acyclic* if it contains no (proper) cycles.[4] Formally,

$$\bar{R} \cap I = \emptyset$$

where $\bar{R} = \cup_{k=1}^{\infty} R^k$ denotes the *transitive closure* of relation R and $I = \{(v, v) : v \in \mathcal{V}\}$.

All the strong components of an acyclic network are trivial: they are single vertices. Therefore, the number of such components is equal to the number of vertices n. To check whether a network is acyclic in Pajek the following command is used:

```
Network/Create Partition/Components/Strong [2]
```

If the graph is acyclic there will be a single component[5] labeled cluster 0.

Real-life acyclic networks usually have a vertex property $p : \mathcal{V} \to \mathbb{R}$ (most often the time stamp of the vertices), that is *compatible* with arcs

$$(u, v) \in R \Rightarrow p(u) < p(v).$$

When p maps vertices to time points, this means that the vertex u appeared before the vertex v.

3.1.1 Some basic properties of acyclic networks

Let $\mathcal{G} = (\mathcal{V}, R)$ be acyclic and $\mathcal{U} \subseteq \mathcal{V}$ and $Q \subseteq R$, then $\mathcal{G}|\mathcal{U} = (\mathcal{U}, R|\mathcal{U})$, $R|\mathcal{U} = R \cap \mathcal{U} \times \mathcal{U}$ and $(\mathcal{V}, Q)$ are also acyclic. A subgraph of an acyclic graph is acyclic.

The inverse relation R^{inv}, $(u, v) \in R^{\mathrm{inv}} \Leftrightarrow (v, u) \in R$, describes a graph $\mathcal{G}' = (\mathcal{V}, R^{\mathrm{inv}})$ which is obtained from the graph $\mathcal{G} = (\mathcal{V}, R)$ by changing the direction of all its arcs. An example of a pair of inverse relations is citations and knowledge flows, as described in the introduction to this chapter. Let $\mathcal{G} = (\mathcal{V}, R)$ be acyclic, then $\mathcal{G}' = (\mathcal{V}, R^{\mathrm{inv}})$ is also acyclic. The transitive closure, $\bar{R}$, of an acyclic relation R is acyclic.

The indegree of a *source* is 0 and the outdegree of a *sink* is 0. For an acyclic network with a finite set of vertices, there is at least one source and at least one sink. They are also called minimal and maximal elements respectively. The set of minimal elements (sources) for a relation, R, over vertices, $\mathcal{V}$, is denoted by $\mathrm{Min}_R(\mathcal{V})$ and the set of maximal elements (sinks) by $\mathrm{Max}_R(\mathcal{V})$. Formally, the set of sources $\mathrm{Min}_R(\mathcal{V}) = \{v : \neg \exists u \in \mathcal{V} : (u, v) \in R\}$ and the set of sinks $\mathrm{Max}_R(\mathcal{V}) = \{v : \neg \exists u \in \mathcal{V} : (v, u) \in R\}$. For all finite graphs these two sets are non-empty. In Figure 3.2, the vertex v_8 and v_{10} are sources and v_3 and v_6 are sinks.

---

[4] For some analyses, loops are allowed.

[5] Strictly, this is not a strong component: every vertex is marked with 0 in Pajek to indicate that there are no non-trivial strong components.

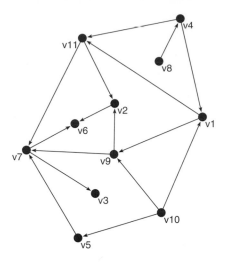

Figure 3.2   An acyclic network.

Every vertex $u \in \mathcal{V}$ and every arc $(u, v) \in R$ belong to at least one path from $\text{Min}_R(\mathcal{V})$ to $\text{Max}_R(\mathcal{V})$. Formally: $\forall u \in \mathcal{V} : \bar{R}(u) \cap \text{Max } R \neq \emptyset$ and $\forall u \in \mathcal{V} : R^{\text{inv}}(u) \cap \text{Min } R \neq \emptyset$

A relation Q is a (transitivity or reachability) *skeleton* of relation $\bar{R}$ iff $Q \subseteq R$, $\bar{Q} = \bar{R}$ and the relation Q is a minimal such relation: no arc can be deleted from it without destroying the second property. A general relation (graph) can have several skeletons. For an acyclic relation it is uniquely determined by $Q = R \setminus R * \bar{R}$.

3.1.2   Compatible numberings: Depth and topological order

Figure 3.1 is drawn in a way to make the ordering of paths from initial vertices to terminal vertices clear. While all acyclic networks can be drawn like this, Figure 3.2 does not have this form. However, it can be redrawn to make this feature clear. Two results of doing this are shown in Figure 3.3. The layouts are identical but the vertices are labeled differently to illustrate two ideas.

A mapping $h : \mathcal{V} \rightarrow \mathbb{N}^+$ is called *depth* or *level* if it is compatible with arcs and all differences on the longest path and the initial value equal to 1. The following algorithm, expressed in pseudo-code,

$\mathcal{U} \leftarrow \mathcal{V}; k \leftarrow 0$
while $\mathcal{U} \neq \emptyset$ **do**
    $\mathcal{T} \leftarrow \text{Min}_R(\mathcal{U}); k \leftarrow k + 1$
    for $v \in \mathcal{T}$ **do** $h(v) \leftarrow k$
    $\mathcal{U} \leftarrow \mathcal{U} \setminus \mathcal{T}$

determines the depth h of a given acyclic graph. The resulting depth for the acyclic graph in Figure 3.2 is presented on the left in Figure 3.3. The depth, in terms of vertex numbering, is not uniquely determined. For example, $h(v_{10})$ can be changed to 2 and $h(v_5)$ can be changed 2 or $h(v_5) = 4$. However, the depth (number of levels) of an acyclic network is unique, as it

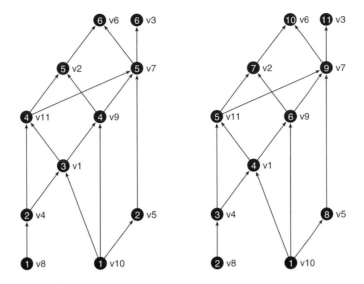

Figure 3.3   Depth and topological ordering in acyclic network.

is determined by the number of arcs in the longest path from an initial vertex to a terminal vertex. There are two longest paths with a length of 5, implying the depth of the network in Figure 3.3 is $h = 6$.

Knowing the depth, h, of a network facilitates drawing acyclic graphs by levels.[6] In Pajek, the macro Layers is available for this task. It is mobilized with:

Macro/Play [Layers]

An injective mapping from one set to another is a one-to-one mapping preserving distinctness: it never maps distinct elements of the first set to the same element of the second set. An injective mapping $i : \mathcal{V} \rightarrow 1..|\mathcal{V}|$ compatible with the relation R is called a *topological ordering*. It can be produced using the 'topological sort' algorithm expressed in pseudo-code as:

$\mathcal{U} \leftarrow \mathcal{V}; k \leftarrow 0$
while $\mathcal{U} \neq \emptyset$ **do**
    select $v \in \mathrm{Min}_R(\mathcal{U}); k \leftarrow k + 1$
    $i(v) \leftarrow k$
    $\mathcal{U} \leftarrow \mathcal{U} \setminus \{v\}$

In the step, 'select $v \in \mathrm{Min}_R(\mathcal{U})$,' there are usually multiple choices for selecting the next vertex: the resulting topological sort is not uniquely determined. However, the maximum numerical label is unique and equals n. On the right side of Figure 3.3 we show a topological order for acyclic network in Figure 3.2. The numbering obtained by changing $i(v_8) = 1$ and $i(v_{10}) = 2$ is also a topological order.

---

[6] Since in such pictures all the arcs are pointing in the same direction – from a lower level to a higher level – the arrows on arcs can be omitted. We prefer to keep them.

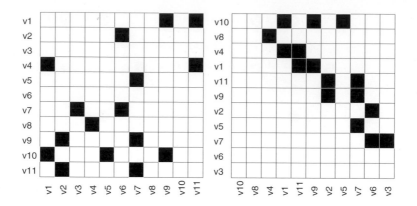

Figure 3.4 Matrix display.

The matrix display of an acyclic network with its vertices reordered according to a topological order has a zero lower triangle. On the left side of Figure 3.4, the matrix display of our example acyclic graph is shown. A matrix reordered according to the topological order is shown on the right of Figure 3.4. Note, the matrix is null below the main diagonal.

3.1.3 Topological orderings and functions on acyclic networks

One advantage of establishing a topological order of an acyclic network is that it sets up a useful efficient computation.

Let the function $f : \mathcal{V} \rightarrow \mathbb{R}$ be defined in the following way:

- $f(v)$ is known in sources $v \in \text{Min}_R(\mathcal{V})$

- $f(v) = F(\{f(u) : uRv\})$

Computing the values of the function f in a sequence determined by a topological ordering can be done in one pass since, for each vertex $v \in \mathcal{V}$, the values of f are known when needed.

3.1.3.1 Topological orderings and CPM

The *critical path method* (CPM) is a method created in operations research. A project consists of tasks. The vertices of a project network represent states of the project, and arcs represent tasks. Such networks are acyclic. For each task, (u, v), the time to complete it, $t(u, v)$, is known. A task can start only when all its preceding tasks are finished. A manager's goal is sequencing tasks to create the shortest project completion time.

Let $T(v)$ denote the earliest time of completion of all tasks entering the state v, then,

$$T(v) = 0, \qquad v \in \text{Min}_R(\mathcal{V})$$

$$T(v) = \max_{u:uRv} (T(u) + t(u, v)).$$

The function T has the form assumed for the function f at the beginning of this subsection.

The path (subgraph) consisting of arcs (u, v) for which $t(u, v) = T(v) - T(u)$ is called a *critical path*. Despite its narrowly defined origins of task sequencing, CPM is another available method for studying acyclic citation networks.

In Pajek, the command for obtaining a CPM path is:

```
Network/Acyclic Network/Critical Path Method-CPM
```

3.2 SPC weights in acyclic networks

An important step was made by Hummon and Doreian (1989). They proposed three indices (NPPC, SPLC, and SPNP) as weights of arcs, which can be used to provide an automatic way of identifying the (most) important part(s) of the citation network. This is known as *main path analysis*. This idea was applied in Hummon and Doreian (1990) and in Hummon et al. (1990) to discern main paths in rather small citation networks. Batagelj (1991, 2003) showed how to compute efficiently the Hummon and Doreian's weights. This implied that the weights can also be used for analysis of very large citation networks with several thousands or millions of vertices.

This section is structured as follows. First, a formal definition of citation networks is presented in Section 3.2.1. A short outline of the main path method is provided in Section 3.2.2. Details of the related methods follow. Section 3.2.3 presents the search path count method, and the computation of the SPLC and SPNP weights follows in Section 3.2.4. Some comments on implementation follow in Section 3.2.5. Thus far, attention has been on defining arc weights. Their usefulness can be extended to constructing vertex weights, which is described in Section 3.2.6. General properties of weights are stated in Section 3.2.7. This section concludes with some formal properties of the SPC weights in Section 3.2.8.

3.2.1 Citation networks

In a given set of units $\mathcal{U}$ (representing articles, books, works, ...) we introduce a *citing* relation. $R \subseteq \mathcal{U} \times \mathcal{U}$. For $u, v \in \mathcal{U}$,

$$uRv \equiv u \text{ cites } v \equiv v \text{ is cited by } u$$

which determines[7] a *citation network* $\mathcal{N} = (\mathcal{U}, R)$.

A citing relation is usually *irreflexive*, $\forall u \in \mathcal{U} : \neg uRu$ (documents do not cite themselves), and (almost) *acyclic* – no vertex is reachable from itself by a non-trivial path, or formally $\forall u \in \mathcal{U} \, \forall k \in \mathbb{N}^+ : \neg uR^k u$. In the following we shall assume that a citation network has this property. Methods for dealing with the problems posed by non-acyclic citation networks are presented in Section 3.4.

For a relation $R \subseteq \mathcal{U} \times \mathcal{U}$ we denote by

$$R(u) \equiv \{v \in \mathcal{U} : uRv\}$$

the set of successors of unit $u \in \mathcal{U}$. If R is acyclic then also R^{inv} is acyclic. This means that the network $\mathcal{N}^{\text{inv}} = (\mathcal{U}, R^{\text{inv}})$, $uR^{\text{inv}}v \equiv u$ is cited by v, is a network of the same type as the

---

[7] The symbol, $\equiv$, stands 'is the same as'.

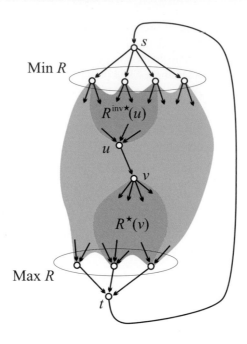

Figure 3.5   Citation network in standard form.

original citation network $\mathcal{N} = (\mathcal{U}, R)$. Therefore it is just a matter of 'taste' which relation, $R \equiv$ cites or $R^{\text{inv}} \equiv$ is cited by , is selected.

Many citation networks, including the one shown in Figure 3.1, have multiple sources and sinks as well as multiple weak components. To simplify the presentation we transform a citation network $\mathcal{N} = (\mathcal{U}, R)$ to its *standard form*, $\mathcal{N}' = (\mathcal{U}', R')$, by extending the set of units $\mathcal{U}' := \mathcal{U} \cup \{s, t\}$, $s, t \notin \mathcal{U}$ with a common *source* (initial unit) s and a common *sink* (terminal, target unit) t, and by adding the corresponding arcs to the relation R:

$$R' := R \cup \{s\} \times \text{Min } R \cup \text{Max } R \times \{t\} \cup \{(t, s)\}$$

The standard form is shown in Figure 3.5. This eliminates problems with networks with several connected components and/or several initial/terminal units. In the following we shall assume that the citation network $\mathcal{N} = (\mathcal{U}, R)$ is in the standard form. Note that, to make the theory smoother, we added to R' also the 'feedback' arc (t, s), thus destroying its acyclicity.

3.2.2   Analysis of citation networks

An approach to the analysis of citation network is to determine for each unit its *importance* or *weight*. These values are used afterwards to determine the essential substructures in the network. In the following we shall focus on the methods of assigning weights $w : \mathbb{R} \to \mathbb{R}_0^+$ to arcs proposed by Hummon and Doreian (1990). They pointed out:

- For the *search path link count* (SPLC) method, $w_l(u, v)$ equals the number of '*all possible search paths through the network emanating from an origin node*' through the arc $(u, v) \in R$.

- In the *search path node pair* (SPNP) method, $w_p(u, v)$ 'accounts for all connected vertex pairs along the paths through the arc $(u, v) \in R$.'

The next two subsections detail the computation of these weights.

3.2.3 Search path count method

To compute the SPLC and SPNP weights we introduce a related *search path count* (SPC) method for which the weight $N(u, v)$ for uRv counts the number of *different* paths from s to t (or from Min R to Max R) with the arc (u, v) on them. The path from a vertex, s, to a vertex, t, is denoted by s-t. To compute $N(u, v)$ we introduce two auxiliary quantities: 1) $N^-(v)$ denotes the number of different s-v paths, and 2) $N^+(v)$ denotes the number of different v-t paths.

Every s-t path, π, containing the arc $(u, v) \in R$ can be uniquely expressed in the form

$$\pi = \sigma \circ (u, v) \circ \tau$$

where σ is an s-u path, τ is a v-t path, and $\circ$ denotes 'followed by'. Since every pair (σ, τ) of s-u / v-t paths gives a corresponding s-t path it follows:

$$N(u, v) = N^-(u) \cdot N^+(v), \qquad (u, v) \in R$$

where

$$N^-(u) = \begin{cases} 1 & u = s \\ \sum_{v:vRu} N^-(v) & \text{otherwise} \end{cases}$$

and

$$N^+(u) = \begin{cases} 1 & u = t \\ \sum_{v:uRv} N^+(v) & \text{otherwise.} \end{cases}$$

This is the basis of an efficient algorithm for computing the weights $N(u, v)$. After establishing a topological sort of the network Cormen et al. (2001), using the above formulae in topological order, implies that these weights can be computed in time of order $O(m)$ where m is the number of arcs in the network. Again, having a sparse (acyclic) network is critical for efficient computation. The counters $N(u, v)$ are used as SPC weights $w_c(u, v) = N(u, v)$ for computing the weights of Hummon and Doreian (1989, 1990).

3.2.4 Computing SPLC and SPNP weights

From the description of the SPLC method, $w_l(u, v)$ equals the number of 'all possible search paths through the network emanating from an origin node' through the arc $(u, v) \in R$. This implies that we have to consider each vertex as an origin of search paths. This is equivalent to applying the SPC method to the extended network $\mathcal{N}_l = (\mathcal{U}', R_l)$:

$$R_l := R' \cup \{s\} \times (\mathcal{U}' \setminus \cup R(s))$$

From the definition of the SPNP weights, $w_p(u, v)$ 'accounts for all connected vertex pairs along the paths through the arc $(u, v) \in R$.' Again, the computation of these arc weights can be completed by applying the SPC method to an extended network: $\mathcal{N}_p = (\mathcal{U}', R_p)$

$$R_p := R \cup \{s\} \times \mathcal{U} \cup \mathcal{U} \times \{t\} \cup \{(t, s)\}$$

in which every unit $u \in \mathcal{U}$ is additionally linked from the source s and to the sink t.

It holds $R' \subseteq R_l \subseteq R_p$.

3.2.5   Implementation details

In our first implementation of the SPC method the values of $N^-(u)$ and $N^+(u)$ for some large networks exceeded the range of Delphi's `LargeInt` (20 decimal places). We decided to use the `Extended` real numbers (range $= 3.6 \times 10^{-4951} .. 1.1 \times 10^{4932}$, 19–20 significant digits) for the counters. This range is safe also for very large citation networks.

Some citation networks can have a wide range for the values of the arc weights. Having very large/small numbers among the weights in such large networks creates a potential problem. One option for dealing with this problem is to use transformations to reduce the range of the weights. For example, using square roots or taking logarithms of the obtained weights can be used. Both transformations are monotone and therefore preserve the ordering of weights (indicating the importance of vertices and arcs). The transformed values are also more convenient for visualizing networks when arcs have different line thicknesses depending on their weights.

In Pajek, the transformation of line weights is done with:

```
Network/Create New Network/Transform/Line Values/ [Ln]
```
or[8]
```
Network/Create New Network/Transform/Line Values/ [Absolute +
  Sqrt]
```

Implementing varying line thicknesses in Pajek is done within the `Draw` window with:

```
Options/Size /of Arrows/Size of Arrow [value]
```

3.2.6   Vertex weights

Given the computation of arc weights, w, these quantities can be used to define and compute vertex weights, t, for the vertices linked by the arcs. These include:

$$t_c(u) = N^-(u) \cdot N^+(u)$$
$$t_l(u) = N_l^-(u) \cdot N_l^+(u)$$
$$t_p(u) = N_p^-(u) \cdot N_p^+(u)$$

They are counting the number of s-t paths of selected type through the vertex u. More specifically: $t_c(u)$ is the product of the number of paths coming into u and the number of paths going out from u; $t_l(u)$ is the same product computed in the extended network, $\mathcal{N}_l = (\mathcal{U}', R_l)$; and $t_p(u)$ is for the extended network, $\mathcal{N}_p = (\mathcal{U}', R_p)$.

---

[8] As negative line values are possible in some networks, taking absolute values of arcs has to precede taking square roots. This part of the general command is irrelevant for citation networks where all arcs are positive.

The relation $R^\star = \bar{R} \cup I$ is the *transitive and reflexive closure* of the relation R. In essence, for the corresponding network, loops on all vertices are joined with the transitive closure of R. Another vertex weight can be computed for $R^\star$:

$$t_d(u) = |R^{\text{inv}\star}(u)| \cdot |R^\star(u)|$$

3.2.7 General properties of weights

There are general properties of weights that are useful. As a general notation, we use $w_k(u, v; R_j)$ to denote the weight, w, on the arc, (u, v), in the relation, R_j, for the weight type, k. Let $\mathcal{N}_1 = (\mathcal{U}, R_1)$ and $\mathcal{N}_2 = (\mathcal{U}, R_2)$ be two citation networks over the same set of units $\mathcal{U}$ and $R_1 \subseteq R_2$. Then

$$w_k(u, v; R_1) \leq w_k(u, v; R_2), \qquad k = c, p.$$

Directly from the definitions of weights we also get

$$w_k(u, v; R) = w_k(v, u; R^{\text{inv}}), \qquad k = c, p$$

and

$$w_c(u, v) \leq w_l(u, v) \leq w_p(u, v).$$

These are formal properties of arc weights. One potential immediate application is when there are two variants of a citation network where citation arcs present in R_2 are not present in R_1. For electronically gathered citation networks this could be due to data errors. In the DNA network studied by Garfield et al. (1964), one citation network was constructed from Asimov's count in *The Gentic Code* with some 'inferred' citations, about which there could be legitimate disagreements. These two properties show constraints on the implications of such omissions. However, this applies only if $R_1 \subset R_2$.

Sometimes, it is useful to extract segments from a citation network for further study. The following general property is useful when this is done. Let $\mathcal{N}_A = (\mathcal{U}_A, R_A)$ and $\mathcal{N}_B = (\mathcal{U}_B, R_B)$ with, $\mathcal{U}_A \cap \mathcal{U}_B = \emptyset$, be two citation networks. We denote the standardized network of $\mathcal{N}_A$ by $\mathcal{N}_1 = (\mathcal{U}'_A, R'_A)$. A second network can be defined as the union of $\mathcal{N}_A$ and $\mathcal{N}_B$. Let $\mathcal{N}_2 = ((\mathcal{U}_A \cup \mathcal{U}_B)', (R_A \cup R_B)')$. Then it holds $\forall u, v \in \mathcal{U}_A$ and $\forall p, q \in R_A$

$$\frac{t_k^{(1)}(u)}{t_k^{(1)}(v)} = \frac{t_k^{(2)}(u)}{t_k^{(2)}(v)}, \qquad \text{and} \qquad \frac{w_k^{(1)}(p)}{w_k^{(1)}(q)} = \frac{w_k^{(2)}(p)}{w_k^{(2)}(q)}, \qquad k = c, l, p$$

where $t^{(1)}$ and $w^{(1)}$ are weights in the network $\mathcal{N}_1$, and $t^{(2)}$ and $w^{(2)}$ are weights in the network $\mathcal{N}_2$. The important implication of this property is that removing or adding segments in a citation network does not change the ratios (ordering) of the weights inside the segments. Extracting segments does not 'distort' important properties of weights when attention is focused on weights.

3.2.8 SPC weights

This section draws upon the concepts and measures introduced in Section 3.2.3 for which the concept of a *network flow*, $N(u, v)$, is defined. For $N(u, v)$, the *Kirchhoff's vertex law* holds:[9]
For every vertex v in a citation network in the standard form it holds

$$\text{incoming flow} = \text{outgoing flow} = t_c(v)$$

Proof:

$$\sum_{x:xRv} N(x, v) = \sum_{x:xRv} N^-(x) \cdot N^+(v) = (\sum_{x:xRv} N^-(x)) \cdot N^+(v) = N^-(v) \cdot N^+(v)$$

$$\sum_{y:vRy} N(v, y) = \sum_{y:vRy} N^-(v) \cdot N^+(y) = N^-(v) \cdot \sum_{y:vRy} N^+(y) = N^-(v) \cdot N^+(v)$$

$\square$

From Kirchhoff's vertex law it follows that the *total flow* through the citation network equals $N(t, s)$. This gives us a natural way to normalize the weights:

$$w(u, v) = \frac{N(u, v)}{N(t, s)} \quad \Rightarrow \quad 0 < w(u, v) \le 1$$

If C is a minimal arc-cut set, then

$$\sum_{(u,v) \in C} w(u, v) = 1.$$

Therefore, $w(u, v)$ can be interpreted also as the probability that an s-t path is passing through the arc (u, v).

Let $\vec{\mathbf{K}}_n = \{(u, v) : u, v \in 1..n \wedge u < v\}$ be the complete acyclic directed graph on n vertices then, since $R \subseteq \vec{\mathbf{K}}_n$, the value of $N(u, v; \vec{\mathbf{K}}_n)$ is maximum over all citation networks on n units. It is easy to verify that

$$N(1, n; \vec{\mathbf{K}}_n) = 2^{n-2}$$

and in general

$$N(i, j; \vec{\mathbf{K}}_n) = 2^{j-i-1}, i < j.$$

From this result we see that the exhaustive search algorithm proposed in Hummon and Doreian (1989, 1990) can require exponential time to compute the arc weights w.

---

[9] While the origins of Kirchhoff's law are found in the flow of electricity over electrical circuits, the flows defined here does not rely on this imagery and depend only on the values of the weights we have defined.

Since

$$N^-(u; R) = N^+(u; R^{\text{inv}}) \quad \text{and} \quad N^+(v; R) = N^-(v; R^{\text{inv}})$$

we have for $(u, v) \in R$ (and therefore $(v, u) \in R^{\text{inv}}$):

$$N(u, v; R) = N^-(u; R) \cdot N^+(v; R) = N^-(v; R^{\text{inv}}) \cdot N^+(u; R^{\text{inv}}) = N(v, u; R^{\text{inv}})$$

Each arc (u, v) in the citation network $(\mathcal{U}, R)$ gets the same SPC weight as its inverted arc (v, u) in the inverted network $(\mathcal{U}, R^{\text{inv}})$. The same holds also for SPNP weights.

Suppose that in an acyclic citation network $N = (U, R)$ the vertices u and v are directly linked with the arc $(u, v) \in R$ and also with a path $(u = t_0, t_1, t_2, \ldots, t_{k-1}, t_k = v)$ of length $k \geq 2$. Then each arc on the path has the SPC weight larger than or equal to the SPC weight of the arc (u, v).

To see this, observe that by definition for an arc $(u, v) \in R$

$$N^-(v) = N^-(u) + N^-(\text{other predecessors of } v)$$

implying $N^-(v) \geq N^-(u)$. Therefore we have $N^-(t_{i+1}) \geq N^-(t_i) \geq N^-(u)$; and similarly $N^+(v) \leq N^+(t_{i+1}) \leq N^+(t_i)$. This finally gives

$$N(t_i, t_{i+1}) = N^-(t_i) \cdot N^+(t_{i+1}) \geq N^-(u) \cdot N^+(v) = N(u, v),$$

proving the claim.

This result tells us that the strongest SPC weights belong to arcs belonging to the transitive reduction (skeleton) of a given citation network.

The algorithmic complexity of transitive reduction is $O(nm)$. This bounds its possible applications to networks with up to some tens of thousands of vertices.

3.3 Probabilistic flow in acyclic network

Another way to measure the importance of vertices and arcs in acyclic networks is the following. Let $\mathcal{N} = (\mathcal{V}, \mathcal{A})$ be a standardized acyclic network with source $s \in \mathcal{V}$ and sink $t \in \mathcal{V}$. The vertex potential, $p(v)$, is defined by:

- $p(s) = 1$ and

- $p(v) = \displaystyle\sum_{u:(u,v)\in\mathcal{A}} \frac{p(u)}{\text{outdeg}(u)}$

The flow on the arc (u, v) is defined as

$$\varphi(u, v) = \frac{p(u)}{\text{outdeg}(u)}.$$

It follows immediately that

$$p(v) = \sum_{u:(u,v)\in\mathcal{A}} \varphi(u, v).$$

Also,

$$\sum_{u:(v,u)\in A} \varphi(v,u) = \sum_{u:(v,u)\in A} \frac{p(v)}{\text{outdeg}(v)} = \frac{p(v)}{\text{outdeg}(v)} \sum_{u:(v,u)\in A} 1 = p(v).$$

Therefore, for each $v \in \mathcal{V}$

$$\sum_{u:(u,v)\in A} \varphi(u,v) = \sum_{u:(v,u)\in A} \varphi(v,u) = p(v),$$

which states that Kirchhoff's law holds for the flow φ.

The probabilistic interpretation of the flow φ has two parts:

1. The vertex potential of v, $p(v)$, is equal to the probability that a random walk starting in the source s goes through the vertex v.

2. The arc flow on (u, v), $\varphi(u, v)$, is equal to the probability that a random walk starting in the source s goes through the arc (u, v).

Note that the measures p and φ consider only 'users' (future) and do not depend on the past.

3.4 Nonacyclic citation networks

The problem with cycles is that even one cycle in a network creates an infinite number of trails between some units. The counts described in Section 3.2 are useless and cannot be used because all of them are massively inflated. This is a serious problem. Fortunately, there are two broad approaches for overcoming it. One focuses attention on the weights:

- introduce some 'aging' factor to force the total weight of all trails to converge to some finite value; or

- restrict the definition of weights to some finite subset of trails, for example, paths or geodesics.

Alas, this approach does not work well because two new problems arise: 1) How do we establish the right value for the aging (down-weighting) factor? 2) Is there an efficient algorithm to count the restricted trails? Most likely, using different down-weighting factors will lead to different orderings of weights which destroys a critical property of the weights established in Section 3.2. To our knowledge, regarding the second problem, no efficient algorithm exists. For our purposes, this approach is a dead end.

The second broad approach involves dealing with the cycles directly. Since a citation network is usually almost acyclic, this approach transforms a non-acyclic network into an acyclic network. There are at least three ways of doing this:

1. since the presence of cycles creates non-trivial strong components, identifying and removing them by shrinking (see Section 2.7);

2. strategically deleting some arcs also creates acyclic networks; and

3. using what we call a `preprint` transformation.

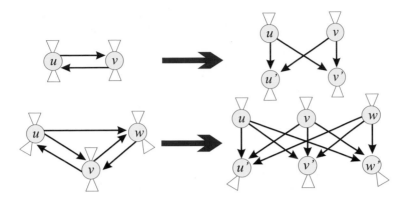

Figure 3.6 Preprint transformation.

Figure 3.6 illustrates these ideas. There are two cyclic configurations on the left. For the top 2-cycle, shrinking simply collapses u and v into a single vertex in a new network. Similarly, for the second configuration, shrinking collapses u, v, and w. If there is no need to preserve the identities of the vertices, this is a very simple way of forming an acyclic network. The other two operations are applicable when there is a need to preserve the identities of the vertices. Deleting arcs to create an acyclic network, while effective, is truly problematic because selecting arcs to delete is arbitrary and the choices change the meaning of specific citations. For the lower cyclic configuration in Figure 3.6, one option is the removal of (u, w), (w, v), and (v, u) which creates a different fragment than removing (u, v) and (w, v). We do not recommend this tactic.

The `preprint` transformation is based on the following idea: Each paper, u, from a strong component is *duplicated* with its 'preprint' version, u'. Under this transformation, the scientific productions, mainly papers, inside a strong component are treated as 'citing' their preprints, and the preprints cite the papers outside the strong component.

Table 3.1 presents some characteristics for some of the citation networks we have considered. The labels for the columns include n and m, the number of vertices and arcs respectively. The next three columns are the size of the largest weakly connected component (n_C), the number of non-trivial weakly connected components (k_C), and the depth of network, the minimum number of levels (h). The last three columns contain the numbers of strongly connected components (cyclic parts) of sizes 2, 3, and 4. In citation networks, large strong components are highly unlikely.[10] This is made clear in Table 3.1, where there are no strong components of size greater than 4. Indeed, the presence of large strong components usually indicates at least one error in the data. In general, shrinking and using the `preprint` transformation are very useful ways of converting a non-acyclic citation network into an acyclic network in order to mobilize the arc counts introduced in the previous section.

---

[10] An exception to this rule is a citation network of the high energy particle physics literature KDD Cup (2003) from *arXiv*. In it, different versions of the same paper are treated as separate units. Doing this leads to large strongly connected components. The idea of preprint transformation can be used in this case to eliminate cycles if necessary.

Table 3.1 Some citation network characteristics.

network	n	m	n_C	k_C	h	2	3	4
DNA	40	60	35	3	11	0	0	0
Coupling	223	657	218	1	16	0	0	0
Small world	396	1988	233	1	16	0	0	0
Small & Griffith	1059	4922	1024	1	28	2	0	0
Co-citation	1059	4929	1024	1	28	2	0	0
Scientometrics	3084	10416	2678	21	32	5	2	1
Kroto	3244	31950	3244	1	32	6	0	0
SOM	4470	12731	3704	27	24	11	0	0
Zewail	6752	54253	6640	5	75	38	1	2
Lederberg	8843	41609	8212	35	63	54	4	0
Desalination	8851	25751	7143	115	27	12	0	1
Clustering	72281	123514	72108	60	24	2	0	0
SNA	193376	324616	190848	174	35	10	2	0
US patents	3774768	16522438	3764117	3627	32	0	0	0

3.5 Two-mode networks from data tables

Thus far, we have treated bibliographic networks only in terms of citations from one document to other documents. However, there is far more information potentially available about these documents. This can include the author(s) of publications, the content of the publication (usually expressed in terms of keywords provided by their authors or extracted from titles and abstracts), the organizational affiliations of the authors, the years that productions appeared, and the journals in which articles are published.[11] Coupling these attributes to scientific productions is to mobilize two-mode networks (Batagelj and Cerinšek 2013).

The sources for the additional attribute data for scientific documents are secondary sources. Two-mode networks were introduced in Section 2.2.3. They can be used effectively to couple the additional information to documents. Most often, these primary data are available in the form of tables. Every such table can be transformed in a collection of compatible two-mode networks from which a variety of *derived networks* can be obtained.

Formally, a *data table* $\mathcal{T}$ is a set of *records* $\mathcal{T} = \{T_k : k \in \mathcal{K}\}$, where $\mathcal{K}$ is the set of *keys*. A record has the form $T_k = (k, q_1(k), q_2(k), \ldots, q_r(k))$ where $q_i(k)$ is the value of the *property* (attribute) $\mathbf{q}_i$ for the key k.

Suppose the property, $\mathbf{q}$, has, as values, subsets of the set Q. In the case of ordinary variables, the subset contains a single element. For example, describing bibliographies usually the properties include Authors, Keywords, and publication year (PubYear). For example, Wasserman and Faust (1994) is a book for which the set of keywords includes:

Authors(SNA) = { S. Wasserman, K. Faust },
PubYear(SNA) = { 1994},
Keywords(SNA) = { network, centrality, matrix, … }, …

---

[11] Overwhelmingly, the publications in scientific citation networks are journal articles, with journals being a primary institution of science.

Table 3.2 An example of bibliography descriptions.

Work	Authors	PubYear
...		
SNA	S. Wasserman, K. Faust	1994
Network Evolution	P. Doreian, F. Stokman	1997
Islands	M. Zaveršnik, V. Batagelj	2004
Blockmodeling	P. Doreian, V. Batagelj, A. Ferligoj	2005
IFCS09	N. Kejžar, S. Korenjak, V. Batagelj	2010
GenCores	V. Batagelj, M. Zaveršnik	2011
ESNA2	W. de Nooy, A. Mrvar, V. Batagelj	2012
...		

The list of publications can be expanded to incorporate a broader literature. Table 3.2 contains seven publications as an initial part of such an expansion. Here, Work is a key, with Authors and PubYear as properties.[12]

If Q is finite, we can assign to the property $\mathbf{q}$ a two-mode network $\mathcal{K} \times \mathbf{q} = (\mathcal{K}, Q, \mathcal{A}, w)$ where $(k, v) \in \mathcal{A}$ iff $v \in q(k)$, and $w(k, v) = 1$. Note that the set Q can always be transformed into a finite set by partitioning it and recoding the values.

3.5.1 Multiplication of two-mode networks

The different two-mode networks are obtained from bibliographic data, for example, include works by authors, works by journals, authors by institutional locations, and institutional locations by countries. Combining these discrete two-mode arrays is straightforward with matrix multiplication. 'Multiplying networks' is done by multiplying their matrices. Expressed more precisely, the product of two *compatible* networks is the network corresponding to the product of matrices corresponding to the given networks. Formally, for a simple (having no parallel arcs) two-mode *network* $\mathcal{N} = ((\mathcal{I}, \mathcal{J}), \mathcal{A}, w)$; where $\mathcal{I}$ and $\mathcal{J}$ are sets of *vertices*, $\mathcal{A}$ is a set of *arcs* linking $\mathcal{I}$ and $\mathcal{J}$, and $w : \mathcal{A} \to \mathbb{R}$ is a *weight*; a *network matrix* $\mathbf{W} = [w_{i,j}]$ is assigned with elements: $w_{i,j} = w(i, j)$ for $(i, j) \in \mathcal{A}$ and $w_{i,j} = 0$ otherwise.

Two two-mode networks are compatible (for multiplication) iff the second set of vertices in the first network is equal to the first set of vertices in the second network. Given a pair of compatible two-mode networks, $\mathcal{N}_A = ((\mathcal{I}, \mathcal{K}), \mathcal{A}_A, w_A)$ and $\mathcal{N}_B = ((\mathcal{K}, \mathcal{J}), \mathcal{A}_B, w_B)$ with corresponding matrices $\mathbf{A}_{\mathcal{I} \times \mathcal{K}}$ and $\mathbf{B}_{\mathcal{K} \times \mathcal{J}}$, the *product of networks* $\mathcal{N}_A$ and $\mathcal{N}_B$ is a network $\mathcal{N}_C = ((\mathcal{I}, \mathcal{J}), \mathcal{A}_C, w_C)$, where $\mathcal{A}_C = \{(i, j) : i \in \mathcal{I}, j \in \mathcal{J}, c_{i,j} \neq 0\}$ and $w_C(i, j) = c_{i,j}$ for $(i, j) \in \mathcal{A}_C$. The product matrix $\mathbf{C} = [c_{i,j}]_{\mathcal{I} \times \mathcal{J}} = \mathbf{A} * \mathbf{B}$ is defined in the standard way:

$$c_{i,j} = \sum_{k \in \mathcal{K}} a_{i,k} \cdot b_{k,j}$$

The general scheme for multiplying two-mode networks is shown in Figure 3.7 where $\mathbf{A}$ represents the matrix for $\mathcal{N}_A$ and $\mathbf{B}$ is the the matrix for $\mathcal{N}_B$. (When $\mathcal{I} = \mathcal{K} = \mathcal{J}$, the multiplication is of ordinary one-mode networks with square matrices.)

---

[12] Single-valued properties can be represented more compactly by a partition. This is more easily done for publication years.

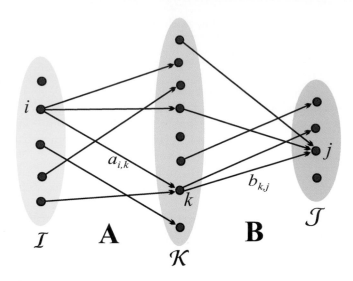

Figure 3.7   A visualization of two-mode network multiplication.

In the expression for $c_{i,j}$, the only terms $a_{i,k} \cdot b_{k,j}$ contributing to its value occur when both $a_{i,k}$ and $b_{k,j}$ are non-zero. The set $N_A(i)$ is made up of the *successors* of vertex i in network $\mathcal{N}_A$, and $N_B^-(j)$ is the set of *predecessors* of vertex j in network $\mathcal{N}_B$. For $N_A(i) \cup N_B^-(j) \neq \emptyset$,

$$c_{i,j} = \sum_{k \in N_A(i) \cup N_B^-(j)} a_{i,k} \cdot b_{k,j}.$$

Therefore, if all weights in networks $\mathcal{N}_A$ and $\mathcal{N}_B$ equal 1, then the product $a_{i,k} \cdot b_{k,j} = 1$ and the value of $c_{i,j}$ counts the number of ways we can go from $i \in \mathcal{I}$ to $j \in \mathcal{J}$ passing through $\mathcal{K}$. As shown in Figure 3.7, the element $c_{i,j}$ has the value of 2.

The standard matrix multiplication has complexity $O(|\mathcal{I}| \cdot |\mathcal{K}| \cdot |\mathcal{J}|)$. As noted in Section 2.3, it is too slow to be practical for large networks. Most of large networks are *sparse*: their matrices contain many more zero elements than non-zero elements. For sparse large networks, a much faster multiplication is possible by considering only the non-zero elements. In pseudo-code:

```
for k in K do
    for (i, j) in N_A^-(k) × N_B(k) do
        if ∃c_{i,j} then c_{i,j} := c_{i,j} + a_{i,k} · b_{k,j}
        else new c_{i,j} := a_{i,k} · b_{k,j}
```

The multiplication of large sparse networks can be a 'dangerous' operation since the result *need not* be a sparse network. When this occurs, the resulting matrix 'explodes'. And when a sequence of compatible networks is multiplied, having one of them not be sparse raises a serious computational problem that must be addressed. Doing this takes the form of establishing conditions under which this explosion does not occur.

For a two-mode network, let $\mathcal{U}$ and $\mathcal{V}$ be the two sets of vertices for the two modes. Further, let $n_1 = |\mathcal{U}|$ and $n_2 = |\mathcal{V}|$. Let $\mathcal{L}$ denote the set of edges. The corresponding complete two-mode network, $K_{n_1 n_2}$ is defined by: $\forall u \in \mathcal{U}$ and $\forall v \in \mathcal{V}$, $(u, v) \in \mathcal{L}$. In words, there are arcs between all pairs of vertices in the two modes.

From the network multiplication algorithm, each intermediate vertex $k \in \mathcal{K}$ adds to a product network a complete two-mode subgraph $K_{N_A^-(k), N_B(k)}$ (or, in the case $\mathcal{I} = \mathcal{J}$, a complete subgraph $K_{N(k)}$). If both degrees $\deg_A(k) = |N_A^-(k)|$ and $\deg_B(k) = |N_B(k)|$ are large then already the computation of this complete subgraph has a quadratic (time and space) complexity. This multiplication ceases to be a practical option.

However, if at least one of the sparse networks $\mathcal{N}_A$ and $\mathcal{N}_B$ has small maximal degree on $\mathcal{K}$, the resulting product network $\mathcal{N}_C$ is sparse. Indeed, there is a stronger result: if for the sparse networks $\mathcal{N}_A$ and $\mathcal{N}_B$, there are, in $\mathcal{K}$, only some vertices with large degree and none of them have large degree in both networks then also the resulting product network $\mathcal{N}_C$ is sparse. A proof follows:

Let

$$d_{min}(k) = \min(\deg_A(k), \deg_B(k)) \quad \text{and} \quad d_{max}(k) = \max(\deg_A(k), \deg_B(k)).$$

Then

$$\deg_A(k) \cdot \deg_B(k) = d_{min}(k) \cdot d_{max}(k).$$

Define also $\Delta_{min} = \max_{k \in \mathcal{K}} d_{min}(k)$ and

$$\mathcal{K}(d) = \{k \in \mathcal{K} : d_{max}(k) \geq d\},$$

let $d^* = \operatorname{argmin}_d(|\mathcal{K}(d)| \leq d)$ and $\mathcal{K}^* = \mathcal{K}(d^*)$. Then $|\mathcal{K}^*| \leq d^*$ and the number of non-zero elements in the product:

$$C \leq \sum_{k \in \mathcal{K}} \deg_A(k) \cdot \deg_B(k) = \sum_{k \in \mathcal{K}} d_{min}(k) \cdot d_{max}(k)$$

$$= \sum_{k \in \mathcal{K}^*} d_{min}(k) \cdot d_{max}(k) + \sum_{k \in \mathcal{K} \backslash \mathcal{K}^*} d_{min}(k) \cdot d_{max}(k)$$

$$\leq \Delta_{min} \cdot \sum_{k \in \mathcal{K}^*} d_{max}(k) + d^* \cdot \sum_{k \in \mathcal{K} \backslash \mathcal{K}^*} d_{min}(k)$$

$$\leq d^* \cdot (\Delta_{min} \cdot \max(|\mathcal{I}|, |\mathcal{J}|) + \min(|\mathcal{A}_A|, |\mathcal{A}_B|))$$

$$\square$$

Therefore: if, for the sparse networks $\mathcal{N}_A$ and $\mathcal{N}_B$, the quantities Δ_{min} and d^* are small then also the resulting product network, $\mathcal{N}_C$, is sparse. This result is equivalent to the above claim. If these conditions are satisfied, the matrix multiplication yields a matrix that does not explode. These conditions are satisfied for all of the citation networks in Table 3.1 and all of the bibliographic networks studied in Chapters 4–6.

A nice application of network multiplication is the computation of other kinship relations (is sister of, is uncle of, is cousin of, etc.) from the basic genealogical relations (is child of, is married to, is female) (Batagelj and Mrvar 2008).

3.6   Bibliographic networks

In Appendix A1 we describe how we can transform a bibliography into a collection of compatible two-mode networks. For a given set of works, besides the two-mode network **WA** on works × authors, we can construct other two-mode networks such as **WK** on works × keywords, **WC** on works × classifications, **WJ** on works × journals, etc.

A *collaboration network* is usually defined in the following way. The set of a network's *vertices* consists of *authors*. There exists an *edge* between authors u and v iff they produced a joint *work* (paper, book, report, etc.). Its *weight* $w(u, v)$ is equal to the number of works to which u and v both contributed.

In this case a more basic network is a *two-mode* network **WA** linking the set of works with the set of authors. There is an *arc* from the work p to the author u iff u is an author of the work p. It is well known that we can compute the corresponding collaboration network as $\mathbf{WA}^T * \mathbf{WA}$ using network multiplication.

Since these networks have the same first set – the set of works – we can obtain from them using multiplication different *derived* networks. For example $\mathbf{WA}^T * \mathbf{WK} = \mathbf{AK}$ gives us the two-mode network **AK** on authors × keywords with the weight of the arc (u, k) counting in how many works the author u used the keyword k. Additional derived networks can be produced considering also the one-mode citation network **Ci** between works.

3.6.1   Co-authorship networks

Let **WA** be the works × authors two-mode co-authorship network; $wa_{pi} \in \{0, 1\}$ is describing the authorship of author i of work p. Then for each work $p \in W$:

$$\sum_{i \in A} wa_{pi} = \text{outdeg}(p)$$

The $\text{outdeg}(p)$ is equal to the number of authors of work p. Similarly

$$\sum_{p \in W} wa_{pi} = \text{indeg}(i).$$

The $\text{indeg}(i)$ is equal to the number of works to which author i contributed.

Let **N** be its normalized version, with n_{pi} describing the share of contribution of author i to work p such that for each work $p \in W$:

$$\sum_{i \in A} n_{pi} \in \{0, 1\}$$

The sum has value 0 for works without authors.

The contributions n_{pi} can be determined by some rules or, assuming that each author contributed equally to the work, it can be computed from **WA** as

$$n_{pi} = \frac{wa_{pi}}{\max(1, \text{outdeg}(p))}.$$

A similar normalization of collaboration links, but with $\text{outdeg}(p) - 1$ instead of $\text{outdeg}(p)$, was proposed by Newman (2001). He is interpreting the weight as a proportion of time spent for the collaboration with each co-author.

Row-normalization $n(\mathcal{N})$ is a network obtained from $\mathcal{N}$ in which the weight of each arc a is divided by the sum of weights of all arcs having the same initial vertex as the arc a. For binary network $\mathbf{A}$ on $\mathcal{I} \times \mathcal{J}$

$$n(\mathbf{A}) = \text{diag}\left(\frac{1}{\max(1, \text{outdeg}(i))}\right)_{i \in \mathcal{I}} * \mathbf{A}.$$

Therefore we can obtain the normalized co-authorship network as

$$\mathbf{N} = n(\mathbf{WA}).$$

In some sense reverse transformation is the *binarization* $b(\mathcal{N})$ of the $\mathcal{N}$: it is the original network in which all weights are set to 1. It holds

$$\mathbf{WA} = b(\mathbf{N})$$

and if $\mathbf{N}$ was obtained from $\mathbf{WA}$ also $\mathbf{WA} = b(n(\mathbf{WA}))$.

Another useful transformation is the transposition. *Transposition* $\mathcal{N}^T$ or $t(\mathcal{N})$ is a network obtained from $\mathcal{N}$ in which to all arcs their direction is reversed. For bibliographic networks we introduce the abbreviations $\mathbf{AW} = \mathbf{WA}^T$, $\mathbf{KW} = \mathbf{WK}^T$, etc.

3.6.2 Collaboration networks

A standard way to obtain the *(first) collaboration network* $\mathbf{Co}$ from the co-authorship network using network multiplication is

$$\mathbf{Co} = \mathbf{AW} * \mathbf{WA}.$$

From

$$co_{ij} = \sum_{p \in W} wa_{pi} wa_{pj} = \sum_{p \in N(i) \cap N(j)} 1$$

we see that co_{ij} is equal to the number of works that authors i and j wrote together.

The weights in the first collaboration network are symmetric:

$$co_{ij} = \sum_{p \in W} wa_{pi} wa_{pj} = \sum_{p \in W} wa_{pj} wa_{pi} = co_{ji}$$

In $\mathbf{Co}$ one can search for authors with most collaborators. The output degree (without loops) of each author is equal to the number of his/her co-authors.

The obvious question is: who are the most collaborative authors? The standard answer is provided by k-cores. In a collaboration network a k-core is the largest subnetwork with the property that each of its author wrote a joint work with at least k other authors from the core. But, because works with many co-authors contribute large complete subgraphs to the collaboration network, some problems can appear as we will show in Chapter 4.

To neutralize the overrating of the contribution of works with many authors we can try alternative definitions of collaboration networks using the normalized co-authorship network. The structure (graph) of the collaboration network remains the same, but the weights change. The *second collaboration network* is defined as

$$\mathbf{Cn} = \mathbf{AW} * \mathbf{N}.$$

The value of the weight cn_{ij}

$$cn_{ij} = \sum_{p \in W} wa_{pi} n_{pj} = \sum_{p \in N(i) \cap N(j)} n_{pj}$$

is equal to the contribution of author j to works, that he/she wrote together with the author i.

In general the entries of $\mathbf{Cn}$ need not to be symmetric. In the case when $n_{pi} = \dfrac{wa_{pi}}{\text{outdeg}_{\mathbf{WA}}(p)}$ they are symmetric, $cn_{ij} = cn_{ji}$.

The total contribution for a work p of terms $wa_{pi} n_{pj}$ from the definition of cn_{ij}

$$\sum_{j \in A} \sum_{j \in A} wa_{pi} n_{pj} = \text{outdeg}_{\mathbf{WA}}(p)$$

is equal to the number of authors of the work p.

Similarly, for an author i the total contribution of entries cn_{ij}

$$\sum_{j \in A} cn_{ij} = \text{indeg}_{\mathbf{WA}}(i)$$

is equal to the number of works that the author i co-authored; and the (diagonal) entry

$$cn_{ii} = \sum_{p \in N(i)} n_{pi}$$

is equal to the total contribution of author i to his/her works.

Therefore we can base on the entries of network $\mathbf{Cn}$ the *self-sufficiency* index

$$S_i = \frac{cn_{ii}}{\text{indeg}_{\mathbf{WA}}(i)}$$

as the proportion of an author's contribution to his/her works and the total number of works that he/she co-authored. The *collaborativeness* index is complementary to it:

$$K_i = 1 - S_i$$

All cn_{ij} values

$$\sum_{i\in A}\sum_{j\in A} cn_{ij} = m_{\mathbf{WA}}$$

sum up to the number of all lines in the network **WA**.

The *third collaboration network* is defined as

$$\mathbf{Ct} = \mathbf{N}^T * \mathbf{N}.$$

The weight ct_{ij} is equal to the total contribution of collaboration of authors i and j to works.

The total contribution of a complete subgraph corresponding to the authors of a work is p:

$$\sum_{i\in A}\sum_{j\in A} n_{ip}^T n_{pj} = 1$$

The weights ct_{ij} are symmetric, $ct_{ij} = ct_{ji}$, and the sum

$$\sum_{j\in A} ct_{ij} = \sum_{p\in W} n_{pi}$$

is equal to the total contribution of author i to works from W.

The sum of all weights ct_{ij}

$$\sum_{i\in A}\sum_{j\in A} ct_{ij} = |W|$$

is equal to the number of all works.

We can also introduce the *author's contribution to the field* as

$$ac_i = \frac{|A|}{|W|}\sum_{p\in W} n_{pi}$$

with the property

$$\sum_{i\in A} ac_i = |A|.$$

Therefore the average ac is 1.

Note also that

$$b(\mathbf{Co}) = b(\mathbf{Cn}) = b(\mathbf{Ct}).$$

3.6.3   Other derived networks

In **WoS2Pajek** the citation relation $p \, \mathbf{Ci} \, q$ means p cites q. Therefore the *bibliographic coupling* network **biCo** can be determined as (Kessler 1963)

$$\mathbf{biCo} = \mathbf{Ci} * \mathbf{Ci}^T.$$

The corresponding weight

$$bico_{pq} = \sum_{s \in W} ci_{ps} ci_{qs} = \sum_{s \in N(p) \cap N(q)} 1$$

is equal to the number of works cited by both works p and q. It is symmetric $bico_{pq} = bico_{qp}$.

Again we have problems with works with many citations, especially with review papers. To neutralize their impact we can introduce a normalized measure such as

$$\mathbf{biCon} = \frac{1}{2}(n(\mathbf{Ci}) * \mathbf{Ci}^T + \mathbf{Ci} * n(\mathbf{Ci})^T).$$

It is easy to verify that $bicon_{pq} \in [0, 1]$ and $bicon_{pq} = bicon_{qp}$ (symmetry). It also holds: $bicon_{pq} = 1$ iff the works p and q are referencing the same works. Note that

$$b(n(\mathbf{Ci}) * \mathbf{Ci}^T) = b(\mathbf{Ci} * n(\mathbf{Ci})^T).$$

The cC_{pq} element of the first term represents the 'importance' of common (p, q)-citations for the work p; and the Cc_{pq} element of the second term represents the 'importance' of common (p, q)-citations for the work q:

$$bicon_{pq} = \frac{1}{2}(cC_{pq} + Cc_{pq})$$

Note that the first term in the definition of **biCon** is equal to the transpose of the second term

$$(\mathbf{Ci} * n(\mathbf{Ci})^T))^T = n(\mathbf{Ci}) * \mathbf{Ci}^T$$

and therefore $Cc_{pq} = cC_{qp}$. This can be used for more efficient computation of **biCon**. We only need to compute the first term **cC**. Then

$$bicon_{pq} = \frac{1}{2}(cC_{pq} + cC_{qp}).$$

Similarly the *document co-citation* network **coCi** can be determined as (Rosengren 1968; Small 1973)

$$\mathbf{coCi} = \mathbf{Ci}^T * \mathbf{Ci}.$$

The corresponding weight

$$coci_{pq} = \sum_{s \in W} ci_{sp} ci_{sq} = \sum_{s \in N^-(p) \cap N^-(q)} 1$$

is equal to the number of works citing both works p and q. $N^-(p)$ denotes the set of neighbors from which the vertex p can be entered.

It holds that $\mathbf{coCi}(\mathcal{N}) = \mathbf{biCo}(\mathcal{N}^T)$ and also for corresponding normalized networks $\mathbf{coCin}(\mathcal{N}) = \mathbf{biCon}(\mathcal{N}^T)$.

The weight aci_{ip} in the *author citation* network

$$\mathbf{ACi} = \mathbf{AW} * \mathbf{Ci}$$

counts the number of times that author i cited work p.

The *author co-citation* network can be obtained as

$$\mathbf{ACo} = b(\mathbf{ACi}) * b(\mathbf{ACi})^T.$$

The weight aco_{ij} counts the number of works cited by both authors i and j.

The weight ak_{ik} in the *authors using keywords* network

$$\mathbf{AK} = \mathbf{AW} * \mathbf{WK}$$

counts the number of works in which the author i used a keyword k.

The network of citations between authors can be obtained as

$$\mathbf{Ca} = \mathbf{AW} * \mathbf{Ci} * \mathbf{WA}$$

The weight ca_{ij} counts the number of times a work co-authored by i is citing a work co-authored by j (see Figure 3.8).

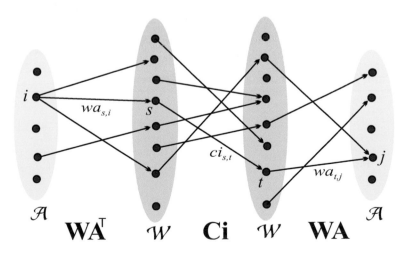

Figure 3.8 Authors' citations network.

3.7 Weights

The *normalization* of weights (line values) approach outlined in Section 2.11.1 was developed for quick inspections of one-mode networks regardless of whether these networks were observed networks or obtained from two-mode networks by matrix multiplication. However, there are often huge differences in the weights on lines in networks obtained from large two-mode networks by matrix multiplication. In these instances, directly comparing vertices according to the raw data is not a 'fair' comparison. In order to make line weights comparable, some normalization is required. Some ways of doing this are described in Section 3.7.1.

Note that most of the obtained derived networks are one-mode networks. To analyse them standard SNA methods can be used. For analysis of two-mode networks we can use direct methods such as (generalized) two-mode cores, two-mode hubs and authorities, and 4-ring islands (Ahmed et al. 2007). The edited collection, Agneessens and Everett (2013) contains recently developed approaches for analyzing two-mode networks.

3.7.1 Normalizations of weights

For networks without loops, diagonal weights for undirected networks can be defined as the sum of the out-diagonal elements in the row (or column): $w_{vv} = \sum_{u \neq v} w_{vu}$ For directed networks, the values of the row and column sums may differ. In this case, the diagonal elements can be defined as $w_{vv} = \frac{1}{2}(\sum_u w_{vu} + \sum_u w_{uv})$. Usually, the network is assumed to not have any isolates.

Some normalizations based on these ideas are presented in the Table 3.3. These normalizations can also be used for other networks.

The general Pajek command for doing this (with a normalization chosen in [method]) is:

```
Network/2-Mode Network/2-Mode to 1-Mode/Normalize 1-Mode/ [method]
```

3.7.2 *k*-rings

In Chapter 2, we described some methods for identifying dense parts of networks. Another method for doing this exploits the idea of a *k*-ring. The materials in this section build upon the idea of a *triangular network* introduced in Section 2.8.

Table 3.3 Some normalizations of network weights.

$$\text{Geo}_{uv} = \frac{w_{uv}}{\sqrt{w_{uu}w_{vv}}}$$

$$\text{GeoDeg}_{uv} = \frac{w_{uv}}{\sqrt{\deg_u \deg_v}}$$

$$\text{Input}_{uv} = \frac{w_{uv}}{w_{vv}}$$

$$\text{Output}_{uv} = \frac{w_{uv}}{w_{uu}}$$

$$\text{Min}_{uv} = \frac{w_{uv}}{\min(w_{uu}, w_{vv})}$$

$$\text{Max}_{uv} = \frac{w_{uv}}{\max(w_{uu}, w_{vv})}$$

$$\text{MinDir}_{uv} = \frac{w_{uv}}{w_{uu}}$$

$$\text{MaxDir}_{uv} = \frac{w_{uv}}{w_{vv}}$$

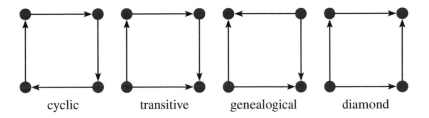

cyclic   transitive   genealogical   diamond

Figure 3.9   4-rings in directed graphs.

A *k-ring* is a simple closed chain of length k. By using k-rings we can define weights for edges as $w_k(e) = $ # of different k-rings containing the edge $e \in \mathcal{E}$

Since for each edge, e, of a complete graph, K_r, $r \geq k \geq 3$, it follows, $w_k(e) = (r - 2)!/(r - k)!$ the edges belonging to cliques have large weights. Therefore, these weights can be used to identify the dense parts of a network. Unfortunately, even in sparse networks the k-rings can be efficiently determined only for small values of $k = 3, 4$, and 5. Even so, these short k-rings have value when coupled to the notion of short cycle connectivity to provide another method for decomposition of networks.

Figure 2.35 shows two types of triangles in directed networks. The transitive ring is on the right.[13] For transitive rings, Pajek provides a special weight obtained by counting the number of times an arc is located on transitive rings as a *shortcut*. We turn our attention to *4-rings*. There are four types of *directed 4-rings* as shown in Figure 3.9.

3.7.3   4-rings and analysis of two-mode networks

In two-mode networks there are no 3-rings. The densest substructures are complete bipartite subgraphs K_{n_1,n_2}. Given n_1 vertices in one set, and n_2 vertices in the second set, K_{n_1,n_2} contains $n_1 \times n_2$ lines. The two-mode complete graphs contain many 4-rings. This is shown in the following example displaying $K_{5,4}$. The general formula for the number 4-rings in K_{n_1,n_2} is on the right together with a formula for a weight, $w_4(e)$ for every edge. There are 30 4-rings in $K_{5,4}$ with each edge having a 4-ring weight of 12.

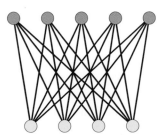

There are

$$\binom{n_1}{2}\binom{n_2}{2} = \frac{1}{4}n_1(n_1 - 1)n_2(n_2 - 1)$$

4-rings in K_{n_1,n_2}; and each of its edges e has the 4-ring weight of

$$w_4(e) = (n_1 - 1)(n_2 - 1).$$

Using 4-rings is useful for extending the notion of cores to two-mode cores.

---

[13] The cyclic triangle on the left of that figure is a 3-ring.

3.7.4 Two-mode cores

The subset of vertices $C \subseteq \mathcal{V}$ is a (p, q)-*core* in a two-mode network $\mathcal{N} = (\mathcal{V}_1, \mathcal{V}_2; \mathcal{L})$, $\mathcal{V} = \mathcal{V}_1 \cup \mathcal{V}_2$ iff

    a. in the induced subnetwork $\mathcal{K} = (C_1, C_2; \mathcal{L}(C))$, $C_1 = C \cap \mathcal{V}_1$, $C_2 = C \cap \mathcal{V}_2$ it holds $\forall v \in C_1 : \deg_{\mathcal{K}}(v) \geq p$ and $\forall v \in C_2 : \deg_{\mathcal{K}}(v) \geq q$;

    b. C is the maximal subset of $\mathcal{V}$ satisfying condition **a**.

Properties of two-mode cores:

- $C(0, 0) = \mathcal{V}$

- $\mathcal{K}(p, q)$ is not always connected

- $(p_1 \leq p_2) \wedge (q_1 \leq q_2) \Rightarrow C(p_1, q_1) \subseteq C(p_2, q_2)$

- $\mathcal{C} = \{C(p, q) : p, q \in \mathbb{N}\}$. If all non-empty elements of $\mathcal{C}$ are different it is a lattice.

To determine a (p, q)-core the procedure similar to the ordinary core procedure can be used:

repeat
    remove from the first set all vertices of degree less than p,
        and from the second set all vertices of degree less than q
until no vertex was deleted

It can be implemented to run in $O(m)$ time.

    How can we select parameters p and q giving interesting (p, q)-cores? We compute a table of cores' characteristics $n_1 = |C_1(p, q)|$, $n_2 = |C_2(p, q)|$ and k – number of components in $\mathcal{K}(p, q)$. The interesting (p, q)-cores are

- $n_1 + n_2 \leq$ selected threshold

- they belong to a 'border line' in the (p, q)-table.

Also in the two-mode case the generalized cores can be defined.

3.8 Pathfinder

Visualization of networks is important. However, in most cases, the drawn pictures of large or dense networks are unreadable: they contain too many elements (vertices and/or lines). The divide and conquer approach outlined in Chapter 2 is one way of handling this problem by selecting parts of networks, analyzing and visualizing them before combining the results. An alternative approach for obtaining visualizations of such a network is to transform it to its *skeleton* by removing less important lines. The most often used skeleton is the minimal spanning tree. We can also preserve only k strongest lines at each vertex. Another option is skeletons produced by the Pathfinder algorithm (Vavpetič et al. 2009).

    This algorithm was proposed in the 1980s (Schvaneveldt 1990; Interlink 1990; Schvaneveldt et al. 1989) for simplifying *weighted* networks, where the weight measures

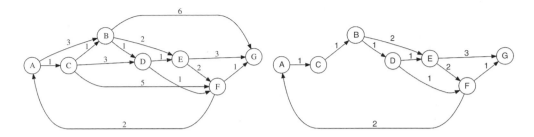

Figure 3.10   An original network (left) and its PFnet$(1, \infty)$ skeleton (right).

a *dissimilarity* between vertices. It removes from a network all lines not satisfying the triangle inequality: if, given a line, there is a path with two lines connecting its endpoints with lower weights sum then the line is removed. Let **W** denote a network *dissimilarity* matrix and $\mathbf{W}^{(q)}$ be the matrix of values of all walks of length at most q computed over the semiring $(\mathbb{R}_0^+, \oplus, \boxed{r}, \infty, 0)$ with $a \boxed{r} b = \sqrt[r]{a^r + b^r}$ (known as the *Minkowski operation*) and $a \oplus b = \min(a, b)$. Some special cases of the Minkowski operation are:

$r = 1 \Rightarrow a \boxed{r} b = a + b$,

$r = 2 \Rightarrow a \boxed{r} b = \sqrt{a^2 + b^2}$,

$r = \infty \Rightarrow a \boxed{r} b = \max(a, b)$.

With this notation, the basic idea of the Pathfinder algorithm is to produce a network denoted by PFnet$(\mathbf{W}, r, q) = (\mathcal{V}, \mathcal{L}_{PF})$ where $\mathcal{L}_{PF}$ are the arcs remaining after applying this algorithm.

Figure 3.10 illustrates some of these ideas. The path $A \rightarrow C \rightarrow B$ has two arcs each with a weight of 1. The arc $A \rightarrow B$ has a weight of 3. As $1 + 1 < 3$, $A \rightarrow C$ is removed in the Pathfinder skeleton. The values of 2 and 3 on $B \rightarrow E \rightarrow G$ and the value of 6 on $B \rightarrow G$ leads to the removal of $B \rightarrow G$. For $D \rightarrow E \rightarrow F$ with weights of 1 and 2 on the arcs and with $D \rightarrow F$ having a value of 1, the triangle inequality is not satisfied, so $D \rightarrow F$ remains in the Pathfinder skeleton. Note that for $C \rightarrow D \rightarrow E \rightarrow F$ and $C \rightarrow F$, the triangle inequality is not applicable. However, the weights on the arcs in $C \rightarrow D \rightarrow F$ imply the removal of $C \rightarrow F$. Note that no vertices are removed and that the connectivity is preserved.

The network on the left of Figure 3.10 is completely readable. In contrast, the networks on the left of Figure 3.11 are not. The results of applying the Pathfinder algorithm is shown on the right of this figure. The same holds for the World Trade 1999 network in Figure 3.12 and its PF skeleton in Figure 3.13. These skeletons show the essential connective structure of the two networks.

One version of this algorithm is expressed as:

compute $\mathbf{W}^{(q)}$;
$\mathcal{L}_{PF} := \emptyset$;
for $e(u, v) \in \mathcal{L}$ **do begin**
    if $\mathbf{W}^{(q)}[u, v] = \mathbf{W}[u, v]$ **then** $\mathcal{L}_{PF} := \mathcal{L}_{PF} \cup \{e\}$
end;

The Pathfinder approach can be based on a Joly and Le Calvé theorem (Joly and Le Calvé 1986): For any even dissimilarity measure d there is a unique number $p \geq 0$, called its *metric index*, such that: d^r is metric for all $r \leq p$, and d^r is not metric for all $r > p$.

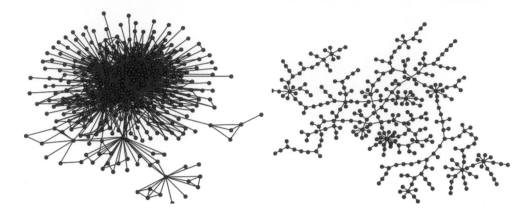

Figure 3.11   A second example of a network and its Pathfinder skeleton.

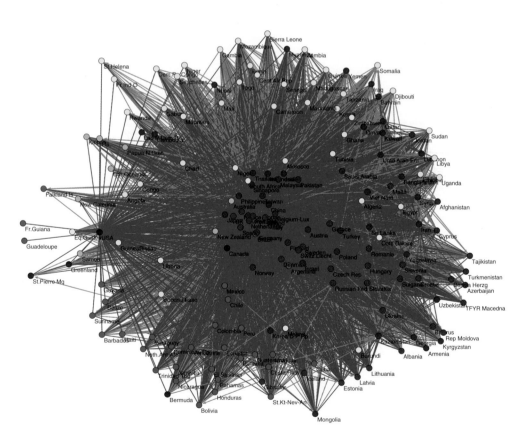

Figure 3.12   A display of World Trade 1999 network.

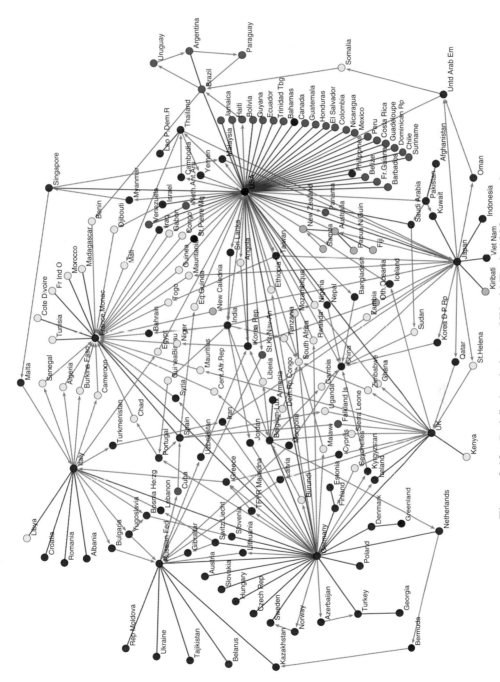

Figure 3.13 Pathfinder skeleton of World Trade 1999 network.

Therefore, let d be a dissimilarity and for x, y, and z we have $d(x, z) + d(z, y) \geq d(x, y)$ and $d(x, y) > \max(d(x, z), d(z, y))$ then there exists a unique number $p \geq 0$ such that for all $r > p$

$$d^r(x, z) + d^r(z, y) < d^r(x, y)$$

or equivalently

$$d(x, z) \boxed{r} d(z, y) < d(x, y).$$

In the transformed space, the largest side of a triangle can be eliminated.

The structure $(\mathbb{R}_0^+, \oplus, \boxed{r}, \infty, 0)$ is a complete semiring with $a^\star = 0$. It is also called a *Pathfinder* semiring. There are some useful theoretical results for the Pathfinder algorithm (Schvaneveldt (1990); Guerrero-Bote et al. (2006); Interlink (1990); Schvaneveldt et al. (1989)). For a given dissimilarity matrix, $\mathbf{W}$, the PFnet($\mathbf{W}, r, q$)

- is unique;

- preserves geodetic distances;

- links nearest neighbors;

- contains the same information as the minimum method of hierarchical clustering; and

- PFnet($\mathbf{W}, r = \infty, q = n - 1$) is the union of all MINTREES.

Some additional results include:

- graph PFnet($\mathbf{W}, r_2, q$) is a spanning subgraph of graph PFnet($\mathbf{W}, r_1, q$) iff $r_1 < r_2$;

- graph PFnet($\mathbf{W}, r, q_2$) is a spanning subgraph of graph PFnet($\mathbf{W}, r, q_1$) iff $q_1 < q_2$;

- similarity transformations preserve structure: the graph PFnet($\mathbf{W}, r, q$) is equal to the graph PFnet($\alpha\mathbf{W}, r, q$) for $\alpha > 0$;

- monotonic transformations preserve structure for $r = \infty$: the graph PFnet($\mathbf{W}, \infty, q$) is equal to the graph PFnet($f(\mathbf{W}), \infty, q$) where $f : \mathbb{R} \to \mathbb{R}$ is strictly increasing mapping and $f(\mathbf{W}) = [f(w_{ij})]$.

3.8.1 Pathfinder algorithms

In the original pathfinder algorithm, the matrix $\mathbf{W}^{(q)}$ is computed by using its definition

$$\mathbf{W}^{(q)} = \sum_{i=0}^{q} \mathbf{W}^i$$

by computing all its powers $\mathbf{W}^i$, $i = 1, \ldots, q$. The complexity of the algorithm is $O(qn^3)$, therefore $O(n^4)$, for $q \geq n - 1$. Therefore, it can be applied only to relatively small (up to some hundreds of vertices) networks. This limits is utility. However, interest in the Pathfinder transformation was renewed by Chen (1998).

3.8.2   Computing the closure over the Pathfinder semiring

In a Pathfinder semiring, the absorption property, $a \oplus (a \boxed{r} b) = a \oplus (b \boxed{r} a) = a$, holds. This property, coupled to this semiring being complete leads to a useful result. If the set of vertices, $\mathcal{V}$, is finite and the set of all paths $\mathcal{E}_{uv}$ is also finite, the value of all walks $w(\mathcal{S}_{uv}^{\star}) = w(\mathcal{E}_{uv})$ can be computed. One option for doing this is to use, for large enough k, the equality

$$\mathbf{W}^{\star} = \mathbf{W}^{(k)} = (\mathbf{1} \oplus \mathbf{W})^{k}.$$

Computation can be made much faster by using the sequence $(\mathbf{1} \oplus \mathbf{W})^{2^{i}}, i = 1, .., s$.

However, this is not the fastest way of computing $\mathbf{W}^{\star}$. Kleene, Warshall, Floyd, and Roy contributed to the development of the procedure whose final form over a closed semiring $(A, +, \cdot, {}^{*})$ is given by Fletcher (1980).

```
C_0 := W ;
for k := 1 to n do begin
    for i := 1 to n do for j := 1 to n do
        c_k[i, j] := c_{k-1}[i, j] + c_{k-1}[i, k] · (c_{k-1}[k, k])* · c_{k-1}[k, j] ;
    c_k[k, k] := 1 + c_k[k, k] ;
end;
W* := C_n ;
```

If we delete the statement $c_k[k, k] := 1 + c_k[k, k]$, the algorithm for computing the strict closure $\bar{\mathbf{W}} = \mathbf{W}\mathbf{W}^{\star}$ results.

3.8.3   Spanish algorithms

Since the Pathfinder semiring is idempotent,

$$\mathbf{W}^{(q)} = (\mathbf{1} \oplus \mathbf{W})^{q}.$$

This power can be computed more quickly by using a *binary algorithm* (for example, to compute $a^{57} = a^{32} \cdot a^{16} \cdot a^{8} \cdot a^{1}$ we need only eight multiplications instead of 56). This improvement was proposed by Guerrero-Bote et al. (2006) and reduces its complexity to $O(n^3 \log q)$.

When $q \geq n - 1$, $\mathbf{W}^{(q)} = \mathbf{W}^{\star}$, it can be determined by using Fletcher's algorithm over the Pathfinder semiring. This improvement was proposed by Quirin et al. (2008b) and reduces complexity to $O(n^3)$.

An additional improvement can be made for undirected networks in the case $q \geq n - 1$ and $r = \infty$. In this case, the network PF is the union of all minimal spanning trees of N. It can be obtained using an adapted version of Kruskal's minimal spanning tree algorithm as described in Quirin et al. (2008a). The complexity of this algorithm is $O(m \log n)$ where m is the number of edges.

3.8.4   A sparse network algorithm

In general, for sparse networks, further improvements are possible. The basic Pathfinder algorithm can be rewritten in the following form.

$\mathcal{L}_{PF} := \emptyset;$
for $v \in \mathcal{V}$ **do begin**
    compute the list $S = ((u, d_u) : u \in N(v))$, where $d_u = \mathbf{W}^{(q)}[v, u]$;
    for $(u, d_u) \in S$ **do**
        if $d_u = \mathbf{W}[v, u]$ **then** $\mathcal{L}_{PF} := \mathcal{L}_{PF} \cup \{(v, u)\}$
end;

where $N(v)$ denotes the set of successors of vertex v.

For determining the values $d_u = \mathbf{W}^{(q)}[v, u]$ for $q = n - 1$ an adapted version of Dijkstra's algorithm determines the list S in a single run. The job is completed when all values of vertices from $N(v)$ are determined. Only a (small) portion of a network has to be inspected for each vertex v. To efficiently implement this algorithm, a special data structure, the *indexed priority queue*, can be used.

3.9   Clustering, blockmodeling, and community detection

The subnetworks obtained as cuts or islands are usually dense and the standard 'dots' and 'lines' representation is unreadable. In such cases a much better visualization is using the matrix representation for the 'right' permutation of vertices. The permutation can be obtained using seriation algorithms, clustering, blockmodeling, or community detection algorithms (Louvain method, VOS clustering).

```
Network/Create Permutation/
Operations/Network+Cluster/Dissimilarity*/Network based/
Network/Create Hierarchy/Clustering*/
Network/Create Partition/Blockmodeling*/
```

3.9.1   The Louvain method and VOS

One of the approaches to determine the clustering $\mathbf{C}$ of a network is to maximize its *modularity*

$$Q(\mathbf{C}) = \sum_{C \in \mathbf{C}} (\frac{l(C)}{m} - (\frac{d(C)}{2m})^2)$$

where $l(C)$ is the number of edges between vertices belonging to cluster C, and $d(C)$ is the sum of the degrees of vertices from C.

The modularity maximization problem is NP-complete.

The Louvain method and VOS are available in Pajek using the commands:

```
Network/Create Partition/Communities/Louvain Method
Network/Create Partition/Communities/VOS Clustering
[Draw] Layout/VOS Mapping
```

3.10   Clustering symbolic data

Given a set of units, $\mathcal{U}$, clustering is a process of organizing these units into clusters where the clusters contain units more similar to each other than they are to units from other clusters, according to some well-specified criteria. In real-life (empirical) clustering problems, we have to deal with their different characteristics:

- descriptions of units in the form of vectors (capturing types of measurement scales, numbers of variables, missing values . . .) or structured units;

- sizes of the resulting clusters of units;

- structure of units in a 'space' (including features such as density and the 'shapes' of clusters).

A recent survey on clustering is given in Gan et al. (2007). Here, we present an approach to clustering (very) large datasets of *mixed* units, that is, units measured in different scales.

The approach is based on representation of units by *symbolic objects* (SOs) (Billard and Diday 2006). The SOs can describe either single units or groups of initial units condensed into SOs in a pre-processing step.

For clustering SOs, we adapted two classical clustering methods:

1. the *leaders method* – a generalization of the *k-means* method (Hartigan 1975) and dynamic clouds (Diday 1979)

2. Ward's *hierarchical clustering method* (Ward 1963)

Both adapted methods are based on the *same* criterion function – they are solving the *same* clustering problem.

With the leaders method, the size of the sets of units is reduced to a manageable number of leaders. The obtained leaders can be further clustered with the compatible agglomerative hierarchical clustering method to reveal relations among them and using the resulting dendrogram also to decide upon the appropriate number of clusters.

3.10.1   Symbolic objects described with distributions

An SO, X, is described by using a list $X = [\mathbf{x_i}]$ of descriptions of m variables $V_i \in V$. In our model, each variable is described with frequency distribution (*bar chart*) of its k_i values:

$$\mathbf{f_{x_i}} = [f_{x_i 1}, f_{x_i 2}, \ldots, f_{x_i k_i}]$$

with

$$\mathbf{x_i} = [p_{x_i 1}, p_{x_i 2}, \ldots, p_{x_i k_i}]$$

we denote the corresponding probability distribution by

$$\sum_{j=1}^{k_i} p_{x_i j} = 1, \quad i = 1, \dots, m.$$

We approach the clustering problem as an optimization problem over the set of *feasible* clusterings, Φ_k, the partitions of units into k clusters. The *criterion function* has the form

$$P(\mathbf{C}) = \sum_{C \in \mathbf{C}} p(C).$$

The *total error*, $P(\mathbf{C})$, of the clustering, $\mathbf{C}$, is a sum of the *cluster errors*, $p(C)$.

There are many possibilities for expressing the cluster error, $p(C)$. We assume a model in which the error of a cluster is a sum of differences of its units from the cluster's *representative*, T,

$$p(C, T) = \sum_{X \in C} d(X, T).$$

Note, in general, the representative need not be from the same 'space' (set) as units.

The *best* representative is called a *leader*:

$$T_C = \mathrm{argmin}_T \, p(C, T).$$

Then we define

$$p(C) = p(C, T_C) = \min_T \sum_{X \in C} d(X, T).$$

The SO, X, is described by a list $X = [\mathbf{x_i}]$. Assume also the representatives are described in the same way $T = [\mathbf{t_i}]$, $\mathbf{t_i} = [t_{i1}, t_{i2}, \dots, t_{ik_i}]$.

We introduce a dissimilarity measure between SOs with

$$d(X, T) = \sum_i \alpha_i d(\mathbf{x_i}, \mathbf{t_i}), \quad \alpha_i \geq 0, \quad \sum_i \alpha_i = 1,$$

where

$$d(\mathbf{x_i}, \mathbf{t_i}) = \sum_{j=1}^{k_i} w_{x_i j} \delta(p_{x_i j}, t_{ij}), \quad w_{x_i j} \geq 0.$$

This one is a generalization of the squared Euclidean distance. Using a more appropriate type of dissimilarity, δ, it is possible to consider the problem of squared Euclidean distance being responsive to the largest values. Some alternatives for $\delta(x, t)$ are shown in the following table:

	$\delta(x,t)$	t^*_{ij}
1	$(p_x - t)^2$	$\frac{P_{ij}}{A_{ij}}$
2	$(\frac{p_x-t}{t})^2$	$\frac{Q_{ij}}{P_{ij}}$
3	$\frac{(p_x-t)^2}{t}$	$\sqrt{\frac{Q_{ij}}{A_{ij}}}$
4	$(\frac{p_x-t}{p_x})^2$	$\frac{H_{ij}}{F_{ij}}$
5	$\frac{(p_x-t)^2}{p_x}$	$\frac{A_{ij}}{H_{ij}}$
6	$\frac{(p_x-t)^2}{p_x t}$	$\sqrt{\frac{P_{ij}}{H_{ij}}}$

$$A_{ij} = \sum_{X \in C} w_{x_i j}$$

$$P_{ij} = \sum_{X \in C} w_{x_i j} p_{x_i j} \qquad Q_{ij} = \sum_{X \in C} w_{x_i j} p^2_{x_i j}$$

$$H_{ij} = \sum_{X \in C} \frac{w_{x_i j}}{p_{x_i j}} \qquad F_{ij} = \sum_{X \in C} \frac{w_{x_i j}}{p^2_{x_i j}}$$

The weight, $w_{x_i j}$, can be different for the same unit X for each variable V_i (as is needed in descriptions of egocentric networks, population pyramids, etc.).

3.10.2   The leaders method

The *leaders method* is a generalization of a popular non-hierarchical clustering *k-means method*. The idea is to get an 'optimal' clustering into a *pre-specified* number of clusters with the following iterative procedure:

determine an initial clustering
repeat
    determine leaders of the clusters in the current clustering;
    assign each unit to the nearest new leader – producing a
      new clustering
until the leaders stabilize.

Given a cluster, C, the corresponding leader, T_C, is the solution of the problem

$$T_C = \text{argmin}_T \sum_{X \in C} d(X, T) = \left[\text{argmin}_{\mathbf{t_i}} \sum_{X \in C} d(\mathbf{x_i}, \mathbf{t_i})\right]^m_{i=1}.$$

Therefore $T_C = [\mathbf{t^*_i}]$ and $\mathbf{t^*_i} = \text{argmin}_{\mathbf{t_i}} \sum_{X \in C} d(\mathbf{x_i}, \mathbf{t_i})$. And again (to simplify the notation we omit the index i):

$$\mathbf{t^*} = \text{argmin}_{\mathbf{t}} \sum_{X \in C} d(\mathbf{x}, \mathbf{t}) = \left[\text{argmin}_{t_j \in \mathbb{R}} \sum_{X \in C} w_{xj} \delta(p_{xj}, t_j)\right]^k_{j=1}$$

and omitting the index j

$$\mathbf{t}^* = \operatorname{argmin}_{t \in \mathbb{R}} \sum_{X \in C} w_x \delta(p_x, t).$$

This is a standard optimization problem with one real variable. The solution has to satisfy the condition

$$\frac{\partial}{\partial t} \sum_{X \in C} w_x \delta(p_x, t) = 0.$$

For $\delta_1(p_x, t) = (p_x - t)^2$, we get

$$t^* = \frac{\sum_{X \in C} w_x p_x}{\sum_{X \in C} w_x} = \frac{P}{A}.$$

The leaders for δ_1 have two important properties. If $w_{x_ij} = w_{x_i}$ then for each $i = 1, \ldots, m$:

$$\sum_{j=1}^{k_i} t_{ij}^* = \frac{1}{A_i} \sum_{j=1}^{k_i} \sum_{X \in C} w_{x_i} p_{x_ij}$$

$$= \frac{1}{A_i} \sum_{X \in C} w_{x_i} \sum_{j=1}^{k_i} p_{x_ij} = \frac{1}{A_i} \sum_{X \in C} w_{x_i} = 1$$

The leaders' components are *distributions*.

Further, let $w_{x_ij} = n_{x_i}$ then for each $i = 1, \ldots, m$:

$$t_{Cij}^* = \frac{\sum_{X \in C} n_{x_i} p_{x_ij}}{\sum_{X \in C} n_{x_i}} = p_{Cij}$$

The representative of a cluster is again a distribution over this cluster.

Given the leaders, $\mathbf{T}$, the corresponding optimal clustering, $\mathbf{C}^*$, is determined from

$$P(\mathbf{C}^*) = \sum_{X \in U} \min_{T \in \mathbf{T}} d(X, T) = \sum_{X \in U} d(X, T_{c^*(X)})$$

where

$$c^*(X) = \operatorname{argmin}_k d(X, T_k).$$

We assign each unit, X, to the closest leader, $T_k \in \mathbf{T}$.

3.10.3   An agglomerative method

The hierarchical agglomerative clustering procedure is based on a step-by-step merging of the two closest clusters.

> each unit forms a cluster: $\mathbf{C}_n = \{\{X\} : X \in \mathcal{U}\}$;
> they are at level 0: $h(\{X\}) = 0,\ X \in \mathcal{U}$;
> **for** $k = n - 1$ **to** 1 **do**
>> determine the closest pair of clusters
>>> $(u, v) = \mathrm{argmin}_{i,j:i\neq j}\{D(C_i, C_j) : C_i, C_j \in \mathbf{C}_{k+1}\}$;
>>
>> join the closest pair of clusters $C_{(uv)} = C_u \cup C_v$
>>> $\mathbf{C}_k = (\mathbf{C}_{k+1} \setminus \{C_u, C_v\}) \cup \{C_{(uv)}\}$;
>>> $h(C_{(uv)}) = D(C_u, C_v)$
>>
>> determine the dissimilarities $D(C_{(uv)}, C_s), C_s \in \mathbf{C}_k$
>
> **endfor**

$\mathbf{C}_k$ is a partition of the finite set of units, $\mathcal{U}$, into k clusters. The level, $h(C)$, of the cluster $C_{(uv)} = C_u \cup C_v$.

Therefore, the computation of dissimilarities between the new (merged) cluster and the rest has to be specified. To obtain the compatibility with the adapted leaders method, we define the dissimilarity between clusters C_u and C_v, $C_u \cap C_v = \emptyset$, as

$$D(C_u, C_v) = p(C_u \cup C_v) - p(C_u) - p(C_v).$$

In an initial general computation, $\mathbf{u}_i$ and $\mathbf{v}_i$ are components of the leaders of clusters, C_u and C_v, and $\mathbf{z_i}$ is a component of the leader of the cluster, $C_u \cup C_v$. It can be shown that a *generalized Ward's relation* holds:

$$D(C_u, C_v) = \sum_i \alpha_i \sum_j \frac{A_{u_i j} \cdot A_{v_i j}}{A_{u_i j} + A_{v_i j}} (u_{ij} - v_{ij})^2$$

and

$$z_{ij} = \frac{A_{u_i j} u_{ij} + A_{v_i j} v_{ij}}{A_{u_i j} + A_{v_i j}}.$$

Instead of the squared Euclidean distance other dissimilarity measures $\delta(x, t)$ can be used – see Kejžar et al. (2011). Relations similar to Ward's can be derived for them.

The proposed approach to clustering symbolic data is implemented in the R-package *clamix* for dissimilarity measures discussed in Section 3.10.1, currently available upon request; and for clustering without weights $w_{x_i j}$ is implemented in R-package *clustDDist* (Kejžar et al. 2009).

3.11   Approaches to temporal networks

In the last two decades, the increased interest in temporal networks has been generated by many different sources. This triggered a need for developing methods for handling such networks. Particularly important for analyzing temporal networks were the study of travel-support services and the analysis of *sequences of events* including email traffic, news events, and phone calls. These approaches and results are surveyed by Holme and Saramäki (2013, 2012). See also Nicosia et al. (2013) and Kempe et al. (2000). Another overview can be found in Casteigts et al. (2012) and Casteigts and Flocchini (2013) based on their formalization of temporal networks using time-varying graphs (TVGs).

In this section we build upon the foundations established earlier in this chapter by incorporating ideas from these literatures to create a broader framework for analyzing temporal networks. The idea of adding the time dimension to networks is not new and has been formulated in many disciplines. Some of the earliest efforts are in the transport(ation) network analysis (Bell and Iida 1997; Correa and Stier-Moses 20ll), project scheduling (CPM, Pert) in operations research (Moder and Phillips 1970), and constraints networks in artificial intelligence (Dechter 2003).

While there are also qualitative approaches to temporal networks, for example Allen (1983) and Vilain et al. (1990), in this book we build on the quantitative approach presented in Section 2.5. For statistical approaches, see Kolaczyk (2009) and the Siena page of Snijders (Ripley et al. 2013).

Based on our survey of these sources, there are two useful views for temporal networks:

1. Networks can be seen as providing constraints on the activities of actors in them or be determined by constraints (for example, the networks linking airports as determined by airline time tables).

2. Networks can represent interactions of events among actors (for example, the KEDS networks, one of which was examined in Section 2.2.2 and citation networks of the sort studied in Chapters 4–6).

It is useful to think of processes on networks in terms of 'what moves' over the network ties. Examples of this include:

- travel (where people and goods move between places) and transmission networks such as electrical grids and postal systems;

- diffusion (of cultural ideas, styles, information, and technologies), broadcasting (of news, entertainment, and advertisements), and election results stemming for network processes (funding candidates, mobilizing grass-root actions, facilitating voting, and preventing voting);

- flows of, for example, people commuting, water delivered over pipes or oil and gas being delivered.

Thinking of what moves or flows over network ties is helped by using some additional concepts.

3.11.1 Journeys – Walks in temporal networks

When dealing with walks in temporal networks we can consider additional information regarding weights on lines capturing aspects of flows:

- The *transition time* $\tau \in \mathcal{W}$; $\tau: \mathcal{L} \to \mathbb{R}_0^+$. $\tau(l)$ is equal to the time needed to traverse the line l. If the function τ is not given we can assume $\tau(l) = 0$ for all lines l.

- The *value* (length, cost, etc.) $w \in \mathcal{W}$; $w: \mathcal{L} \to \mathbb{R}$. If the function w is not given we can assume $w(l) = 1$ for all lines l.

In applications related to flows in networks we need an additional weight:

- The *capacity* $c \in \mathcal{W}$; $c: \mathcal{L} \to \mathbb{R}_0^+$. $c(l)$ is equal to the maximum of quantity of items transferred in a time unit over line l. If the function c is not given we can assume $c(l) = \infty$ (representing no flow limits) for all lines l.

In many real networks the values of functions $\tau(l)$, $w(l)$, and $c(l)$ can also vary through time. For example, $\tau(l)$ can depend on the overall traffic, for example telephone calls, email volumes, vehicles on roads in the network. Similarly, $w(l)$ for trade flows between nations can vary depending on currency exchange rates. However, in the following we assume that they are constant on each line.

A *temporal walk* or *journey* $\sigma(v_0, v_k; t_0)$ from (source) vertex v_0 to (destination) vertex v_k starting at time $t_0 \in T(v_0)$ is a finite sequence

$$(t_0, v_0, (t_1, l_1), v_1, (t_2, l_2), v_2, \dots, v_{k-2}, (t_{k-1}, l_{k-1}), v_{k-1}, (t_k, l_k), v_k)$$

where $l_i \in \mathcal{L}$, $t_i \in T$, $i = 1, 2, \dots k$. The triples $v_{i-1}, (t_i, l_i)$ state for vertex v_{i-1} in time t_i the line l_i was selected for the next transition. The sequence σ has to satisfy the following conditions: the line l_i links vertex v_{i-1} to vertex v_i and is active during the transition:

- $l_i(v_{i-1}, v_i)$

- $t'_{i-1} \le t_i$

- $[t_i, t'_i] \subseteq T(l_i)$

for $i = 1, 2, \dots k$; where $t'_i = t_i + \tau(l_i)$ and $t'_0 = t_0$.

The number k is called a *length* of the walk σ. The *time used* by the walk σ is equal to

$$t(\sigma) = t'_k - t_0$$

and its *value* is

$$w(\sigma) = \sum_{i=1}^{k} w(l_i).$$

Note: by the consistency from the third condition it follows that $t_i \in T(v_{i-1})$ and $t'_i \in T(v_i)$.

Departure time: $dep(\sigma) = t_1$
Arrival time: $arr(\sigma) = t'_k$
Duration: $dur(\sigma) = arr(\sigma) - dep(\sigma)$

A temporal walk is *regular* if also

$$[t'_{i-1}, t_i] \subseteq T(v_i), \text{ for } i = 1, 2, \ldots k - 1.$$

While waiting for the next step (transition) in vertex v_{i-1} this vertex should be active all the time.

In the following we shall observe the temporal network inside a *time window* $[a, b] \subseteq T$.

Some special classes of temporal networks can be important in solving the problems of different types of temporal networks:

a) networks with *fixed vertices*: $T(v) = T$ for each $v \in V$

b) networks where time is *discrete*: T is a finite set

c) networks where time is *integer*: $T \subseteq Z$ and $\tau : L \to N$

d) case where there are *single intervals*: $T(v)$s and $T(l)$s consist of single intervals

Networks with fixed vertices or single interval temporal networks are regular.

3.11.2   Measures

Using quantities such as shortest durations, earliest arrivals, etc., the standard network measures such as degrees, betweenness, closeness, and clustering indices can be extended (albeit in different ways) to temporal networks. See for example Nicosia et al. (2013). Some of these measures are essentially time functions. In developing such temporal measures some kind of averaging over the time (window) is needed.

3.11.2.1   Measures based on time slices

The approach, supported by Pajek, based on slices is valid only in the case when $\tau(l) = 0$ for each line l. The slices are considered as static networks on which different structural properties such as *degrees*, *closeness*, *betweenness*, *clustering coefficients*, and *flow volumes* can be computed, thus producing different time series describing the evolution of a network. This approach is used extensively in Chapter 5.

It is useful to define the activity of a network. Formally,

Activity:

$$T(v, t) = \begin{cases} 1 & t \in T(v) \\ 0 & t \notin T(v) \end{cases}$$

$$act(v) = \int_{T} T(v, t) dt$$

There are some corresponding optimization problems including identifying the fastest walk for a given amount of available time and establishing the minimal cost (or shortest) walk for the value flowing. Such optimization problems were studied in different fields for special cases of temporal networks in *artificial intelligence* and *operations research*. See, for example, some adapted methods in Nicosia et al. (2013), Holme and Saramäki (2012), Casteigts and Flocchini (2013), and Xuan et al. (2003).

3.11.3 Problems and algorithms

A set of labels can be created to define desirable outcomes when solving problems into a set of labels, Q, for example,

$$Q \in \{\text{short}, \text{fast}, \text{early}, \dots\}.$$

Using Q, it is possible to consider the Q-est journey from vertex u to vertex v for a given starting time t. This would include the shortest journey in terms of the length of temporal walks or the fastest journey in time. Journeys can be constrained, for example, finding the Q-est journey from vertex u to vertex v for any starting time $t \in T(u)$.

One approach to creating and using algorithms is to reduce the problem to traditional problems on static networks. Another option is to develop new algorithms. See for example, Casteigts and Flocchini (2013), George et al. (2007), and Xuan et al. (2003).

3.11.3.1 Properties of journeys

We now consider some useful properties for journeys, starting with the extension of reachability to a temporal context.

Reachability relation: vertex v is reachable from vertex u (in time window W), $u \leadsto_W v$, iff a journey exists from vertex u to vertex v in time window W.

$$\text{InHor}(v; W) = \{u \in \mathcal{V} : u \leadsto_W v\}$$

$$\text{OutHor}(v; W) = \{u \in \mathcal{V} : v \leadsto_W u\}$$

In static networks the reachability relation is transitive. In contrast, this is not true for temporal networks. In the realm of political negotiations, for example, the creation of a peace deal or forming a coalition to pass a piece of legislation, the expression of a 'short window' is used often to capture the notion of temporal constraints under which negotiations take place. When the window 'closes' no deal can be reached because the passage of time implies that opportunities can change. In transportation networks, the closure of bridges, roads, or ports changes the networks and disrupts flows and calculations concerning optimization problems for traffic flows.

This concept is illustrated in the following simple example where, in general, $W = [t_s, t_f]$ denotes a time window from t_s to t_f with integer time. Their activity sets are marked on the arcs.

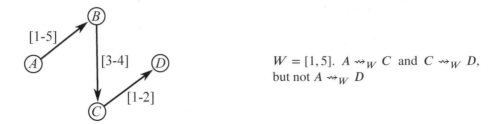

$W = [1, 5]$. $A \leadsto_W C$ and $C \leadsto_W D$, but not $A \leadsto_W D$

For Xuan et al. (2003): the fundamental property for Dijkstra's shortest path algorithm is the property that the prefix paths of the shortest paths are the shortest paths themselves. In general, this is not true for temporal networks with varying time windows. Consider the following simple example.

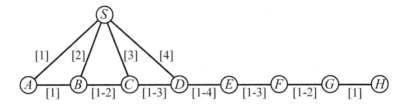

For both $W = [1, 4]$ and $W = [1, 1]$ the shortest journey from S to H is

$$\sigma(S, H) = \{S, A, B, C, D, E, F, G, H\}.$$

Also, the shortest journey from S to E is

$$\sigma(S, E) = \{S, D, E\}.$$

But in $W = [3, 4]$, none of A, B, and F through H can be reached from S. However, the shortest journey from S to E is still $\sigma(S, E) = \{S, D, E\}$.

3.11.3.2   Semirings

For networks with $\tau = 0$ we can determine the reachability time sets

$$T(u, v) = \{t \in \mathcal{T} : u \leadsto_{[t]} v\}$$

using the concept of a semiring, Batagelj (1994). There are two cases to consider: 1) parallel arcs between vertices u and v and 2) a sequence of two arcs from u to v via a third vertex, z, Formally, they are represented by parallel: $T(u, v) = T_1(u, v) \cup T_2(u, v)$ and sequential: $T(u, v) = T_1(u, z) \cap T_2(z, v)$. Reachability time sets form a semiring with these operations. In early 2014 we extended this approach to temporal networks with $\tau = 0$ described by temporal quantities.

Operations $\cup$ and $\cap$ are illustrated by the following pairs:

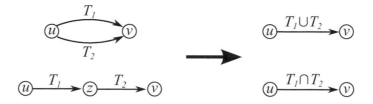

3.11.3.3   Transformation of temporal networks to static networks

In operations research, a temporal network $\mathcal{N} = (\mathcal{V}, \mathcal{L}, T, \{\tau, w\})$ with integer time $\mathcal{T} = \{0, 1, 2, \ldots, T\}$ is often transformed into a traditional network $\mathcal{N}' = (\mathcal{V}', \mathcal{L}', w)$ which is determined as follows:

1) For each active time point, $t, \in T(v)$ of vertex, v, we produce a copy (v, t) as

$$\mathcal{V}' = \bigcup_{v \in V} \{v\} \times T(v)$$

2) where the line $l' \in \mathcal{L}'$ links the vertices $(v_1, t_1), (v_2, t_2) \in \mathcal{V}'$, $l'((v_1, t_1), (v_2, t_2))$, iff

$$\exists l \in \mathcal{L} : (l(v_1, v_2) \wedge [t_1, t_2] \subseteq T(l) \wedge t_2 - t_1 = \tau(l)$$

3) with $w(l') = w(l)$.

By using this transformation, some traditional network problems for this type of temporal network can be transformed to larger static networks for which traditional problems can be solved using methods appropriate for static networks.

An example of this kind of transformation of a temporal network to a static network from George et al. (2007) is the following, see Figure 3.14. On the left is a time aggregated network with four vertices $\{A, B, C, D\}$. The time $\mathcal{T} = [1, 7]$ is integer. The vertices are active all the time. The labels on the arcs describe the transition times τ and have the form $(t_1 : \tau_1, \ldots, t_k : \tau_k)$. The term $t_i : \tau_i$ says: if the transition starts in time interval t_i its transition time is τ_i; if the time point t doesn't belong to any of the listed time intervals the transition is not possible. On the right is the corresponding temporally ordered network with time-referenced 'columns' containing each vertex for all time points $t = 1$ through $t = 7$ with the detailed arcs linking them through time.

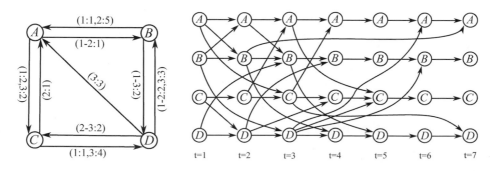

Figure 3.14   Time-aggregated and time expanded graph.

Some useful algorithms for dealing with these transformed networks include the SP-TAG algorithm (George et al. 2007) and the BEST algorithm (George et al. 2007). See also Casteigts and Flocchini (2013, 31–38), for a discussion of several special classes of TVGs.

3.11.4 Evolution

Network evolution has become a topic of interest since Doreian and Stokman (1997), but the development of tools for studying network evolution, especially for large temporal networks, has been lacking. One fundamental paper discussing methods for studying network evolution is Palla et al. (2007). Additional useful ideas can be found in Greene et al. (2010). Figure 3.15 illustrates some of the changes that can be captured.

One approach is to identify groups, or communities, inside the temporal network and analyze their changes. These changes include 'births' of new vertices joining the network, 'deaths' of vertices leaving the network, the overall expansion (growth) of the network, the overall contraction (decline) of the network, the merging of disjoint networks or disjoint parts of networks, networks splitting up, and the structural reorganization of networks. The groups involved can be determined by using a variety of different procedures including identifying leaders and islands, as described earlier.

The evolution of (selected and interesting) disjoint groups can be described by a *temporal partition* $C(v,t)$ with values

$$C(v,t) = \begin{cases} 0 & v \text{ is active in time } t \text{ but not a member of any group} \\ i & v \text{ is active in time } t \text{ and a member of group } i > 0 \\ NA & v \text{ is not active in time } t. \end{cases}$$

Then the i-th temporal cluster $C(i,t)$ is

$$C(i,t) = \{v \in \mathcal{V} : C(v,t) = i\}.$$

In applications, some additional conditions have to be imposed on groups to express a form of stability. One example of doing this for an integer single interval temporal network comes from Palla et al. (2007):

$$G(i,t) = \frac{C(i,t-1) \cap C(i,t)}{C(i,t-1) \cup C(i,t)}, \quad t \in \mathcal{T} \setminus \{t_0\}$$

if $C(i,t-1) \cup C(i,t) \neq \emptyset$, otherwise $G(i,t) = 0$.

By using the temporal clusters, it is possible to characterize the above transition changes, including network expansion, network contraction, and network reorganization. The idea of deaths can be expanded to include the demise of entire networks as well as the creation of new networks or parts of networks (see Figure 3.13).

3.12 Levels of analysis

Pajek allows analyses on different levels specified by a partition of the corresponding set of units. For example: partition of authors by institutions, or partition of institutions by countries, . . .

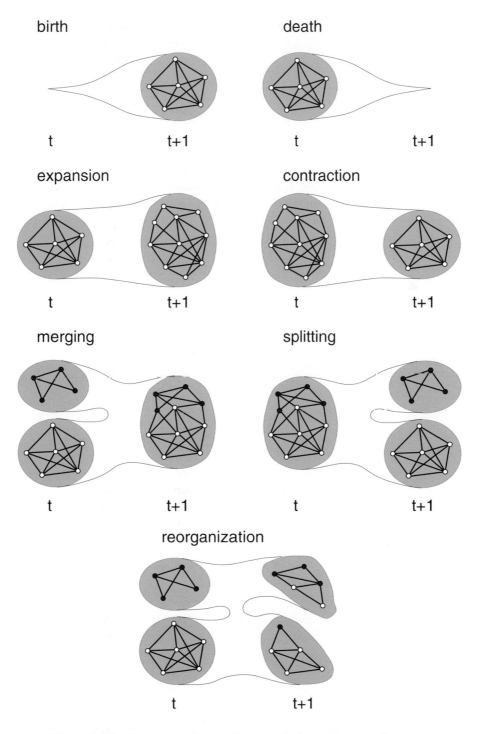

Figure 3.15   Examples of network events in the evolutions of groups.

The main tools are the *shrinking* of classes:

```
Operations/Shrink Network/Partition
```

and for reducing the area of interest the *extraction* of selected classes:

```
Operations/Extract from Network/Partition
```

3.13   Transition to substantive topics

In Chapters 2 and 3 the network analysis approaches used in the following chapters are presented. In Chapter 2 the basic network concepts are given and in Chapter 3 some advanced approaches with emphasis on large and temporal networks. The symbolic clustering was also described as it is used in some chapters.

4

Scientific citation and other bibliographic networks

The literature covering all of the many scientific disciplines is vast. As a result, its citation network is enormous. It can be viewed as one whole network, or attention can be focused on coherent parts of it. Our strong preference is for the latter strategy. For the substantive reasons laid out in Section 1.3, we pursue only two of these scientific citation networks here. One is for the centrality literature[1] with the other featuring network analysis as a whole.

4.1 The centrality citation network

As we note in Section 1.3.1, the centrality literature was started by the Bavelas (1948) paper reporting on experiments for task-oriented groups. The experimental control variable was the communication structure of a task-oriented group charged with distributing information to all of its members. The outcome variable was the time it took the group to complete its task. While centrality did not appear explicitly in this paper, the experimental outcomes can be expressed as: 1) centrality in the network is predictive of actor outcomes, and 2) network centralization is predictive of the task completion times. A research literature was spawned featuring social psychology, small group processes, experimental design, social organization, business administration, and measurement as substantive and technical concerns. Centrality was seen as a useful concept linking these diverse topics, one with great implications for the study of social network processes. Indeed, the above two basic findings morphed into two mantras for social network analysts: 1) actor locations in networks affect actor outcomes, and 2) group network structures matter for collective outcomes.

---

[1] This relatively small part of the network literature, as shown below, has almost a million works and close to two million citation arcs.

Understanding Large Temporal Networks and Spatial Networks: Exploration, Pattern Searching, Visualization and Network Evolution, First Edition. Vladimir Batagelj, Patrick Doreian, Anuška Ferligoj and Nataša Kejžar.
© 2014 John Wiley & Sons, Ltd. Published 2014 by John Wiley & Sons, Ltd.

The experimentally varied communication structures differed in terms of what became recognized as the centralization of networks. While both the idea of group (network) centralization and task-oriented small group interests were present within the SNA literature, attempts were made to formalize this concept, albeit with mixed results. Freeman (1977, 1979) laid foundations for using centrality ideas with a clear formalization of three alternative conceptualizations of centrality for the vertices in a network and the corresponding centralization of a network as a whole. One was based on degree centrality, a concept presented in Sections 2.1.2 and 2.11.1. In Section 2.11.1, we provided statements of closeness centrality and betweenness centrality, the other two conceptualizations of Freeman. We note that both degree centrality and closeness centrality were present in the literature before his foundational papers. Betweenness centrality was genuinely new, although we note that this was also (independently) defined by Anthonisse (1971). Of greater importance, Freeman resolved the problem of comparing centrality measures across networks of different sizes by proposing normalizations of the three measures controlling for network size. These normalizations are laid out in Section 2.11.1.

Hummon and Doreian (1989, 1990) proposed a (then) new method for analyzing the structure of citation networks, calling it *main path analysis* (see Section 3.2). The primary objective was the delineation of main paths of scientific productions featuring important works. Hummon et al. (1990) used main path analysis to study the complete network for the centrality literature between the Bavelas (1948) and Freeman (1979) papers. There were 119 articles, technical reports, and books created in this period. The main path they identified is shown in Figure 4.1.[2] The structure of this main path is instructive:

- The sequence of publications at the beginning of the main path dealt with both centrality and group productivity.

- After 1956, the main path split into two parts. In the papers of the left branch, attention was focused on centrality and its operationalization. It culminated in the Freeman (1979) paper. The substantive concern with group productivity was abandoned.

- Papers in the right branch focused on group productivity but dropped centrality as a structural principle. They feature publications involving one author.

Having the primary focus diverge makes it clear that, when studying citation networks, attention has to be paid to the content of the works as well as to network structures. Hummon et al. (1990) suggested that an appropriate next step was to consider the centrality literature published after 1979. This is undertaken here. However, given the large size of the citation network, as shown in Table 1.1, this task is far more complicated than was envisioned initially.

4.2 Preliminary data analyses

Before examining the citation structure of the centrality literature, some preliminary results about the network as a whole are presented. The distribution of publications through time is in Section 4.2.1. Some simple results based on the degree concept are in Section 4.2.2. Works are classified into conceptual types in Section 4.2.3. Finally, the network boundary

---

[2] Although authors are shown in this figure, not all are relevant for our discussion and are omitted from the references. However, if needed, they are all listed in the appendix of Hummon et al. (1990).

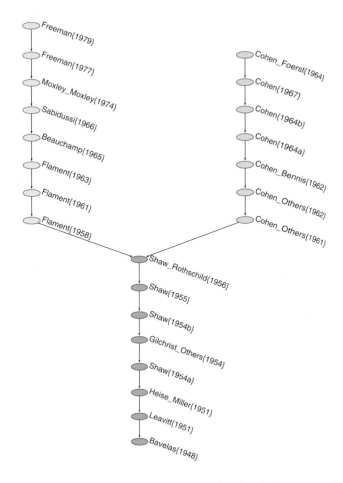

Figure 4.1   The main path structure of early citation network.
Color codes: initial path (dandelion); centrality (light faded green); and group productivity (light sky blue).

problem is tackled in Section 4.2.4 so as to define a *useful* citation network for detailed analyses. Throughout this chapter, we maintain the convention of using a different typeface for Pajek items: specific Pajek commands and Pajek objects will be written as `commands` and `objects`. This applies also for the R commands we used.

4.2.1   Temporal distribution of publications

The complications mentioned at the end of Section 4.1 stem from the size of the centrality citation network in 2013: there are 995,783 scientific productions with 1,856,102 citation links. By comparison, the network studied by Hummon et al. (1990) was minute. The size of the later centrality citation network is implied by the distribution shown in Figure 4.2 for the number of publications featuring centrality.[3] In the first half decade of the 21st century

---

[3] Included are publications citing publications that can deal with other topics. Complicating things further, users can cite papers for other reasons that are likely to be extraneous as far as centrality is concerned.

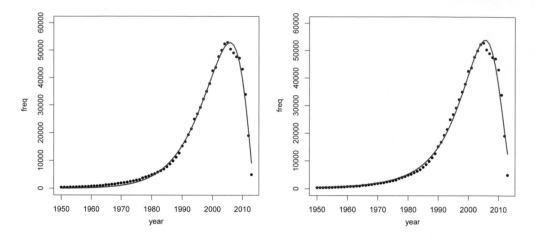

Figure 4.2   The number of works featuring centrality by publication years.
Left: the fitted gamma distribution, right: the fitted generalized reciprocal power exponential curve.

alone, more than 250,000 works appear on this topic. The rate of publication appears to have dropped – but the data are incomplete for 2013 (and cited works by publications yet to appear).

In addition to having a visual impression of this temporal distribution, it is often useful to fit an estimated parameterized curve. To do this, a file `Year.clu` was used where each publication is marked by the year it appeared. The corresponding distribution of *papers by years* has a shape similar to the curves such as the gamma distribution, the lognormal distribution and the generalized reciprocal power exponential curve: $c * (x + d)^{\frac{a}{b+x}}$.

The first part of the following R code constructs the empirical distribution for time, t. The nonlinear (weighted) least-squares procedure `nls` was used to fit the gamma distribution and the generalized reciprocal power exponential curves. The rest of the R code below was used for fitting these curves to the data.

```
> # read data and tabulate
> setwd("D:/Data/Centrality/net")
> years <- read.table(file="Year.clu",header=FALSE,skip=2)$V1
> t <- table(years)
> year <- as.integer(names(t))
> freq <- as.vector(t[1950<=year & year<=2013])
> y <- 1950:2013; x <- as.vector(2014-y)
> # fitting gamma
> c <- 900000; a <- 2.98; b <- 0.23
> modelG <- nls(freq ~ c*dgamma(x,a,b), start=list(c=c,a=a,b=b))
> modelG
Nonlinear regression model
  model: freq ~ c * dgamma(x, a, b)
   data: parent.frame()
         c          a          b
9.595e+05 2.428e+00 1.747e-01
  residual sum-of-squares: 84883808
Number of iterations to convergence: 5
```

```
Achieved convergence tolerance: 3.725e-07
> plot(y,freq,pch=20,ylim=c(0,60000),xlab="year")
> lines(y,predict(modelG,list(x=2014-y)),col='red',lw=2)
> # fitting  generalized reciprocal power exponential
> c <- 3; a <- 105; b <- 17; d <- 2.4
> modelP <- nls(freq ~ c*(x+d)^(a/(b+x)), start=list(c=c,a=a,b=b,d=d))
> modelP
Nonlinear regression model
  model: freq ~ c * (x + d)^(a/(b + x))
   data: parent.frame()
        c        a        b        d
  0.1517 175.5650  27.5712   5.3230
 residual sum-of-squares: 173666241
Number of iterations to convergence: 31
Achieved convergence tolerance: 1.466e-06
> plot(y,freq,pch=20,ylim=c(0,60000),xlab="year")
> lines(y,predict(modelP,list(x=2014-y)),col='red',lw=2)
```

The red lines mark the fitted curves in Figure 4.2. The fitted gamma distribution is on the left and the fitted generalized reciprocal power exponential curve is on the right. The empirical distribution is well approximated by these two distributions.[4]

4.2.2   Degree distributions of the centrality literature

Examining the degree distributions is a useful first step for studying citation networks. The indegree (received citations) and outdegree (citations made) data are obtained in Pajek as partitions with the commands

    Network/Create Partition/Degree/Input
    Network/Create Partition/Degree/Output

or as vectors with

    Network/Create Vector/Centrality/Degree/Input
    Network/Create Vector/Centrality/Degree/Output

After exporting the input and output vectors to R, the corresponding frequency distributions were plotted as described in Section 2.5.1. They are shown in Figure 4.3 using log-log scales. The non-zero indegree distribution follows the characteristic power-law for large citation networks.

Interest can center on identifying the most heavily cited works in the citation network. This is obtained by selecting the indegree vector, clicking on the vector info button and entering +20 (for a list of the 20 most-cited articles). The result is shown in Table 4.1. Although the focus in this section is restricted to centrality, our interest in the role of physicists in studying social networks appears here also with the heavy presence of physicists (including Barabási, Watts, and Newman) in these top 20 cited works. The foundational Freeman papers starting the new centrality literature (after 1979) appear at ranks 6 and 14. Granovetter (1973, 1985) are 9th and 16th with Burt (1992) occupying the 8th rank. The classic Wasserman and Faust (1994) textbook on SNA is 3rd in the ranking.

---

[4] In some other similar cases the lognormal distribution gives the best fit.

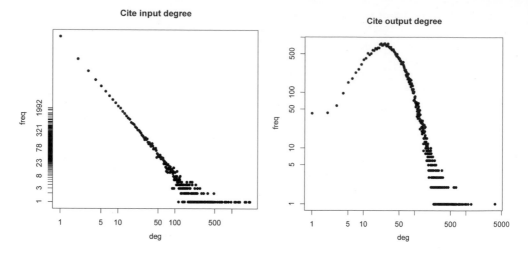

Figure 4.3   Non-zero indegree and outdegree distributions.

The other end of the cited distribution is also useful. To list the indegree frequency distribution, clicking the `partition info` button provides a list of all these values. The initial segment of the resulting frequency distribution is in Table 4.2 where the representative works have little significance beyond marking the cluster to which they belong. (In its terminal segment we get the list of the most frequently cited works from Table 4.1.) We used the list

Table 4.1   The works with the largest indegrees.

Rank	Vertex	Value	Id
1	223	2047.0000	BARABASI_A(1999)286:509
2	1114	1992.0000	WATTS_D(1998)393:440
3	305	1754.0000	WASSERMA_S(1994):
4	1078	1692.0000	ALBERT_R(2002)74:47
5	1099	1365.0000	NEWMAN_M(2003)45:167
6	74	1285.0000	FREEMAN_L(1979)1:215
7	1700	1164.0000	JEONG_H(2001)411:41
8	1253	1117.0000	BURT_R(1992):
9	81	966.0000	GRANOVET(1973)78:1360
10	62084	869.0000	HOLBEN_B(1998)66:1
11	2633	797.0000	ALBERT_R(2000)406:378
12	647	768.0000	JEONG_H(2000)407:651
13	641	759.0000	GIRVAN_M(2002)99:7821
14	464	754.0000	FREEMAN_L(1977)40:35
15	1082	673.0000	BOCCALET_S(2006)424:175
16	82	659.0000	GRANOVET_M(1985)91:481
17	1108	654.0000	STROGATZ_S(2001)410:268
18	57	632.0000	BORGATTI_S(2002):
19	897	603.0000	BARABASI_A(2004)5:101
20	96895	554.0000	HAGMANN_P(2008)6:1479

Table 4.2 Initial part of frequency distribution of indegrees (by cluster numbers).

Cluster	Freq	Freq%	CumFreq%	Representative
0	21512	2.1603	2.1603	VILLANUE_S(2008)28:613
1	744571	74.7724	76.9327	ALONSO_J(2004)4:2392
2	115311	11.5799	88.5127	ALONSO_J(2005):
3	42499	4.2679	92.7806	WEI_D(2005):
4	20729	2.0817	94.8622	CHANDRA_A(2000):
5	12370	1.2422	96.1045	PAYTON_D(2001)11:319
6	7997	0.8031	96.9076	EVANS_M(2001)49:542
7	5515	0.5538	97.4614	BROOKS_R(1991)47:139
8	4044	0.4061	97.8675	FLAP_H(1998)4:109
9	3122	0.3135	98.1810	LEYDESDO_L(1987)11:295
10	2393	0.2403	98.4213	CHISHOLM_D(1989):

of the largest vector values because of the possibility that some works have the same number of citations.

There are several noteworthy features in this partial distribution:

- There are 21,512 works with indegree = 0. For our purposes, these are *not* unimportant productions. They represent pure users and can be viewed as 'starting articles' for identifying parts of the overall centrality literature by providing citations having relevance for using centrality ideas.

- There are 744,571 articles cited only once, another 115,311 are cited twice and so on.

- Slightly over 98% of these scientific productions received ten or fewer citations.

- If the 'terminal vertices' (source articles of indegree smaller than 4) were removed from the network, only slightly more than 10% of the original vertices would remain. This is fully consistent with the left-hand distribution in Figure 4.3.

The same treatment can be applied to the outdegree of works. It yields a list of the 20 articles with the longest lists of references. The results are shown in Table 4.3. These works are overview papers. For example, the highest outdegree work in Table 4.3 appeared in *Pharmacology and Theraputics*, Volume 110(2) in 2006 with 246 pages. The second listed paper appeared in *Current Protein and Peptide Science*, Volume 138(3) in 2013. It was a 'comprehensive review' on the structure and dynamics of molecular networks. The third was published in *Physiological Reviews*, Volume 86(1), in 2006 and had 89 pages.

While all of these papers are reviews with many citations, only one of the items in Table 4.3 is present in the main path (Figure 4.6) or the CPM path (Figure 4.7). It appeared in *Physics Reports*, Volume 424(4-5) in 2006 and was another long production with 124 pages. Its title is 'Complex networks: Structure and dynamics': clearly it is network related. This suggests that review articles, while playing an important role in science, do not necessarily get cited enough, or in ways to appear on main paths (see below) in the literature, especially if they are tangential.

Part of the full distribution is shown in Table 4.4 starting from the lowest counts. Note that the list finishes with the high outdegree items shown in Table 4.3. A total of 957,349

Table 4.3 Works with the largest outdegrees.

Rank	Vertex	Value	Id
1	318269	3729.0000	MILIAN_M(2006)110:135
2	806124	1268.0000	CSERMELY_P(2013)138:333
3	150886	1109.0000	MEHTA_D(2006)86:279
4	629249	1104.0000	WANG_X(2010)90:1195
5	538665	1009.0000	MAYER_A(1995)67:629
6	115288	964.0000	RANDIC_M(2003)103:3449
7	782567	873.0000	LINDQUIS_K(2012)35:172
8	1082	867.0000	BOCCALET_S(2006)424:175
9	481228	816.0000	RUSKOLA_T(2000)52:1599
10	793396	768.0000	SAGER_T(2011)76:147
11	755324	740.0000	STYNEN_B(2012)76:331
12	692148	725.0000	DEFEE_C(2010)21:404
13	511076	707.0000	HUTTO_D(1998)25:459
14	163982	650.0000	ALVES_R(2008)3:98
15	99171	639.0000	TUMA_P(2003)83:871
16	375844	639.0000	SCHROEDE_S(2004)12:311
17	474850	635.0000	GINI_F(2001)37:329
18	286341	624.0000	ASHBY_D(2006)25:3589
19	170589	618.0000	KOWALEWS_S(2008)16:225
20	683625	613.0000	HAGEN_M(2012)46:89

(96%) works in this network have outdegree 0: they are only cited (sources). The modal value of citations made is around 26 citations.

4.2.3 Types of works

We distinguished three broad kinds of works in this citation network: sources, base works, and user works. The *base* works were obtained as hits to a WoS query. The works citing the base works are the *user* works. For base and user works, we obtained their description records from WoS. The only cited works contained in these descriptions in the CR field are called the *source* works having outdegrees equal to 0 in this citation network. Sources can be further divided into *basic sources* and *secondary sources*.

The works in the base are devoted to the centrality concept, not unlike the works in the left-hand branch in Figure 4.1. Works in this base cite works among the basic sources and user works. Works using centrality ideas cite works from the base and both types of sources. This conceptual partition of works is shown in Figure 4.4.

We focus initially on obtaining the sizes of the three broad types of works. After reading the DC (described, cited only) partition file, the corresponding distribution is obtained by clicking on the `partition info` button. The resulting size distribution of these parts of the centrality citation network (0 – sources, 1 – base, and 2 – users) is 957,248 sources, 6951 base works and 34,584 user works.

The source vertices can be further split into those works (to go into class 3) cited by the base vertices and those cited by the remaining user vertices (to go into class 4). The Pajek commands for doing this are:

Table 4.4 Part of the outdegree distribution for works.

Cluster	Freq	Freq%	CumFreq%	Representative
0	957349	96.1403	96.1403	*IEEE(2003):
1	42	0.0042	96.1445	CASTRO_M(2008)15:32
2	43	0.0043	96.1489	KANNAN_G(2008)54:1087
3	58	0.0058	96.1547	ADEANE_J(2007)43:39
4	97	0.0097	96.1644	PETTET_M(2008)91:289
5	152	0.0153	96.1797	EVERETT_M(2005)27:31
6	179	0.0180	96.1977	FUJIWARA_M(2008)20:1612
7	229	0.0230	96.2207	KUO_W(1998)76:287
8	264	0.0265	96.2472	BULUSU_N(2000)7:28
9	324	0.0325	96.2797	ASHTON_D(2005)94:058701
10	395	0.0397	96.3194	ALONSO-Z_J(2008)84:619
. . . .				
20	644	0.0647	96.8860	KIM_J(2004)82:238
21	747	0.0750	96.9610	GOH_K(2003)67:017101
22	772	0.0775	97.0385	FRASER_H(2002)296:750
23	718	0.0721	97.1106	ECKHARDT_S(2008)407:1
24	722	0.0725	97.1831	SERRANO_J(2008)24:1629
. . . .				
30	767	0.0770	97.6306	GUIMERA_R(2005)433:895
31	679	0.0682	97.6988	GUTIERRE_A(2008)8:7545
32	659	0.0662	97.7650	NG_D(2007)46:9107
. . . .				
964	1	0.0001	99.9995	RANDIC_M(2003)103:3449
1009	1	0.0001	99.9996	MAYER_A(1995)67:629
1104	1	0.0001	99.9997	WANG_X(2010)90:1195
1109	1	0.0001	99.9998	MEHTA_D(2006)86:279
1268	1	0.0001	99.9999	CSERMELY_P(2013)138:333
3729	1	0.0001	100.0000	MILIAN_M(2006)110:135
Sum	995783	100.0000		

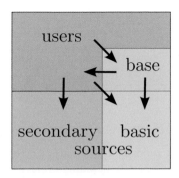

Figure 4.4 Three broad types of works: sources, base, and users.

Read Cite.net
Read DC.clu (0-source,1-base,2-user)

```
Partition/Binarize Partition [1]
File/Partition/Change Label [base]
Partition/Copy to Vector
Network/Create New Network/Transform/Transpose 1-mode [Yes]
Operations/Network+Vector/Network*Vector [1]
Vector/Copy to Partition by Truncating
Partition/Binarize Partition [1-*]
File/Partition/Change Label [base neighbors]
```

Select DC partition

```
Partition/Binarize Partition [0]
File/Partition/Change Label [sources]
```

Select base neighbors as Second partition

```
Partitions/Min (First,Second)
File/Partition/Change Label [basic sources]
Partition/Binarize Partition [0] -> complement
```

Select sources as Second partition

```
Partitions/Min (First,Second)
File/Partition/Change Label [secondary sources]
```

Select DC partition

```
Partition/Binarize Partition [2]
File/Partition/Change Label [users]
```

Save partitions base, users, basic and secondary to files

Applying the `partition info` button on the obtained partitions reveals that class 0 was split into 158,102 basic source vertices and 799,146 secondary source vertices. Sometimes it is handy to have all four types in a single partition: 1 – base; 2 – users; 3 – basic sources; and 4 – secondary sources. Such a partition can be produced as follows:

Select DC as First partition
Select basic as Second partition

```
Partitions/Intersection of partitions
```

In the obtained partition we have: 1 – basic; 2 – base; 3 – users; and 4 – secondary. To get the desired numbering of types we made a 4-element partition map with values: 3, 1, 2, and 4 by using:

Select the intersection as First partition
Select the map as the Second partition

```
Partitions/Functional Composition First*Second
```

The resulting partition was saved in the file `types.clu`.

4.2.4 The boundary problem

The network boundary problem rears its ugly head here. Recall, there are two parts to this problem: 1) missing relevant units, and 2) including irrelevant units. Networks obtained from the WoS file using the program WoS2Pajek are in a 'raw' form. It is possible that some works may have been missed. Even so, it is highly unlikely that many important units (works) will have been omitted.

However, the other part of the boundary problem for citation networks hinges on a definition of importance for works.[5] Many works receive few citations. Indeed, from Table 4.2, the modal value for incoming citations is only 1. Such works do not qualify as important with regard to *received* citations beyond noting their existence. A richer (more useful) network can be obtained by including only works that are referenced more than some threshold. We decided to omit (consider as irrelevant) all *basic source* works that were cited at most $k_1 = 1$ times, and all *secondary source* works cited at most $k_2 = 3$ times. To obtain the corresponding *boundary* partition (1 – preserve, and 0 – omit) we used the following procedure (available also as a Pajek macro `CiteBoundary`). We assume that we already have the basic and secondary source partitions (to be read or computed):

Select Cite.net
```
Network/Create Partition/Degree/Input
Partition/Binarize Partition [2-*]
```
$(k_1 = 1)$
Select basic sources as Second partition
```
Partitions/Min (First,Second)
File/Partition/Change Label [basic and (indeg>1)]
```
Select indegree partition
```
Partition/Binarize Partition [4-*]
```
$(k_2 = 3)$
Select secondary sources as Second partition
```
Partitions/Min (First,Second)
File/Partition/Change Label [secondary and (indeg>3)]
```
Select DC partition
```
Partition/Binarize Partition [1-*]
```
Select secondary and (indeg>3) as Second partition
```
Partitions/Max (First,Second)
```
Select basic and (indeg>1) as Second partition
```
Partitions/Max (First,Second)
File/Partition/Change Label [boundary]
Operations/Network+Partition/Extract Subnetwork [1]
File/Network/Change Label [citeB]
```
The obtained partition was saved to `boundary.clu` and the corresponding extracted 'bounded network' to `CiteB.net`. The resulting network has 139,505 vertices and 854,032 arcs with neither loops nor multiple lines present. The distributions of types in the original citation network and in the bounded network are presented in Table 4.5. This is to define a meaningful and manageable network focused on the centrality literature: in doing this, tangential and extraneous works have been removed. The commands for doing this are:

Select or read types partition as partition 1
Select boundary partition as partition 2
```
Partitions/Extract Subpartition (Second from First)
File/Partition/Change Label [typesB]
```
Using this typesB partition, the following command produces a 'blockmodel' with classes (1 – base, 2 – users, 3 – basic, and 4 – secondary):
```
Operations/Network+Partition/Shrink Network [1 0]
```

---

[5] One option is to limit the network to the works with complete descriptions (records) from the WoS file. However, this does not deal with the second part of the boundary problem.

Table 4.5 Distributions of works in the citation and bounded citation networks.

Class		cite	citeB
base	1	6951	6951
users	2	31584	31584
basic	3	158102	70291
secondary	4	799146	30679
Totals		995783	139505

Table 4.6 A blockmodel with four types of works.

		1	2	3	4
base	1	8702	13665	131007	0
users	2	38924	80784	386562	194388
basic	3	0	0	0	0
secondary	4	0	0	0	0

By definition, sources cite no other works.

This blockmodel is shown in Table 4.6. The 139,505 works are divided into 100,970 sources, 6951 base productions, and 31,584 works using centrality ideas.

4.3 Transforming a citation network into an acyclic network

For computing the SPC (search path count) weights proposed by Hummon and Doreian (1990), see Section 3.2, the citation network must be acyclic. In principle, citation networks are acyclic because new works cite older works. However, this need not be the case in practice. Usually, the cycles, should they exist, are short and relatively infrequent.

4.3.1 Checking for the presence of cycles

It is necessary to test the acyclicity of a citation network by determining its non-trivial (sized larger than 1) strong components. If there are none, the network is acyclic.[6] Otherwise, all cycles are inside the non-trivial strong components. The Pajek commands for this test are:

```
Network/Create Partition/Components/Strong [2]
Partition/Canonical Partition/With Decreasing frequencies
```

When this was applied to the bounded centrality citation network, citeB (constructed in Section 4.2.4), there were 73 non-trivial strong components. The distribution of the sizes of the non-trivial strong components is: one with 12 works; two of size 8; one of size 7; two of size 6; two with size 5; four components of size 4; ten components of size 3 and 51 components of

---

[6] The patent citation network used in Chapter 5 is acyclic. This is due to the stringent review of patent applications by the United States Patent and Tradmark Office.

size 2. On checking, every work in a non-trivial strong component had the same publication year. To obtain these non-trivial components, the following Pajek commands were used:

```
Operations/Network+Partition/Extract SubNetwork [1-*]
Operations/Network+Partition/Transform/Remove Lines/
   Between Clusters
Draw/Network + First Partition
[Draw] Layout/Energy/Kamada-Kawai/Separate Components
```

Such components can be examined visually. The small strong components with two or three works (and the cycles within them) are not surprising. Of more interest are the larger strong components. To obtain them, the Pajek command for searching strong components is changed to:

```
Network/Create Partition/Components/Strong [4]
```

Figure 4.5 shows the strong components with four or more works. At face value, these strong components and the cycles they contain are simply nuisances: their presence prevents a citation network from being acyclic. Below, we describe two ways of dealing with this problem in order to obtain acyclic citation networks. However, before doing so, there is additional value in looking more closely at these strong components. First, this provides insight into the diverse ways by which cycles are created and, second, it points to a clear need to identify tangential works appearing in a citation network ostensibly devoted to a single topic.

We consider first the largest non-trivial strong component. The publications all come from Baum and Rowley (Eds) *Advances in Strategic Management*, Vol. 25, JAI Press. This volume was the result of a two-day workshop at the University of Toronto.[7] Papers were presented and critiqued. Following the workshop, participants revised their presentations for the volume. In doing so, they were encouraged to cite other papers in the volume.[8]

The second large strong component is made up of eight papers among 20 papers appearing in a supplementary issue of the *Journal of Adolescent Health*, Volume 29(3), 2001. This issue focused on 'HIV research in American youth.' The papers were published as part of the REACH (Reaching for Excellence in Adolescent Care and Health) Project. There was an extensive collaboration with members of The Adolescent Medicine HIV/AIDS Research Network. It is not surprising that there would be citations between these eight publications. Indeed, some authors appear on multiple papers. While most of these papers do not cite any of the well known SNA centrality papers, some do include sexual partners suggesting that SNA issues involve disease transmission over sexual networks.

All of the papers in the second strong component with eight works appeared in a special issue of *Urban Studies* in 2010, Vol. 47(9). It dealt with 'examining changing patterns of transnational intercity connectivity and hierarchy' as described in the introduction to the special issue. Although the papers took different approaches, all dealt with the global city

---

[7] One of the current authors (Doreian) took part in this workshop.

[8] This large strong component was present in the 2010 version of this citation network in which there were many fewer non-trivial strong components. To the extent that special journal issues are appearing based on similar workshops or conferences, such strong components will become more frequent in the literature. This expectation was born out in the 2013 data.

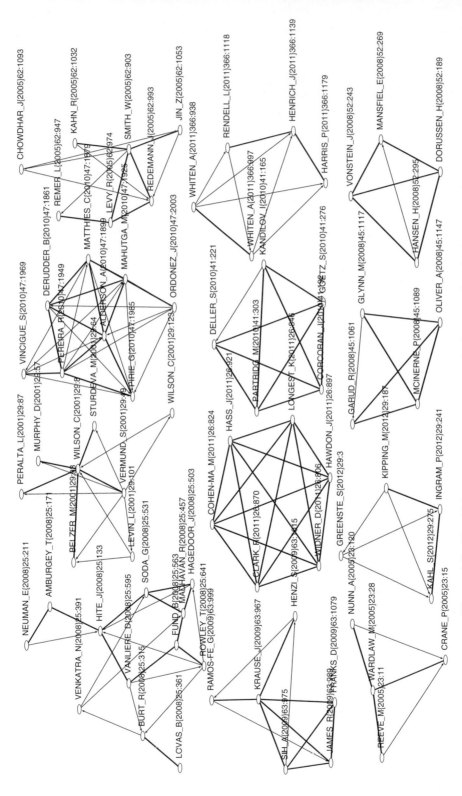

Figure 4.5   The 12 largest strong components in the centrality citation network.
Blue: reciprocated ties, black: directed ties.

network. In most of these papers, centrality was a directly relevant concept. Again, given the common substantive focus of a special issue, it is not surprising to see reciprocal citations and cycles in this set of productions.

The strong component with seven vertices has articles from the *Journal of the Atmospheric Sciences*, Vol. 62(4), 2005. They all feature papers about The Chesapeake Lighthouse and Aircraft Measurements for Satellites (CLAMS) 2001 field campaign to validate some measurement instruments of aerosol properties and radiative flux profiles. These papers cite none of the SNA papers. Indeed, there is no reason for them to do so because they come from an entirely separate literature. The general notion of centrality can appear in many places. These papers do *not* point to an invasion of the physicists.

The papers appearing in the large strong component with seven vertices come from a special issue of *Behavioral Ecology and Sociobiology* (Volume 7, 2009) entitled 'Social Networks: new perspectives.' The specific subject for these papers was primate networks within the broader topic of animal social networks.

The first strong component with six vertices has papers coming from *Sociological Forum*, Volume 26 (4), 2011. The topics include employment discrimination against Arab Americans, neighborhood organizations and resident assistance to police, conflicts in emerging beliefs about religion and science, human rights, survival in wartime, and the American Dream in a multiethnic America. Despite the diversity of topics, there are reciprocated citation ties between all pairs of these papers. With one potential exception, none of them has much to do with SNA: they seem completely tangential for a consideration of centrality.

SNA is highly relevant in the next strong component of five papers coming from *Growth and Change*, Volume 41(2) in 2010. Both commuting networks and social capital were studied and linked. The papers in the next strong component appeared in a special issue 'Culture evolves' of in *Philosophical Transactions of the Royal Society of London, Series B*, 2010. The networks involved link cultural components. Some attention was given also to human relations and culture among non-human primates. Of interest is that one these papers has reciprocal links with a paper appearing in *Trends in Cognitive Sciences*, Volume 15(2), 2011, even though there are no common authors for this pair of papers.

The next strong component, with four vertices, features a journal that was a surprise – the *BT Technology Journal*. These papers came from an issue devoted to the next generation of (telecommunication) services and intelligence. This journal is published by the BT Group plc, a British corporation that evolved from the old Post Office and then British Telecommunications. Its publisher describes the journal as providing in-depth views of technologies at the heart of communications and networked ICT units. Each issue is devoted to a key aspect of the communications business. Items appearing in this journal are written by in-house experts, partners, customers, suppliers, and other 'leaders' of the communications field. Despite appearing to be a self-promotional publication, it has been cited enough, presumably in the context of computer networks. However, both the journal and the articles in this strong component may be tangential regarding centrality as far as SNA is concerned.

Two strong components with four vertices come from the management literature. The first has papers from *Advances in Strategic Management*, Volume 29, 2012 devoted to history and strategy, while the second features papers from the *Journal of Management Studies*, Volume 45(6), 2008. Both sets of papers appear tangential regarding centrality. One of the papers from the second management journal uses the term 'the centrality of fields as a concept in organizational research' suggesting that centrality is used in a general (and meaningless) metaphorical rather than analytical sense.

The final four-vertex strong component has papers from *The Journal of Conflict Resolution*, Volume 52(2), 2008 devoted to 'International Organizations Count.' These papers do feature SNA ideas where centrality could be important.

One of the three-vertex strong components has papers from three different journals: the *International Journal of Primatology*, Volume 32(3), the *American Journal of Primatololgy*, Volume 73(8), and *The Journal of Experimental Biology*, Volume 273(1), all appearing in 2011. One author was involved in producing all three papers, while another author shared in two of them. SNA ideas are used. Some of the three-vertex strong components come from special issues of journals. They include *Social Networks*, Volume 29(2), 2007, devoted to exponential random graph models, clearly featuring SNA concerns, and *Philosophical Transactions of the Royal Society of London, Series B* on a theme issue devoted to evolutionary and ecological approaches to the study of personality where SNA ideas were seen as relevant.

The remaining three-vertex strong components all come from articles linked by cycles within regular journal issues. Five of them are completely outside the realm of SNA: *Atmospheric Chemistry and Physics*, Volume 13, 2013; the *Journal of Geophysical Research: Atmospheres*,[9] Volume 108, Issue D23 2011; *Molecular Ecology*, Volume 19(17), 2010 *Neuroinformatics*, Volume 1(4) and the *International Journal of Radiation Oncology, Biology, Physics. 1995 Dec 1;33(5)*, Volume 33(5), 1995. In contrast, the papers from *Global Networks: A Journal of Transnational Affairs*, Volume 10(1), 2010, and *Journal of Computer Assisted Learning*, Volume 24(2), 2008 explicitly use SNA ideas including centrality.

Most often, cycles are created under one the following circumstances: 1) one author is involved in a pair of (or multiple) publications published at the same time; 2) scholars working at the same place publish related papers on the same topic; 3) scholars working at different places but collaborating on common projects publish related papers (the REACH and CLAMS projects listed above provide examples); 4) small gatherings (in terms of the number of presentations) focusing on specific topics (for example the volumes of *Philosophical Transactions of the Royal Society of London, Series B*); 5) larger (in terms of the number of topics considered) conferences or workshops devoted to a set of related topics (the two issues of *Advances in Strategic Management* are examples); 6) special issues of journals devoted to a specific topic that can be narrowly or broadly focused; and 7) scholars who, while working at different places, are aware of each other's work. Knowing these differences is important for understanding scientific citation networks. Indeed, they are used to help interpret patterns identified in this citation network later in this chapter.

Examining only these strong components still provided a glimpse into the huge diversity of fields where centrality ideas can be relevant. The results reported above came from going to the papers in the strong components and examining them, albeit in a cursory fashion. Even this was time-consuming: it cannot be done for almost one million scientific productions. A way of discerning this diversity is to examine keywords in works as described in Section 4.8. The same can be done with regard to the journals from which papers appear in this network.[10]

---

[9] The papers for these two strong components form a weak component with the papers in the *Journal of Geophysical Research: Atmospheres*, citing papers from *Atmospheric Chemistry and Physics* published two years earlier.

[10] The (bounded) two-mode network (works × keywords) has 1,040,514 units with 995,783 works and 44,731 keywords. The corresponding two-mode network (works × journals) has 49,472 journals and the 995,783 works. Additionally, there are 319,591 authors in the two-mode (works × authors) network.

Looking at papers in these strong components suggested that not all papers are relevant for a sustained analysis of the centrality citation network. Tangential works can be removed without much consequence: this was the rationale motivating the approach to establishing boundaries as described in Section 4.2.4. Certainly, for studying the *genuine use* of analytic centrality ideas, metaphorical uses of the concept must be removed. When this is done the resulting network can still contain cycles. So, we turn to the issue of making citation networks genuinely acyclic.

4.3.2 Dealing with cycles in citation networks

There are different ways to make a citation network acyclic – see the WoS2Pajek manual. We focus on two of them. The simplest solution is to shrink each strong component into a vertex. The huge majority of the strong components in a citation network are trivial (singletons) and are unaffected by this shrinking. However, the differences between works in strong components are lost in this shrinking procedure. This will be inconsequential for most analyses. The Pajek commands for this solution are:

Select the `citeB` network

```
Network/Create Partition/Components/Strong [2]
Operations/Network+Partition/Shrink Network [1 0]
Network/Create New Network/Transform/Line Values/ Set All Line
   Values to 1 [No]
Network/Create New Network/Transform/Remove/Loops [No]
```

In those instances where the identities of the units need to be preserved (usually to focus on content or specific authors), we used a second procedure, called `preprint` and introduced in Section 3.4. The bounded citation network citeB has 139,505 vertices, 854,032 arcs and 246 weak components (made up of 203 isolated vertices and 43 non-trivial components). The main component has 139,133 vertices, the second largest has only 18 vertices and the third largest has 13 vertices. We limit our further analysis to the main component. After extracting it, strong components can be analyzed. There are 72 non-trivial strong components in the main (largest weak) component. To transform the main component network `citeMain` into an acyclic network the preprint transformation was applied to it. The resulting acyclic network `citeAcy` has 139,336 vertices and 854,368 arcs.

To produce the types partition, `typesAcy` for the `citeAcy` network the following commands were used:

Select `citeMain` as the First network

Select `citeAcy` as the Second network

```
Networks/Match Vertex Labels
```

Select typesMain as Second partition

```
Partitions/Functional Composition First*Second
Partition/Create Constant Partition [139336 5]
```

Select `composition` as Second partition

```
Partitions/Min(First,Second)
File/Partition/Change Label [typesAcy]
```

In this partition we have a new class 5 representing the preprint vertices (whose names are preceded by an '=' sign when they appear on network diagrams below).

Table 4.7   The 30 most important vertices.

Rank	Value	Id
1	2481.1796	BARABASI_A(1999)286:509
2	2413.1823	WATTS_D(1998)393:440
3	2099.6951	ALBERT_R(2002)74:47
4	1807.7400	WASSERMA_S(1994):
5	1656.5485	JEONG_H(2001)411:41
6	1559.2715	FREEMAN_L(1979)1:215
7	1521.8437	NEWMAN_M(2003)45:167
8	1171.6774	ALBERT_R(2000)406:378
9	1142.8359	JEONG_H(2000)407:651
10	1083.6487	FREEMAN_L(1977)40:35
11	1055.2631	GIRVAN_M(2002)99:7821
12	785.9130	STROGATZ_S(2001)410:268
13	782.6379	BOCCALET_S(2006)424:175
14	777.3402	GRANOVET(1973)78:1360
15	734.1673	GAREY_M(1979):
16	718.6859	HAGMANN_P(2008)6:1479
17	662.5071	BARABASI_A(2004)5:101
18	659.7386	AMARAL_L(2000)97:11149
19	635.2949	BURT_R(1992):
20	589.2179	ALBERT_R(1999)401:130
21	584.6247	NEWMAN_M(2001)98:404
22	566.9214	SHANNON_P(2003)13:2498
23	558.6965	KATZELA_I(1996)3:10
24	505.0996	DOROGOVT_S(2002)51:1079
25	497.5110	SCOTT_J(2000):
26	496.6379	RAVASZ_E(2002)297:1551
27	496.3690	UETZ_P(2000)403:623
28	483.7304	ERDOS_P(1959)6:290
29	456.5162	ERDOS_P(1960)5:17
30	453.8277	MASLOV_S(2002)296:910

4.4   The most important works

To identify the most important works we computed the probabilistic flow index for works from centrality citation network. In Table 4.7 the 30 vertices with the largest flow value (multiplied by 10^6) are listed.

The table contains mainly papers and four books (Burt, 1992; Garey and Johnson, 1979; Scott, 2000; Wasserman and Faust, 1994). Most of the paper are written by physicists. The highest placed papers written by traditional SNA researchers are Freeman (1979), Freeman (1977), and Granovetter (1973).

4.5   SPC weights

Having created a genuinely acyclic citation network by either of the methods described in Section 4.3.2, the next step is to put weights on arcs. Obtaining them is described in

Section 4.5.1 with the first results of doing so contained in Section 4.5.2. Most often, these analyses are done on the acyclic network produced by the `preprint` transformation. The few exceptions to this are noted.

4.5.1 Obtaining SPC weights and drawing main paths

Hummon and Doreian (1989, 1990) proposed ways of determining main paths through acyclic citation networks based on depth first search and exhaustive search methods. (See the discussion in Section 3.2.) These methods included search path count (SPC), node pair projection count (NPPC), and search path node pair count (SNPC). All have been implemented in Pajek. We focus on SPC here. There is an issue of whether the flows in the citation network are normalized or not. In the following, we assume that a flow-normalization is appropriate.[11]

The relevant commands in Pajek are (with the normalization completed with the first Pajek command):

[Select an acyclic network]

```
Network/Acyclic Networks/Citation Weights/
Normalization of Weights/Normalize-Flow
Network/Acyclic Networks/Citation Weights/Search Path Count (SPC)
```

This results in two networks: 1) the citation weights SPC network, and 2) the main path network.

Since all subnetworks of an acyclic network are also acyclic, their standard visualization is obtained by using the levels of an acyclic network. This can be produced by using a macro `Layers` available in Pajek. The vertices can be colored using the DC partition. More specifically, the following commands were used:

```
Macro / Play [Layers]
[select or read DCA as partition 1]
[select Main Path as partition 2]
Partitions / Extract Subpartition (Second from First)
Draw / Draw-Partition
```

4.5.2 The main path of the centrality citation network

The main path result[12] is shown in Figure 4.6. The initial vertices all preceded the work of Freeman (1977, 1979). They include the Bavelas (1950), Beauchamp (1965), Flament (1963), MacKenzie (1966), Sabidussi (1966), and Harary (1969) papers plus Coleman (1964), a book on mathematical sociology. They are all linked in the main path to Moxley and Moxley (1974). Then come the two foundational Freeman papers ordered in their temporal sequence. After the Freeman (1979) paper comes a sequence of papers by social network analysts: Burt (1980), Mizruchi and Bunting (1981), Cook et al. (1983), Bonacich (1987), Stephenson and Zelen (1989), Marsden (1990), Valente et al. (1997), and Valente and Foreman (1998).

---

[11] The available normalizations are: 1) without normalization; 2) normalize-flow; 3) normalize-max and 4) logarithmic weights. Most often, using normalized links is the most appropriate, although unnormalized links can be used for small networks.

[12] Exactly the same main path was identified with both the NPPC and SNPC weights.

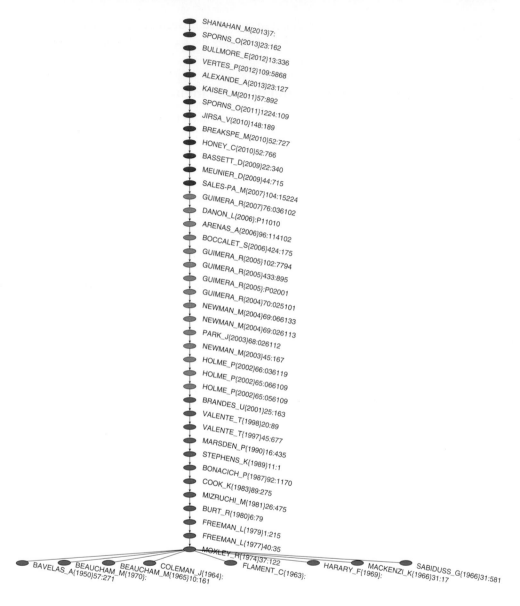

Figure 4.6   The main path of the bounded centrality citation network.

The colors are: blue, SNA; Orange: physics; Salmon. neuroscience.

Next in this main path is Brandes (2001) which appeared in *The Journal of Mathematical Sociology* and represents the last item from the traditional SNA literature on the main path.

The middle part of the main path is dominated by productions contributed by physicists: the journals include *Physical Review E: statistical, nonlinear, and soft matter physics* (seven times), *SIAM Review*, *Nature*, *The Journal of Statistical Mechanics*, *Physics Reports*, and

Physical Review Letters. The invasion of the physicists appears to have resulted in a takeover of the centrality literature as it is represented in the main path through the citation network. However, this is not the end of the story.

The end of the main path is dominated by neuroscience, focused on studies of the brain and mental illness including schizophrenia. The journals featured in this part of the main path include: *Neuroimage* (four times), *Cerebral Cortex*, *Proceedings of the National Academy of Sciences* (twice), *Nature Reviews Neuroscience*, *Current Opinion in Neurobiology* (twice), *European Journal of Neuroscience*, *Archives Italiennes de Biologie*, *Annals of the New York Academy of Science*, and *Frontiers in Computational Neuroscience*. Two very prominent places among the author affiliations in this segment are the Behavioural and Clinical Neuroscience Institute at the University of Cambridge and the Melbourne Neuropsychiatry Centre, The University of Melbourne, Melbourne, Victoria. Having large enough contributions of scientists at specific locations is one of the mechanisms for creating works that become important.

In addition to a clear disciplinary concentration in the middle and late segments of the main path, there are very small numbers of authors that are featured many times in these productions. This productivity by a small number of authors is another mechanism, not unrelated to place, for producing important manuscripts, in contrast to the early part of the main path where many social network analysts contributed.

Focusing on the five largest SPC weights in the main path, there are three small paths of works. One is Cook et al. (1983) → Mizruchi and Bunting (1981) → Burt (1980) in the traditional SNA part of the main path. It contains the largest, by far, two weights on the main path. Another path is Holme et al. (2002) → Brandes (2001) → Valente and Foreman (1998): the first arc is the transition from the traditional SNA literature to the part of the centrality network literature involving physicists. The final heavyweight (third in magnitude) path is Park and Newman (2003) → Newman (2003) within the physicist segment of the main path.

The main path procedure producing Figure 4.6 is a *greedy* algorithm. It is prudent to examine alternative algorithms not having this feature. One such alternative is the critical path method (CPM) introduced in the operations research literature (to design the best sequence of tasks in a project from its start to its finish). As described in Section 3.2.1, an acyclic temporal network, such as a citation network, can be modified by introducing a *source vertex* before all of the early vertices in the network and a *sink vertex* after all of the last vertices in the network. This does not change the essential structure of the temporal citation network. However, it facilitates the use of the CPM method, one determining the path(s) from the source to sink having the largest sum of weights on the critical path. The commands for doing this in Pajek are:

[Select Citation weights network]

```
Network / Critical Path Method CPM
```

Again the relevant produced files can be saved if desired. To draw the CPM path, the following Pajek commands are used:

```
Macro / Play [Layers]
```
[Select DCAr as partition 1]
[Select the CPM partition as partition 2]
```
Partitions / Extract SubPartion (Second from First)
Draw / Draw-Partition
```

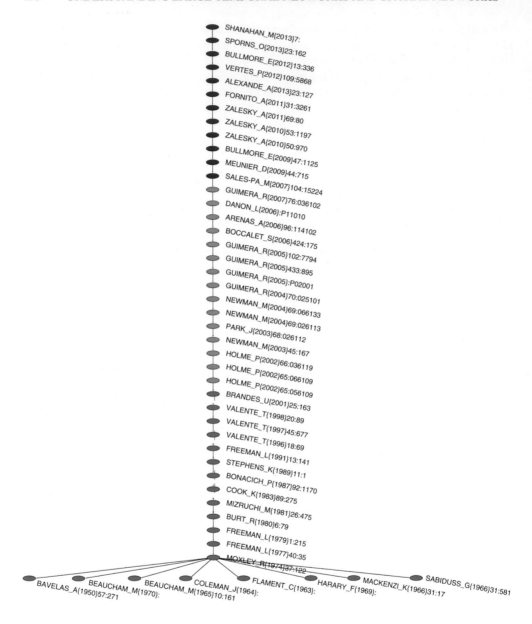

Figure 4.7   The CPM main path of the centrality citation network.

The colors are: blue, SNA; Orange: physics; Salmon. neuroscience.

The result is shown in Figure 4.7. The start of the main path using CPM is very close to the one obtained by SPC. One SNA paper of the SPC main path was dropped with two different SNA papers being included in the CPM main path. The beginnings of the two main paths are almost identical. The physics productions in the middle of these main paths are the same. However, the ends of the two main paths differ regarding papers from neuroscience.

Six of these papers in the SPC main path do not appear in the CPM main path. However, they are replaced by five other such papers in the latter. Three are from *Neuroimage* with one each from *Biological Psychiatry* and *The Journal of Neuroscience*. There is great commonality in the authors for the two different short segments of the main paths (and with the part common to both). Both main paths show the same disciplinary transitions (SNA → physics → neuroscience).

The top five weights in the CPM path are also in three paths. Two, Cook et al. (1983) → Mizruchi and Bunting (1981) → Burt (1980), and Park and Newman (2003) → Newman (2003), are the same as for the SPC main path. The third differs. It is Valente (1996) → Freeman et al. (1991) → Stephenson and Zelen (1989) in the traditional SNA part of the CPM path. This analysis used `CiteAcy` produced by `preprint`. We note that using the acyclic network produced by shrinking `citeB` produced the same SPC and CPM main paths.

4.6 Line cuts

Another useful technique for extracting parts of network is using the line cut where only lines (and their vertices) above a certain value are retained. This introduces a certain amount of judgment: one criterion for this extraction is getting subnetworks that are not too large. Inspecting the distribution of *normalized* SPC weights provides useful information. Taking 0.01 as a threshold for a line cut produced the network shown in Figure 4.8.

The subnetwork produced by this line cut contains all but four of the works in the SPC and CPM main paths shown in Section 4.5. This points to a difference in the treatment of weights on citation arcs. Obtaining a line cut is done by considering *only* the values of the arcs above a specific selected threshold, regardless of any other features of the network. In contrast, for both the SPC and CPM main paths, some weighted arcs are included with lower values because main paths must run through the length of the acyclic citation network. Even so, the subnetwork obtained by using a line cut still provides a wider interpretive context for the main paths by showing works and their citation links 'around' the works on a main path. The early works in Figure 4.8 are base works linked to the foundational Freeman papers on the main path. We note that some of the heavily weighted arcs in Figure 4.8 are on the main paths while others are not. While useful, main paths provide only simple images of the structure of a citation network.

Structurally, one work in the line cut network is particularly salient. The Newman (2003) production links the early and later halves of this subnetwork. Indeed, but for one other tie, it would have been a cut vertex in this subnetwork.[13] This paper on 'the structure and function of complex networks' is a review paper with 429 citations providing a broad overview of network analysis. One concept considered by Newman is the betweenness centrality idea of Freeman.[14]

---

[13] An arc to an earlier Newman paper from a later physicist paper prevents Newman (2003) from being a cut vertex.

[14] One complaint often voiced by social network analysts, especially at conferences, concerns physicists ignoring the earlier work of their field. This cannot be said of Newman (2003) given the number of citations it contains to the traditional SNA literature.

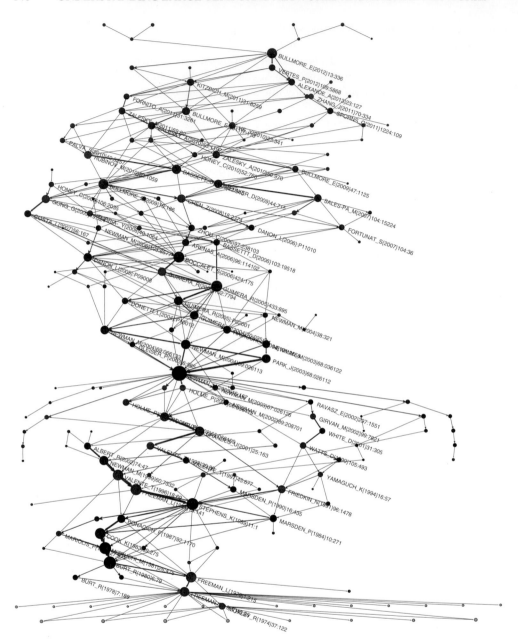

Figure 4.8   The subnetwork produced by a line cut.

Green vertices are basic sources, red vertices belong to the base, and blue vertices are users.

Another near cut-vertex (with only one arc preventing it being a cut-vertex) is the foundational Freeman (1977) paper on centrality. It links the base works (shown in red) to the rest of the line cut network and is cited by Newman (2003). The other arc linking the base part of the centrality literature to the rest of the subnetwork is Freeman (1979) → Moxley and Moxley (1974).

The Pajek commands for extracting this subnetwork based on a line cut are:
[Select SPC network]
```
Network/Create New Network/Transform/Remove/Lines with Value/
   lower than [0.01]
Network / Create Partition/ Components / Weak
Operations / Network+Partion/ Extract SubNetwork
Macro / Play [Layers]
```
[Select weak components partition]
[Select SPC vector]
```
Operations / Vector+Partition / Extract Subvector
```
[Select weak components as partition 2]
[Select DCAr as partition 1]
```
Partitions / Extract SubPartition (Second from First)
Draw/ First Partition + First Vector
```

4.7   Line islands

Line islands, introduced in Section 2.9, can be used to identify other interesting parts of citation networks. Delineating islands is a way of 'filtering' a network to focus on manageable but coherent parts of networks. While it is an important first step, it is only a starting point. Understanding the structure of an island and the substantive content it contains requires looking beyond only the ties in the citation network.

When establishing islands, a choice can be made between identifying simple (single peak) islands or more general (multiple peak) islands. There are 3120 simple line islands identified in bounded citation network when the size bounds are set between 2 and 1000. They are relatively small: the largest has 54 vertices. In general, single peak islands are less 'interesting' structurally by virtue of being narrowly defined (i.e. they have a single peak with works dealing with a single topic). For example, the largest simple island has papers on the topic of diagnosing prostate cancer. Of course, researchers interested in this topic will find such an island of interest.[15] To obtain islands with more interesting structures, delineating multi-peaked line islands is more fruitful. It facilitates exploring the links between multiple topics and even multiple disciplines within an island. The Pajek commands for obtaining general line islands are:
[Select or compute SPC weights network]
```
Network/Create Partition/Islands/Generate Network with Islands
Network/Create Partition/Islands/Line Weights [20 1500] [Yes]
Partition/Canonical Partition/with Decreasing Frequencies
```
The result of using these commands is a set of 42 islands of sizes [20,1500], see Table 4.8. The total flow through network is $5.5147901 \cdot 10^{15}$. The column labeled `Pos` gives the cluster number (position) of a cluster in the Islands [`Line weights`] partition. The column labeled `Freq` gives the number of works in an island. However, `Cluster 0` does not label an island. The largest ten islands range in size from 50 to 1497. While

---

[15] This line island can be identified as part of a broader health-related island with multiple peaks to embed this particular topic in a wider health context.

Table 4.8 Largest line islands of sizes using [20,1500] bounds.

Cluster	Pos	Freq	Freq%	CumFreq%	Representative
0	0	135285	97.0926	97.0926	*IEEE(2003):
1	42	1497	1.0744	98.1670	BONACICH_P(1987)92:1170
2	40	774	0.5555	98.7225	HARRISON_L(1994)33:5118
3	41	296	0.2124	98.9349	ANDERSON_C(1993)3:69
4	12	126	0.0904	99.0254	HENSCHKE_C(1999)354:99
5	35	97	0.0696	99.0950	QIAO_C(1999)8:69
6	33	94	0.0675	99.1625	GORDON_T(1995)221:43
7	23	66	0.0474	99.2098	ALOIA_J(1995)172:896
8	13	55	0.0395	99.2493	SZYMANSK_J(2007)14:137
9	19	52	0.0373	99.2866	ERICKSON_M(1996):
10	38	50	0.0359	99.3225	BAGAJEWI_M(2001)79:600
11	32	46	0.0330	99.3555	*COMM RAD ONC STUD(1981):
...					

each identified line island can be examined in detail, we look closely at only some of them here.

Before such an examination can be undertaken, the first step is to extract the islands subnetwork. This is done with:

[Select Islands [Line weights] network]
[Select Islands [Line weights] partition]
Operations/Network+Partition/Extract SubNetwork [Yes] [1-*]
File/Network/Change Label [Islands]
File/Partition/Change Label [Islands]
[Select Islands [Line weights] partition]
[Select SPC weights vector]
Operations/Vector+Partition/Extract Subvector [1-*]
File/Vector/Change Label [Islands]
[Select or read typesAcy as First partition]
[Select Islands [Line weights] as Second partition]
Partitions/Extract Subpartition (Second from First) [1-*]
File/Partition/Change Label [Islands types]

While the Change Label operations can be skipped, we do not advise doing this because the changed labels facilitate finding the corresponding units in the list more quickly. They also make the descriptions of these procedures easier.[16]

The steps for extracting and analyzing/visualizing a selected island for $Pos = k$ are:

[Select Islands network]
[Select Islands partition]
Operations/Network+Partition/Extract SubNetwork [k]
Macro/Play [layers]

---

[16] Another useful but optional command is:
Network/Create New Network/Transform/Sort Lines/Line Values/Ascending.

```
[Select Islands vector]
[Select Islands partition]
Operations/Vector+Partition/Extract Subvector [k]
[Select Islands types partition]
[Select Islands partition]
Partitions/Extract SubPartition (Second from First) [k]
Network/Create New Network/Transform/Sort Lines/
  Line Values/Ascending
Draw/Network+First Partition+First Vector
[Draw] Move / Fix / y
```

The option to fix y in the draw operation keeps all of the *y* coordinates the same for vertices in the same layer. The resulting figure, if necessary, can be edited using manual picture editing (with neither x nor y fixed). By selecting the appropriate display `options` the picture can be improved and exported to a selected format (including EPS or BMP).[17]

Table 4.8 shows that there are two very large islands of sizes 1497 and 774 plus an additional eight islands of size at least 50. It is reasonable to expect that the main stories will be found in these two islands. Since they are too large for direct visualization and inspection, it is possible to treat these islands as networks and reapply the `islands` procedure to them. We did this for each of the two large islands with limits [50 200].

4.7.1 The main island

The visualized extracted subisland (island from the largest island) is shown in Figure 4.9. It has 200 vertices and normalized weights in the range [0.01047353, 0.32603716]. It tells the same story as the line cut subnetwork; indeed, the 200 vertices in this line island are among the 202 vertices of the line cut subnetwork.

When the contents of the works in this line island were considered, all the scientific productions belonged to the three disciplines identified in the main paths. Works from SNA (or the social sciences more broadly) are colored in red, works by physicists are colored blue and neuroscience papers are colored green. There are three temporarily ordered streams of disciplinary 'dominance,' consistent with the ordering of the main paths. The distribution of publications by discipline is SNA (82), physics (60), and neuroscience (58).

Publications from the SNA literature are at the bottom of Figure 4.9. The initial works in the middle of the SNA swath deal with formalizations of centrality ideas, considerations of alternative measures, and assessments of these measures. The journals primarily featuring these papers are *Social Networks*, *Sociometry*, *The Journal of Mathematical Sociology*, *Sociological Methododogy*, *Psychometrika*, and *Quality & Quantity*. Other journals from psychology and anthropology are also present.

Then there are two broad and separated parts of the SNA literature in this island differing by substance rather than by method. They are user works as defined earlier. One substantive area where social network ideas have been used extensively is business administration. The articles on the left of Figure 4.9 come from this literature: the journals featured include the *Academy of Management Journal*, the *Academy of Management Review*, *Admimistrative*

---

[17] Using IrfanView converts BMP pictures to the PNG format and using GsView converts EPS pictures into the PDF format.

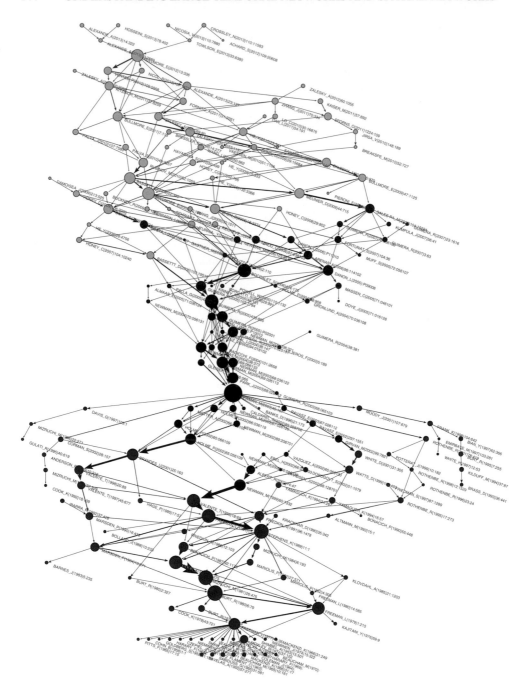

Figure 4.9   The main line island.

The works are colored by discipline: Red, SNA; blue, physics; and green, neuroscience.

Science Quarterly, *The Journal of Marketing*, and *Organizational Science*. There are also papers from the *American Journal of Sociology* and the *American Sociological Review* in this part of the network.

The segment of the literature represented on the right of Figure 4.9 is focused on public health and the sexual transmission of disease. The journals containing these works include *Social Science and Medicine*, *The American Journal of Public Health*, *Sexual Transmission of Disease*, *The Journal of Urban Health*, and *AIDS*.

Overall, the top five social science journals appearing in the SNA part of this line island are, in order, *Social Networks* (15), *The American Journal of Sociology* (8), *Administrative Science Quarterly* (5), *The Journal of Mathematical Sociology* (4), and the *American Sociological Review* (4). The most times an author appears in the SNA publications in this line island is five.

The journal and author distributions differ dramatically for the physicist and neuroscientist productions. For the journals featuring productions by physicists, the top five journal distribution is *Physical Review E* (23), *Physical Review Letters* (6), *European Physical Journal B* (5), *Proceedings of the National Academy of Sciences* (5), and two journals each with a count of 2. Overwhelmingly, it appears that *Physical Review E* is the dominant journal for this part of the citation network. One author appears in at least 12 of these productions. Together, this part of the literature is driven by one journal *Physical Review E* and one author (Newman). This expands the impression obtained from considering main paths.

In the neuroscience part of this line island, the top five journal outlets are *Neuroimage* (18), *Journal of Neuroscience* (8), *Proceedings of the National Academy of Sciences* (8), *Nature Reviews Neuroscience* (3), and *Cerebral Cortex* (3). One author appears on at least 11 productions. As was the case for the main paths, the Behavioural and Clinical Neuroscience Institute at the University of Cambridge dominates the author affiliations in this part of the network. Here, one journal (*Neuroimage*) and one place appear to drive this part of the literature. We note the imprimatur provided by the *Proceedings of the National Academy of Sciences* in the physicist and neuroscientist segments of the main island.

There are two additional minor notes of interest in Figure 4.9. One is the early appearance of Shimbel (1953) in the *Bulletin of Mathematical Biophysics* which we coded as belonging to the physicist part of the literature. He, and Anatol Rapoport, were members of the Committee on Mathematical Biology at the University of Chicago. Rapoport was a major early contributor to the probalistic approach to studying social networks.[18] Another is the appearance of two papers from the SNA literature in the physicist phase of this line island. They were both cited in a large review paper but it was in a secondary fashion more in the style of window dressing to provide motivation for an approach to social networks solely within a claimed 'new' paradigm.

4.7.2   A geophysics and meteorology line island

We extracted a smaller line island for the second largest line island identified in Table 4.8, again using a specification of [20, 200] for the size bounds of islands. The largest

---

[18] As a comment on the change in citation patterns over the years, Shimbel cited only one production of his own work while Rapoport, in the next paper of this issue of *Bulletin of Mathematical Biophysics*, cited only two earlier papers, one of which was by him.

sub-island of the second largest island has 200 vertices with weights in the (normalized) range $[0.01305290 \cdot 10^{-6}, 0.47073834 \cdot 10^{-6}]$. These weights are very much smaller than in the main island. To bring them in the usual range we multiply the weights by 1,000,000:

```
Network/Create New Network/Transform/Line Values/Multiply by
   [1000000]
```

This island is presented in Figure 4.10, one that is dramatically different from the main island. First, its productions are primarily user works using the categorization of Figure 4.4. Second, the substantive focus centers on geophysics, meteorology, geoscience, remote sensing, and earth sciences. Third, both the author participation and journal dominance are dramatically different.

In terms of the substantive foci, in brief: 1) geophysics can be viewed as the physics of the Earth and its environment in space; 2) meteorology is the interdisciplinary scientific study of the atmosphere where attempts are made to understand observed weather events using scientific methods;[19] 3) geoscience involves the prediction of the behavior of the systems of the Earth and the universe, especially in the context of the effects of human behavior; 4) remote sensing is obtaining information about an object without making physical contact with it;[20] and (5) earth science amounts to an all-encompassing term for the sciences related to the Earth. The primary traditional disciplines uniting these diverse areas are chemistry and physics. Even so, we label the topic of this line island as a geophysics and meteorology island even though 'earth science' can be seen as a broader alternative focus.

When the journals featuring the papers in this second line island are considered, the distribution is remarkable. Although 200 scientific productions appear on this line island, only 159 of them, as retrieved from the WoS system, had information about the titles and journals involved. We focus only these productions[21] dating from 1998 through 2013, a mere 15 years. Consistent with our consideration of earlier subnetworks of this citation network, we list the five journals with the most frequent presence. They are the *Journal of Geophysical Research: Atmospheres* (81), the *Journal of Atmospheric Science* (28), *Atmospheric Chemistry and Physics* (10), the *Bulletin of the American Meteorological Society* (8), and *Geophysical Research Letters* (7).

To contrast this distribution with the results from the main island: the top five journals in the SNA segment accounted for 44% of the productions; the top five physicist journals accounted for 72% of the productions in the middle segment, and the top five neuroscience journals accounted for 66% of the productions. In contrast, the top five productions of earth science account for 83% of the productions. The differences are striking also for the most frequent journal: SNA (18%); physics (38%); neuroscience (31%); and earth science (51%).

Seemingly, a very different co-authorship culture exists for earth science (and all of its subparts including geophysics and meteorology). Among the 159 productions we examined, one paper had 32 co-authors. Another had 28 co-authors, with a third having 27 contributors to the production. There was only one production with a single author. From a social

---

[19] This includes examining variables such as temperature, water content, and air pressure plus how they interact over time and across space.

[20] This involves using aerial sensors to detect signals about the Earth and both its atmosphere and oceans. Most often, this involves the use of aircraft and satellites.

[21] The earlier productions have publication years between 1957 and 1998; our interpretations apply only for the recent citation features of this line island.

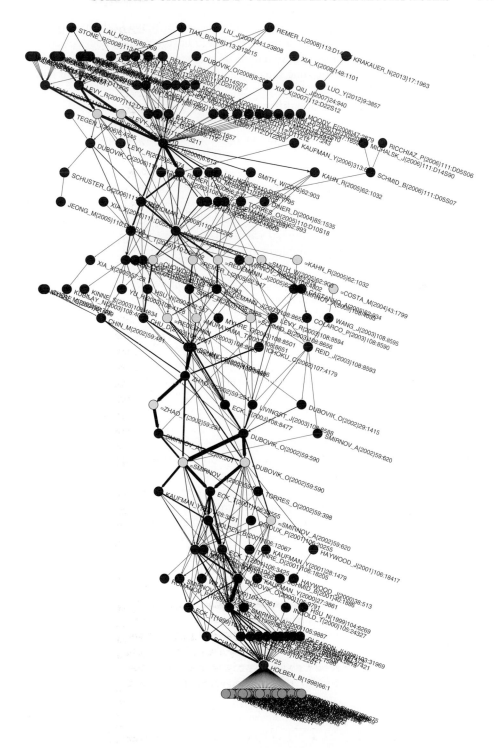

Figure 4.10   The geophysics and meteorology line island.

Colors: green, sources; red, base works; blue, user works; pink, 'duplicate' vertices from using `preprint`.

science perspective, this borders on being incomprehensible. While both the physicist and neuroscientist productions of the main island tended to have larger numbers of co-authors than in the SNA part, their number of co-authors did not remotely approach the numbers for earth science. Consistent with this, one scientist had his name as a co-author on at least[22] 46 productions (29% of them).

There is suggestive evidence of a different kind of institutional support for earth science (or more specifically, geophysics and meteorology) in the form of professional associations. The American Geophysical Union, formed in 1919, publishes the *Journal of Geophysical Research: Atmospheres* and *Geophysical Research Letters* as official journals. The American Meteorological Society publishes the *Journal of Atmospheric Science* and the *Bulletin of the American Meteorological Society*. Finally, one prominent affiliation for authors of productions in the earth science line island is the Laboratory for Terrestrial Physics at the NASA Goddard Space Flight Center located in Greenbelt, MD. This laboratory 'is home to the nation's largest organization of combined scientists, engineers and technologists that build spacecraft, instruments and new technology to study the Earth, the sun, our solar system, and the universe.'[23]

One work is distinctive in Figure 4.10 because it is a cut-vertex separating all that went before from what came after. Holben et al. (1998) has 13 co-authors from nine different affiliations in four countries. The lead author is the scientist with his name on at least 46 productions on this line island. This paper introduced AERONET, A Federated Instrument Network and Data Archive for Aerosol Characterization, and appears to have transformed this subfield. Virtually every paper in this line island has aerosol in its title and it serves as a keyword. An aerosol is a colloid[24] of fine solid particles or liquid droplets in air or another gas. Included among aerosols are clouds, haze, and air pollution including smog and smoke. Given that the science involved in studying aerosols includes the generation and removal of aerosols, and their effects on the environment and people, 'aerosol science' is an alternative – more narrowly defined – name for the core topic of this line island. This suggests a need for using keywords more extensively as a way of defining substantive foci. Also, as useful as using size bounds of [20, 200] was, using a larger upper bound would lead to establishing a larger and potentially more interesting line island.

Two vertices, colored in pink,[25] about two-thirds from the bottom of Figure 4.10 almost form a cut set. Authors from the NASA Goddard Space Flight Center are prominent in these productions. Many of these authors appear on the papers in arcs preventing the two papers from forming a cut set. It seems like this line island 'belongs' to the Goddard Space Flight Center and fits the characterization of scientists working at the same location who have colleagues elsewhere working on common problems. One place providing such collaborators is Laboratoire d'Optique Atmospherique at the University of Lille in France, with another being the Instituto de Pesquisas Espaciais, Sao Jose dos Campos, Brazil. As aerosol properties

---

[22] We suspect that, on searching these publications more closely, his name will appear more often.

[23] Source: http://www.nasa.gov/centers/goddard/about/index.html#.Uio0ZzUNHk accessed September 6, 2013.

[24] A colloid is a substance microscopically dispersed throughout another substance.

[25] Both are 'additional' vertices added by using the preprint method. Immediately above them are the same works in Figure 4.10. Using the simpler transformation for obtaining an acyclic network would mean that these works would have been merged into a single vertex. As their citation patterns differ, this suggests that using the preprint transformation was preferable in this instance. In general, this way of creating an acyclic version of citation network that is not acyclic preserves the identities of works.

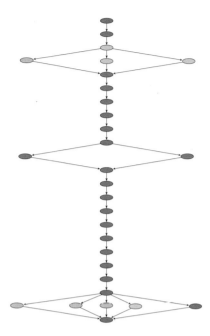

Figure 4.11   The SPC main path through the second line island.

Orange: Goddard Space Flight Center authored; light orange: collaborating authors; gray: other affiliations.

are measured in many places, additional colleagues join who are located at, or closer to, the places where measurements are made. The main path through the aerosol science line island (Figure 4.11) confirms the dominance of Goddard Space Flight Center in this segment of the literature. Vertices colored in orange have authors from this center, those in light orange have authors who collaborate[26] with authors from the Goddard Space Flight Center while four vertices have authors unaffiliated with this center. Three of the early papers in the row of four works near the bottom of Figure 4.11 featured work based on POLDER (POLarization and Directionality of the Earth's Reflectance), a French instrument launched on a Japanese satellite. One of the authors involved in these productions later became a regular co-author with scientists at the Goddard Space Flight Center. Indeed, the first orange-colored vertex above this row of works has him as a co-author.

Our examination of the aerosol science line island is *not* a comprehensive examination of this part of the literature. Even less is it such a study of geophysics and meteorology. The *Journal of Geophysical Research: Atmospheres* alone publishes 24 issues a year with a page count well into the thousands[27] while *Geophysical Research Letters* has 15 issues a year. The works we examine were identified using terms involving centrality. However, searching WoS using 'aerosol' would identify another large citation network that could be studied using the tools used here.

---

[26] This includes two papers with authors from the Goddard Institute for Space Studies in New York City.

[27] Many of the authors of works in the line island we studied have many other publications in this journal.

4.7.3   An optical network line island

We finish our examination of line islands by looking briefly at those identified in Table 4.8 in lines 4–10. An arbitrary outcome of delineating islands is their numbering. As shown in this table, the numbers these islands received when using Pajek were 12, 35, 33, 23, 13, 19, and 38. To obtain these islands, the following commands were used.

[Select Islands network]

[Select Islands partition]

```
Operations/Network+Partition/Extract SubNetwork [12, 35, 33, 23,
    13, 19, 38]
Draw/Network+First Partition
```

With the Draw window, the command for producing the actual layout were

```
[Draw] Layout/Energy/Kamada-Kawai/Separate Components
```

Once obtained, they can be drawn and examined. The result is presented in Figure 4.12. These islands are smaller and differ in their 'shapes.' No doubt, each island will have some interest value. The three in the center row of this figure are centered on particular scientific

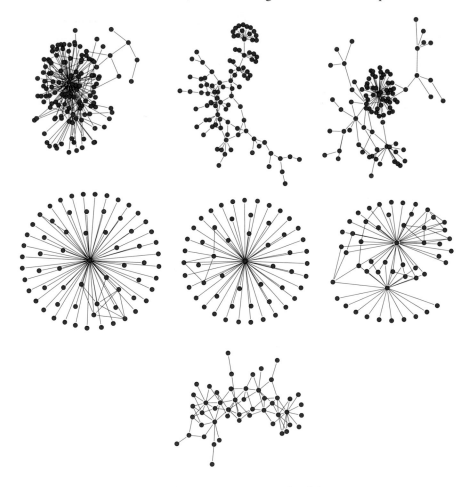

Figure 4.12   Another seven line islands.

productions: the first two have star-like structures while the third seems centered on two productions. These central works are likely to be important in terms of the substantive or technical content of the works in their line islands. The other islands appear to have more interesting structures, in the sense of having multiple levels, and corresponding narratives.

Somewhat arbitrarily for illustrative purposes only, we consider only the second of the islands in the first row of Figure 4.12, as another example of looking at line islands in a citation network. This island has 97 vertices with normalized weights in the range $[0.99115287\cdot10^{-12}, 38.48378589\cdot10^{-12}]$ which are even smaller than in the second (aerosol) island of Figure 4.10. Again, to bring them in the usual range, it is necessary to multiply them by some value, in this case by 10^{12}. The resulting network is presented in Figure 4.13, where the thicker lines have greater weights.

Of the 97 works, the WoS data contained article titles for only 61 of them. Examining them revealed that the subject in this line island is a combination of optics (especially optic fibers), photonics, and optical (transport) networks.[28] The time span covered in this line island is also short (1994–2013) indicative of a rapidly changing scientific, technological, and commercial[29] area. Within this line island, there are multiple initial vertices suggesting a range of relatively narrowly focused works joined together to form the broader literature. Examining the titles of these works reinforces this impression. Multiple cut vertices are also present, with a path through the literature with particularly heavy weights on their arcs. Wu (1994), the lowest red vertex in Figure 4.13, cites all of the yellow vertices below it. Tada et al. (1996) cites the Wu paper and also another row of vertices from the early 1990s. There is another heavy arc from Sharma et al. (1997) to the Tada paper, with Sharma et al. citing another row of initial vertices. It appears that these three papers set the foundations for the core part of this literature by drawing upon a wide set of other earlier papers.

The top five identified journals in this island are *IEEE Journal of Lightwave Technology* (19 times, 31%), *IEEE Photonics Technology Letters* (14, 23%), *Optics Express: The International Online Journal of Optics* (8, 13%), *Journal on Selected Areas in Communications* (5, 8%), and *Journal of Optical Communications and Networking* (4, 7%). The Institute of Electrical and Electronics Engineers (IEEE) publishes seven of the twelve identified journals appearing in this line island, containing 70% of the works.

A clear institutional dominance pervades this part of the literature. The IEEE describes itself[30] as being 'dedicated to advancing innovation and technological excellence for the benefit of humanity, is the world's largest technical professional society. It is designed to serve professionals involved in all aspects of the electrical, electronic and computing fields and related areas of science and technology that underlie modern civilization.' The Optical Society (OSA)[31] 'was organized to increase and diffuse the knowledge of optics, pure and applied; to promote the common interests of investigators of optical problems, of designers and of users of optical apparatus of all kinds; and to encourage cooperation among them.

---

[28] Photonics is a scientific field dealing with the generation, emission, transmission, modulation, signal processing, switching, amplification, and sensing of light. An optical transport network (OTN) is a set of optical network elements connected by optical fiber links that are able to provide the transport, multiplexing, switching, management, supervision, and survivability of optical channels carrying signals.

[29] Among the author affiliations for works in this line island are some of the companies producing the technologies thought to be transforming our communication technologies and lives.

[30] Source: http://www.ieee.org/about/ieee_history.html accessed September 7, 2013.

[31] Source: http://www.osa.org/en-us/about_osa/overview/ accessed September 7, 2013.

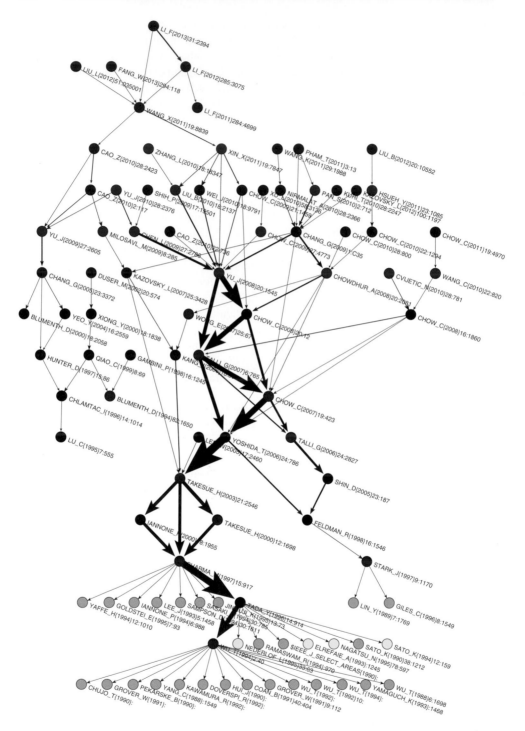

Figure 4.13   The optical network line island.

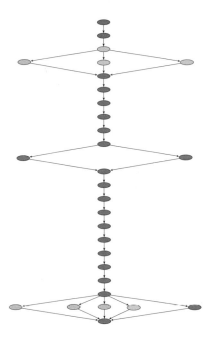

Figure 4.14   The SPC main path through the optical networks line island.

Orange: IEEE journal, salmon: OSA journal, cadet blue gray: IEEE-OSA journal, gray other journal.

The purposes of the Society are scientific, technical and educational.' OTA publishes *Optics Express: The International Online Journal of Optics*. Also, IEEE and OTA collaborate in publishing *Journal of Optical Communications and Networking*. Together, these professional associations published 90% of the identified works in this line island.

Figure 4.14 shows the modified SPC main path through the 'optical networks' line island, one that is identical to the CPM main path. The modification comes from starting the main path with the Wu (1994) paper. The color coding for this main path has orange for IEEE journals, salmon for an OSA journal, cadet blue for the joint IEEE-OSA journal, and gray for the solitary journal not published by these two professional associations. Without surprise, the heavily weighted arcs in Figure 4.13 are on the main path. Looking more closely at these journals reveals that both *IEEE Journal of Lightwave Technology* and *IEEE Photonics Technology Letters* have 24 issues per year, with *Optics Express: The International Online Journal of Optics* having 28 issues a year. This high frequency of publications is one mechanism of the institutional dominance of these professional societies.

Of additional interest is the absence of IEEE's *Journal on Selected Areas in Communications* (J-SAC) on the main path. The aims and scope of this journal state that each issue 'is devoted to a specific technical topic and thus provides readers a collection of up-to-date papers on that topic. These issues are valuable to the research community and become valuable references. The technical topics covered by J-SAC issues span the entire field of communications and networking.' The claim of creating 'valuable references' does not appear to include a major role in creating the literature represented by the works in this line island. This suggests that such review documents provide summaries with little relevance for creating new works in rapidly developing fields.

4.7.4 A partial summary of main path and line island results

The results of the analyses using line islands to identify coherent parts of the centrality citation network and main paths to identify specific works yielded some insights into the processes generating citation networks within science. These include:

1. Some areas, perhaps better viewed as subfields, are dominated by specific authors, journals, research centers, and professional associations. Even though this list expands outwards, these very different objects are best considered in the reverse order.

2. The works in the line island dealing with optical networks, and related topics, were dominated by the Institute of Electrical and Electronics Engineers (IEEE) and the Optical Society of America, two large professional associations. In a similar fashion, works in the geophysics and meteorology line island were dominated by the American Meteorological Society and the American Geophysical Union. While it is obvious that such organizations exist to promote the interests of their members, this includes the management and promotion of publication outlets for research in areas of interest to them. The result is institutional dominance of this field. This represents another mechanism for publishing many publications that become important when they receive citations from other manuscripts within this controlled domain.

3. Specific research centers dominate the production of important works. The most prominent physicist in the main line island studied is at the University of Michigan's Department of Physics, which is also the affiliation of other authors (including Barabási) listed in the table of highly cited works. Both the Behavioural and Clinical Neuroscience Institute at the University of Cambridge and the Melbourne Neuropsychiatry Centre at the University of Melbourne dominated in the neuroscience section of the main line island. Finally, the Laboratory for Terrestrial Physics at the NASA Goddard Space Flight Center dominated the works in the aerosol line island. Larger concentrations of scientists at places with large research budgets, together with a seemingly aggressive self-promotional agenda, drive parts of the literature. A small number of authors are prominent in the publications coming from these research centers.

4. A small number of journals dominate fields and subfields as outlets for relevant work. In part, this reflects a common research focus, which is not surprising. However, this is due also to control by professional associations, especially in the hard sciences. Having annual publications with as many as 24 issues a year, with many thousands of pages confers a considerable advantage. By contrast, social scientific professional associations are smaller and seem like amateurs in terms of creating dominance in their literatures. Size matters for professional associations, research institutes, and journals.

5. When particular authors are promenant, this is not solely due to their own brilliance and productivity. Being located at prominent research centers seems the more important mechanism for prominence.

6. The very different cultures among disciplines concerning the phenomenon of the numbers of co-authors – where the range in the number of authors of the publications we studied went from 1 to 32 – creates a problem for studying the impact of individual

scientists on their fields. Again, affiliations may be more important as a driving mechanism, especially when large research teams work on collective projects.

7. Review articles appear to have a small role in the creation of *new* knowledge. With one exception, the identified works with huge reference lists did not appear on line islands or in main paths. Instead, these works document what has been done rather than create anything new. However, to the extent that researchers cross intellectual boundaries, review works may appear more often as convenient reference or as surrogates for not reading what has been done on different research fronts.

4.8   Other relevant subnetworks for a bounded network

There is far more information than citations between works in the WoS database. This includes the authors of the works, the journals where works in the form of articles appear, and keywords describing each work. Further analyses can include author-to-author citation networks, journal-to-journal citation networks,[32] and linking works to each in terms of their common contents. This was done easily by Hummon et al. (1990) because of the small size of the network they studied. Doing this is far more complicated for the sizes of the citation networks considered here.

We used the term *bounded network* for the kind of network labeled `CiteB` in Section 4.2.4. All of these additional subnetworks must be constructed to be compatible (in sizes) with such a bounded network. We illustrate the operations for creating these subnetworks for a (works × keywords) two-mode network denoted by WK. The following Pajek commands yield some basic information about WK:

```
[Read WK network]
Network / Info / General
[Read boundary partition]
Partition / Info [1] [0]
```

The two-mode network WK has 740,656 vertices (with 548,600 works and 192,056 keywords) and 1643356 arcs. The `boundary.clu` partition of the works has 55,376 vertices in class 0 (inside the imposed boundary) and 493,224 vertices in class 1 (outside the boundary). Clearly, the matrices do not match: it is necessary to construct a partition to extract the relevant data from WK. The Pajek command for this are:

```
Partition / Create Constant Partition [192056] [0]
[Select Constant partition as partition 2]
[Select boundary partition as partition 1]
Partitions / Fuse Partitions
Partition / Info [1] [0]
Operations / Extract SubPartition (second from first)
Network / Info / General
File / Network / Change Label [WKb network]
[Save network to WKb.net]
File / Partition / Change Label [WKb partition]
[Save partition to WKb.clu]
```

---

[32] These two networks are not acyclic.

Table 4.9 Part of the outdegree distribution.

Cluster	Freq	Freq%	CumFreq	CumFreq%	Representative
0	55251	22.3298	55251	22.3298	1
1	2	0.0008	55253	22.3306	171
2	12	0.0048	55265	22.3354	486
46	2	0.0008	55376	22.3803	55363
47	192056	77.6197	247432	100.0000	55377
Sum	247432	100.0000			

We get the (two-mode) partition with classes 0 : 55376, 1 : 493224, 2 : 192056. This partition was used to extract the subnetwork WKb.net with 55,376 works and 192,056 keywords. The next step is examining the reduced network according to the strong components partition established earlier. This is done as follows:

[Read or compute strong components partition]
[Select the constant partition as partition 2]
Partitions / Fuse Partitions
Partition / Info [1 [0]
Partition / Fuse Clusters [47] [0]
[Select WKb network]
Operations / Network+Partition / Shrink Network
[Save network to file WKa.net]

The partition, partially presented in Table 4.9, is the result.

The class 0 corresponds to works in the acyclic part of citation network, classes 1 to 46 correspond to the strong components, and class 47 to the keywords.[33]

There remain some final straightforward cleaning operations for which the commands are:

[Read corrected WKa network]
Network / Transform / Remove / Multiple Lines / Sum Values
Network / Creat Partition / Degree / Input
Partition / Info [1] [0]

The result is a list of the most frequent keywords, as presented in Table 4.10. Most of these keywords are very general: they are unlikely to have much value because of their generality. More useful keyworks are specific substantive terms and technical terms. Using the most frequently used keyworks does not take us very far.

At the other extreme, keywords with degrees of 0 and 1 can be eliminated without loss from the reduced network by using the following commands:

Operations / Network+Partition / Extract 2-Mode Network
Partition / Fuse Clusters [1] [99999]
Partition / Fuse Clusters [2] [0]

---

[33] Currently, the shrink operation does not work correctly for two-mode networks because it fails to produce a two-mode network. However, this is corrected easily by editing the first line in a text editor to include the right sizes. The edited line will read: *Vertices 247353 55297.

Table 4.10   Keywords with frequency at least 3000.

3003	1	0.0004	247323	99.9879	complex
3054	1	0.0004	247324	99.9883	management
3260	1	0.0004	247325	99.9887	number
3337	1	0.0004	247326	99.9891	algorithm
3341	1	0.0004	247327	99.9895	develop
3364	1	0.0004	247328	99.9899	social
3388	1	0.0004	247329	99.9903	control
3416	1	0.0004	247330	99.9907	process
3458	1	0.0004	247331	99.9911	problem
3460	1	0.0004	247332	99.9915	design
3500	1	0.0004	247333	99.9919	method
3535	1	0.0004	247334	99.9923	structure
3575	1	0.0004	247335	99.9927	time
3631	1	0.0004	247336	99.9931	research
3865	1	0.0004	247337	99.9935	different
4013	2	0.0008	247339	99.9943	new
4087	1	0.0004	247340	99.9947	information
4623	1	0.0004	247341	99.9951	performance
4737	1	0.0004	247342	99.9956	propose
4806	1	0.0004	247343	99.9960	provide
4947	1	0.0004	247344	99.9964	present
5007	1	0.0004	247345	99.9968	base
5283	1	0.0004	247346	99.9972	datum
5341	1	0.0004	247347	99.9976	analysis
6681	1	0.0004	247348	99.9980	model
7046	1	0.0004	247349	99.9984	paper
7260	1	0.0004	247350	99.9988	study
7883	1	0.0004	247351	99.9992	result
9971	1	0.0004	247352	99.9996	use
12156	1	0.0004	247353	100.0000	network

```
[select input degree partition as partition 2]
Partitions / Max (First,Second)
Operations / Network+Partition / Extract SubNetwork
File / Network / Change Label [WKar network]
[save network to WKar.net]
```

4.9   Collaboration networks

In Section 3.5 we introduced matrix multiplication of two-mode networks and sketched the use of this in Section 4.8. Further, Section 3.6.2 outlined methods for constructing collaboration networks. Here we apply these methods to look at some collaboration networks providing different views of collaboration. We proposed three collaboration networks (based on the same graph, but having different weights) **Co**, **Cn**, and **Ct**. Their definitions are:

$$\mathbf{Co} = t(\mathbf{WA}) * \mathbf{WA}$$

$$n(\mathbf{WA}) = \mathrm{diag}(1/w) * \mathbf{WA}$$

$$\mathbf{Cn} = t(\mathbf{WA}) * n(\mathbf{WA})$$

$$\mathbf{Ct} = t(n(\mathbf{WA})) * n(\mathbf{WA})$$

In the analysis of collaboration we consider only works with information about all contributing authors in the DC partition discussed earlier that belong to class 1.

In this section we describe how to compute these networks using Pajek. Since the procedures are relatively long we saved them as macros.

4.9.1   Macros for collaboration networks

The macro, normal, transforms a given two-mode network $\mathcal{N}$ into its normalized version $n(\mathcal{N})$ using the following steps:

```
Network/Create Vector/Centrality/Degree/Output
Network/2-Mode Network/Partition into 2 Modes
Operations/Vector+Partition/Extract Subvector [1]
Vector/Transform/Invert
Operations/Network+Vector/Vector#Network/Output
File/Network/Change Label [Normalized]
File/Vector/Dispose [Yes]
File/Vector/Dispose [Yes]
File/Vector/Change Label [# of authors]
```

The macro also returns a vector of number of authors of each work.

The macro collaboration combines the normalization, $n(\mathbf{WA})$ with computation of the first collaboration network, **Co**, with the following commands where the lines in square parentheses indicate choices to be made before invoking the commands:

```
[Select the WA network]
Network/2-Mode Network/2-Mode to 1-Mode/Columns
File/Network/Change Label [Collaboration 1]
```

For the second collaboration network, **Cn**, the commands are

```
[select the WA network as First]
[select the Normalized network as Second]
Network/2-Mode Network/Transpose 2-Mode
Networks/Multiply Networks [Yes]
File/Network/Change Label [Collaboration 2]
[select the Transpose network]
File/Network/Dispose [Yes]
```

And for the third collaboration network, **Ct**:

```
[select the Normalized network]
Network/2-Mode Network/2-Mode to 1-Mode/Columns
File/Network/Change Label [Collaboration 3]
```

And the vectors of self-contribution, total number of works, and collaborativeness:

```
[select the Collaboration 2 network]
Network/Create Vector/Get Loops
File/Vector/Change Label [self-contribution]
[select the WA network]
Network/Create Vector/Centrality/Degree/Input
[select the 2-Mode partition]
Operations/Vector+Partition/Extract SubVector [2]
File/Vector/Change Label [author's total # of works]
[select the self-contribution vector as First]
[select the author's total vector as Second]
Vectors/Divide (First/Second)
File/Vector/Change Label [self-sufficiency]
[select the Collaboration 2 network]
Vector/Create Constant Vector [OK]
[select the self-sufficiency vector as Second]
Vectors/Subtract (First-Second)
File/Vector/Change Label [collaborativeness]
[select the self-contribution vector]
Vector/Make Permutation
Permutation/Mirror Permutation
Operations/Network+Permutation/Reorder Network
File/Network/Change Label [Reordered Collaboration 2]
Operations/Vector+Permutation/Reorder Vector
File/Vector/Change Label [Ordered self-contribution]
[select the author's total vector]
Operations/Vector+Permutation/Reorder Vector
File/Vector/Change Label [Ordered total]
[select the collaborativeness vector]
Operations/Vector+Permutation/Reorder Vector
File/Vector/Change Label [Ordered collaborativeness]
```

4.9.2   An initial attempt of analyses of collaboration networks

The degree of an author in the first collaboration network is equal to the number of his/her different co-authors. In Table 4.11 the authors with the largest number of different co-authors for the centrality collaboration network are listed. Probably, many of these counted item represent more than one author. This is due to the way the names of the authors are currently produced in the program WoS2Pajek and to the lack of the unique identification of authors inside the WoS system.

The Chinese expression 'Some Zhang, some Li' is used to mean 'anyone' or 'everyone' (Wikipedia, 2013). The 100 most common Chinese family names are shared by 85% of the Chinese population. Since many Chinese names emerged on the top level we cannot consider them as noise. Indentifying the individual authors corresponding to the named item task would be an extremely time-consuming task. We decided to abandon the analysis of these collaboration networks until this kind of a task could be completed automatically. This problem lies with the data as they are recorded in WoS. Instead, we switched to the 'Social networks' (SN5) collection of networks to illustrate the creating of the derived networks and the analysis of them.

Table 4.11 Authors with the largest number of co-authors.

rank	deg	name	rank	deg	name	rank	deg	name
1	1018	WANG_J	16	534	ZHANG_P	31	461	ZHANG_J
2	1010	ZHANG_Y	17	527	CHEN_H	32	460	HALL_J
3	795	LIU_J	18	513	WU_G	33	460	KUMAR_S
4	765	LEE_S	19	512	WONG_G	34	454	SMITH_T
5	677	YANG_H	20	503	WANG_X	35	453	YU_Q
6	672	WANG_Y	21	502	LEE_W	36	453	BALDWIN_J
7	671	LIN_C	22	501	EVANS_J	37	449	BOSAK_S
8	641	LIU_X	23	500	JOHNSON_J	38	447	LIN_M
9	603	LI_R	24	500	GARCIA_A	39	447	EDWARDS_K
10	601	LIU_S	25	487	LI_X	40	447	SHI_L
11	585	LI_Y	26	483	LI_J	41	443	ZHANG_X
12	559	SMITH_D	27	481	JONES_C	42	441	NGUYEN_T
13	557	LU_J	28	466	CHANG_J	43	441	WILSON_A
14	549	YANG_S	29	461	SINGH_R	44	440	JAFFE_D
15	546	WU_C	30	461	WANG_L	45	438	PRUITT_K

4.10   A brief look at the SNA literature SN5 networks

4.10.1.1   Tendencies of individual authors to collaborate

One topic of potential interest is identifying those authors with the most collaborators. Such authors in SN5 are listed in Table 4.12 – the authors with the largest input degree in **WA**. The majority of these authors are from the SNA literature but we note the use of a cut-off of 50 collaborators. This threshold could be set at a lower level to include more authors if desired.

4.10.1.2   Participation in co-authored productions

Table 4.13 shows the distribution of output degree of works in **WA**(SN5a) where the output degree of each work equals the number of authors that co-authored the productions. We list

Table 4.12 Authors with the largest number of different collaborators in SN5.

Rank	Author	Collaborators	rank	Author	Collaborators
1	Snijders,T	77	11	Rothenberg,R	58
2	Krackhardt,D	71	12	Doreian,P	56
3	Wasserman,S	65	13	Breiger,R	56
4	Ferligoj,A	63	14	Valente,T	52
5	Berkman,L	63	15	Butts,C	52
6	van Duijn,M	63	16	Goodreau,S	52
7	Donovan,D	62	17	Draper,D	51
8	Friedman,S	60	18	Batagelj,V	51
9	Latkin,C	59	19	Barabasi,A	51
10	Faust,K	59	20	Kelly,J	50

Table 4.13 Outdegree distribution in **WA**(SN5).

Outdeg	Frequency	Outdeg	Frequency	Paper
1	2637	12	8	
2	2143	13	4	
3	1333	14	3	
4	713	15	2	
5	396	21	1	Pierce et al. (2007)
6	206	22	1	Allen et al. (1998)
7	114	23	1	Kelly et al. (1997)
8	65	26	1	Semple et al. (1993)
9	43	41	1	Magliano et al. (2006)
10	24	42	1	Doll et al. (1992)
11	10	48	1	Snijders et al. (2007)

only the names of papers where the outdegree is greater than 20, for these are unique for the degree. For the lower outdegree levels there are multiple papers – so only a representative paper could be listed. The papers can be identified if needed.

While we provide the reference list for the book as a whole at the end of the text, we include a brief reference list at the end of this chapter for the works with large numbers of authors featuring in our analyses of the collaboration networks.

4.10.1.3 *k*-cores in collaboration networks

In a collaboration network a *k-core* is the largest subnetwork with the property that each of its author wrote a joint work with at least *k* other authors from the core.

Figure 4.15 shows the cores of orders (values of *k*) 20–47. The bottom left *k*-core (marked in dark green) shows some heavier levels of collaboration within the traditional SNA community including a trio (Batagelj, Doreian, Ferligoj) and two pairs (Snijders and van Duijn) and (Faust and Wasserman). Larger small configurations are shown in two other cores. Even so, the main feature of this figure reveals a serious drawback of directly applying cores for analysis of collaboration networks. A work with *k* authors contributes a complete subgraph, K_k, on *k* vertices to a collaboration network. For the bibliographies with works with a large number of authors the cores procedure identifies as the highest level cores the complete subgraphs corresponding to these works, and not the groups of really the most collaborative authors, as one would expect.

For the SN5 bibliography the components of the cores of orders 20–47 in **Co**(SN5) (are induced by the papers Allen et al. (1998); Doll et al. (1992); Kelly et al. (1997); Magliano et al. (2006); Pierce et al. (2007); Semple et al. (1993); Snijders et al. (2007), which correspond to the works with the largest number of authors (21–48) – see Table 4.13 and the references at the end of this chapter. In the picture, only the names of authors that are the endnodes of links with weights larger than 1 are displayed. Also high degrees of most of the authors in Table 4.12 result from these papers with very large numbers of co-authors.

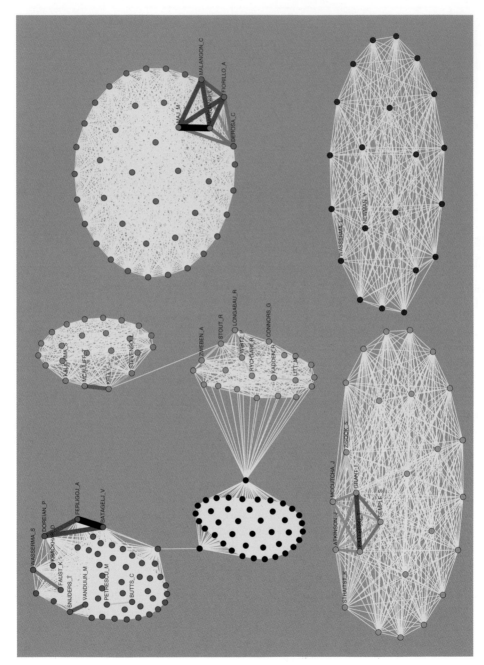

Figure 4.15 Cores of orders 20–47 in **Co**(SN5).

One approach for dealing with this problem would be to remove all links with weight 1 (or up to some other small threshold) and apply cores on the resulting reduced network. Choosing the threshold can be arbitrary. A better solution is to identify the works with (too) many authors, those having very high outdegree in the network, **WA**, for this analysis, and removing them from **WA**. We can always review the removed works separately if needed.

Yet another approach is to apply on the collaboration network **Co** the p_S-cores. In Figure 4.16 the p_S-core at level 20 is presented. Each author belonging to it has at least 20 collaborations with other authors inside the core within which they are located. The thickness of the lines indicates intensity of pairwise collaboration.

Again in the network, **Co**(SN5), the cliques corresponding to papers with the largest number of authors appear in the p_S-core. Even so, some strongly collaborating groups of authors were identified including: {S. Borgatti, M. Everett}, {H. Bernard, P. Killworth, C. McCarty, E. Johnsen, G. Shelley}, {V. Batagelj, A. Ferligoj, P. Doreian}, {R. Rothenberg, S. Muth, J. Potterat, D. Woodhouse}, {L. Magliano, M. Maj, C. Malangon, A. Fiorillo}. The more specific topics focused on by the first four sets of authors are: 1) applying graph theoretic ideas for general analyses of networks; 2) informant accuracy studies and scale-up methods for estimating the size of hidden populations; 3) blockmodeling; 4) networks for medical issues including the sexual transmission of disease; and 5) networks for psychiatric interventions.

Table 4.14, obtained from the second collaboration network, **Cn**(SN5), shows the 50 authors with the largest self-contribution scores, cn_{ii}, to the topic of 'social network analysis' together with the total number of works on the topic that an author co-authored and his/her collaborativeness index, K_i.

Figure 4.17 presents the p_S-core of order 0.75 in the third collaboration network, **Ct**(SN5). In this core the large cliques disappear, a desirable feature given the distorting problem their presence poses. The largest core's component consists of the mainstream social networks researchers with the most intensive collaboration pairs: Borgatti and Everett, Killworth and Bernard, Bonacich and Bienenstock, Ferligoj and Batagelj, Pattison and Robins, etc. The second largest component consists of physicists with more intensive collaboration pairs: Newman and Park, Barabási and Albert, and Masuda and Konno. In the smaller components we find additional three pairs: Leinhardt and Holland (with Leinhardt represented by two vertices), Metzke and Steinhausen, plus Chou and Chi.

In the network **biCon**(SN5) the larger components with edges with *bicon* = 1 correspond to papers with a single reference to a well-known and widely used book (Wasserman, S., Faust, K.: Social network analysis. Cambridge UP, 1994; Taylor, Howard F.: Balance in small groups. Van Nostrand Reinhold, 1970; Belle, D.: Children's social networks and social supports. Wiley, 1989; Gottlieb, B. H.: Social networks and social support. Sage, 1981; Yan, Yunxiang: The flow of gifts. Stanford UP, 1996; Zhang, L.: Strangers in the City. Stanford UP, 2001). There are also several pairs of papers with *bicon* = 1, mostly written by the same author. We can obtain more interesting groups as larger islands with values below 1. We obtain 19 islands of size in [10, 50] on 290 vertices. A selection from this set is displayed in the Figure 4.18.

4.10.1.4 The most important works

In Table 4.15, 50 works with the largest values of probabilistic flow index in CiteAcy(SN5) are listed. The index values are multiplied with 10^6. Most of the works in this list are books

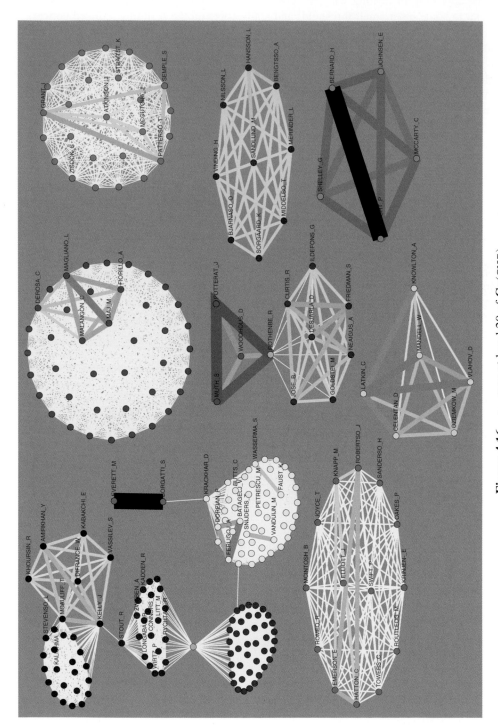

Figure 4.16 p_S-core at level 20 of **Co**(SN5).

Table 4.14 List of the most collaborative authors in SN5.

i	Author	cn_{ii}	Total	K_i
1	Burt,R	43.83	53	0.173
2	Newman,M	36.77	60	0.387
3	Doreian,P	34.44	47	0.267
4	Bonacich,P	30.17	41	0.264
5	Marsden,P	29.42	37	0.205
6	Wellman,B	26.87	41	0.345
7	Leydesdorf,L	24.37	35	0.304
8	White,H	23.50	33	0.288
9	Friedkin,N	20.00	23	0.130
10	Borgatti,S	19.20	41	0.532
11	Everett,M	16.92	31	0.454
12	Litwin,H	16.00	21	0.238
13	Freeman,L	15.53	20	0.223
14	Barabasi,A	14.99	35	0.572
15	Snijders,T	14.99	30	0.500
16	Valente,T	14.80	34	0.565
17	Breiger,R	14.44	20	0.278
18	Skvoretz,J	14.43	27	0.466
19	Krackhardt,D	13.65	25	0.454
20	Carley,K	12.93	28	0.538
21	Pattison,P	12.10	27	0.552
22	Wasserman,S	11.72	26	0.549
23	Berkman,L	11.21	30	0.626
24	Moody,J	10.83	15	0.278
25	Scott,J	10.47	15	0.302
26	Latkin,C	10.14	37	0.726
27	Morris,M	9.98	20	0.501
28	Rothenberg,R	9.82	28	0.649
29	Kadushin,C	9.75	11	0.114
30	Faust,K	9.72	18	0.460
31	Batagelj,V	9.69	20	0.516
32	Mizruchi,M	9.67	15	0.356
33	[Anon]	9.00	9	0.000
34	Johnson,J	8.89	21	0.577
35	Fararo,T	8.83	16	0.448
36	Lazega,E	8.50	12	0.292
37	Knoke,D	8.33	11	0.242
38	Ferligoj,A	8.19	19	0.569
39	Brewer,D	8.03	11	0.270
40	Klovdahl,A	7.96	17	0.532
41	Hammer,M	7.92	10	0.208
42	White,D	7.83	15	0.478
43	Holme,P	7.42	14	0.470

(*continued*)

Table 4.14 *(Continued)*

i	Author	cn_{ii}	Total	K_i
44	Boyd,J	7.37	13	0.433
45	Kilduff,M	7.25	16	0.547
46	Small,H	7.00	7	0.000
47	Iacobucci,D	7.00	12	0.417
48	Pappi,F	6.83	10	0.317
49	Chen,C	6.78	12	0.435
50	Seidman,S	6.75	9	0.250

with Wasserman and Faust (1994) as a clear leader. The most important papers (for SNA literature) turned out to be Watts and Strogatz (1998), Granovetter (1973), Freeman (1979), and Barabási and Albert (1999).

4.10.1.5   SPC weights

In Figure 4.19 the CPM main path through the SN5 citation network with SPC weights is presented.

In Figure 4.20 the main island for SPC weights in $\mathbf{Ci}$(SN5) is presented. The bottom part consists of traditional SNA while the left upper part can be seen as representing the 'invasion of the physicists'. The heavy black line is the citation by Newman (2003) to Valente (1996) identified also in the main path in the centrality literature.

4.10.1.6   Other derived networks

In Figure 4.21 some line islands from $\mathbf{Ca}$(SN5), $\mathbf{Ca} = \mathbf{AW} * \mathbf{Ci} * \mathbf{WA}$, are presented. The weight ca_{ij} counts the number of times a work authored by i is citing a work authored by j. The largest island consists of the mainstream social networks researchers with some subgroups: the star around R. Burt in the top left part; the S. Borgatti and M. Everett tandem in the bottom left part; the probabilistic group in the top right part with G. Robins, P. Pattison, T. Snijders, S. Wasserman, and P. Holland as the most prominent; and others: J. Skvoretz, D. Krackhardt, P. Doreian, R. Breiger, H. White, L. Freeman, and P. Marsden.

Immediately above the mainstream island is a 'scale-free' island (marked in dark red) which consists mainly of physicists M. Newman, A. Barabasi, D. Watts, R. Albert, P. Holme, and others. In the 'medical' island the central authors are J. Potterat, R. Rothenberg, D. Woodhouse, S. Muth, A. Klovdahl, and S. Friedman. There is also an island on 'education and psychology' with T. Farmer, R. Cairns, B. Cairns, H. Xie, and P. Rodkin.

Most of the other islands are star-like, most likely a professor with his/her PhD students.

In Figure 4.22 we present the results of applying condensation on the 'mainstream' (on the left) and 'medical' (on the right) islands. The acyclic component is dominant in each shown part. The shrunken vertices are labeled by names of member researchers.

Other analyses could be done. For example, the entry aj_{ij} of $\mathbf{AJ} = \mathbf{AW} * \mathbf{WJ}$ counts how many works the author i published in the journal j; and the entry $jj_{j_1 j_2}$ of $\mathbf{JJ}_A = b(\mathbf{JA}) * b(\mathbf{AJ})$ counts the number of authors that are publishing in journals j_1 and j_2. Again the normalization problem should be considered.

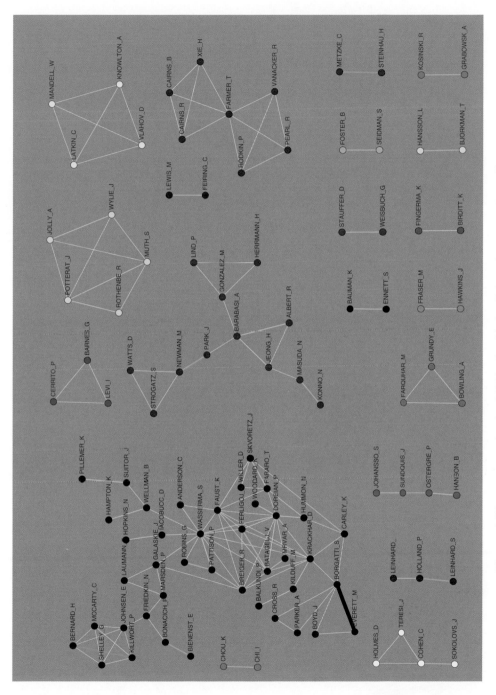

Figure 4.17 p_S-core of order 0.75 in the third collaboration network on **Ct**(SN5).

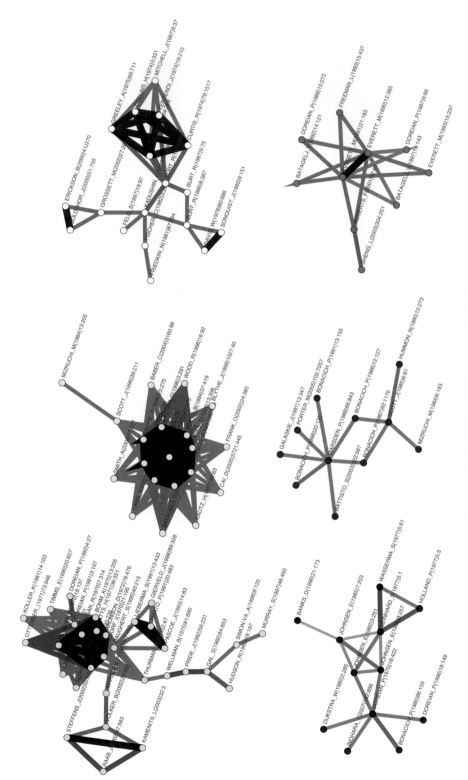

Figure 4.18 Selected **biCon**(SN5) islands.

Table 4.15 Fifty works with the largest value of probabilistic flow index.

Rank	Value	Work	Rank	Value	Work
1	4238.3971	WASSERMA_S(1994):	26	784.2723	FISCHER_C(1982):
2	1993.3290	MITCHELL_J(1969):	27	762.5744	COCHRAN_M(1990):
3	1665.0870	WATTS_D(1998)393:440	28	753.3617	PILISUK_M(1986):
4	1610.8547	GRANOVET(1973)78:1360	29	743.9300	LOURENCO_I(2002):
5	1300.7664	YAN_Y(1996):	30	711.9562	HOLLAND_P(1979):
6	1221.1978	LAUMANN_E(1973):	31	708.2055	FRIEDMAN_S(1999):
7	1215.4193	GOTTLIEB_B(1981):	32	684.1711	MILARDO_R(1988):
8	1139.2607	FREEMAN_L(1979)1:215	33	678.3867	PUTNAM_R(2000):
9	1136.0781	BURT_R(1992):	34	669.8170	GRIECO_M(1987):
10	1131.6018	BARABASI_A(1999)286:509	35	659.6933	MAGUIRE_L(1983):
11	1122.4933	SCOTT_J(1991):	36	656.1357	BOTT_E(1971):
12	1094.2761	BELLE_D(1989):	37	655.7511	LITWIN_H(1995):
13	1071.6445	ZHANG_L(2001):	38	646.8604	HASSINGE_E(1982):
14	925.4650	ROGERS_A(1995):	39	641.1038	GRIECO_M(1996):
15	918.4549	WELLMAN_B(1988):	40	626.8148	WATTS_D(1999):
16	892.2797	PERRUCCI_R(1982):	41	606.7083	COLEMAN_J(1988)94:95
17	888.7358	ANGERMEY_M(1989):	42	573.8576	HEDIN_A(2001):
18	874.9730	ALBERT_R(2002)74:47	43	557.9572	TAVECCHI_L(1987):
19	874.1589	NEWMAN_M(2003)45:167	44	557.0382	COLEMAN_J(1990):
20	865.8406	BIEGEL_D(1985):	45	551.1556	DENOOY_W(2005):
21	854.9899	PHILLIPS_C(2004):	46	535.3469	BERKOWIT_S(1982):
22	851.8199	CLARKE_S(2002):	47	523.1899	MILROY_L(1980):
23	850.1594	RUSSELL_G(2002):	48	519.8359	DEGENNE_A(1999):
24	813.3185	ROGERS_E(1981):	49	517.4263	ALBERT_R(1999)401:130
25	799.1026	BERKMAN_L(1979)109:186	50	504.7977	PATTISON_P(1993):

Figure 4.19 SN5: CPM path for SPC.

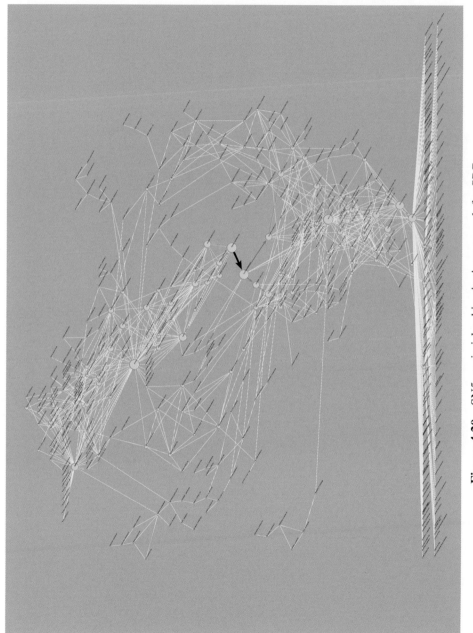

Figure 4.20   SN5: main island in citation network for SPC.

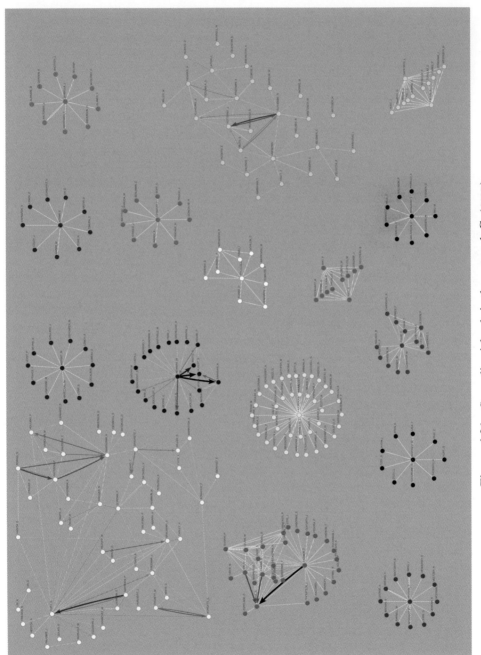

Figure 4.21 Some line islands in the network **Ca**(SN5).

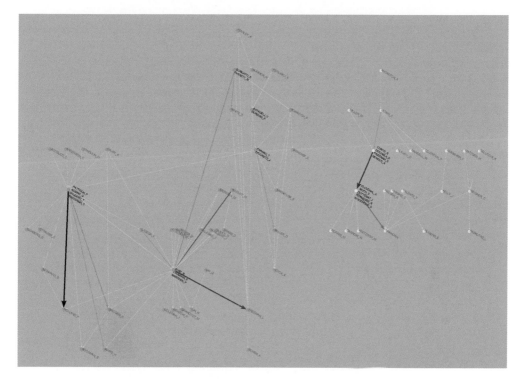

Figure 4.22   Acyclic structure of two main islands of **Ca**(SN5).

4.11   On the centrality and SNA collaboration networks

Based on the tools presented in Chapters 2 and 3, we have presented a series of analyses of the centrality literature and the collaboration network among social network analysts and contributors to network science. The results are readily interpretable and add to our understanding of these networks. They also set foundations for future work on these networks for there is much more to be done. Also, the techniques used here hold great promise for studying other scientific citation networks.

References

Allen, J., Anton, R.F., Babor, T.F., et al. (1998) Matching Alcoholism Treatments to Client Heterogeneity: Project Match Three-Year Drinking Outcomes. *Alcoholism-Clinical and Experimental Research*, **22**(6), 1300–1311.

Doll, L.S., Petersen, L.R., White, C.R., et al. (1992) Homosexually and nonhomosexually identified men who have sex with men – a behavioral comparison. *Journal of Sex Research*, **29**(1), 1–14.

Kelly, J.A., Murphy, D.A., Sikkema, K.J., et al. (1997) Randomised, controlled, community-level HIV-prevention intervention for sexual-risk behaviour among homosexual men in US cities. *Lancet*, **350**(9090), 1500–1505. doi: 10.1016/S0140-6736(97)07439-4

Magliano, L., Fiorillo, A., Malangone, C., et al. (2006) Family psychoeducational interventions for schizophrenia in routine settings: impact on patients' clinical status and social functioning and on relatives' burden and resources. *Epidemiologia e Psichiatria Sociale*, **15**(3), 219–27.

Pierce, J.P., Natarajan, L., Caan, B.J., et al. (2007) Influence of a Diet Very High in Vegetables, Fruit, and Fiber and Low in Fat on Prognosis Following Treatment for Breast Cancer. *JAMA*, **298**(3), 289–298. doi: 10.1001/jama.298.3.289

Semple, S.J., Patterson, T.L., Temoshok, L.R., et al. (1993) Identification of psychological stressors among HIV-positive women. *Women & Health*, **20**(4), 15–36. doi: 10.1300/J013v20n04_02

Snijders, T.A.B., Robinson, T., Atkinson, A.C., et al. (2007) Discussion on the paper by Handcock, Raftery and Tantrum. *Journal of the Royal Statistical Society: Series A – Statistics in Society*, **170**(2), 322–354. doi: 10.1111/j.1467-985X.2007.00471.x

5

Citation patterns in temporal United States patent data

Schumpeter (1942) identified innovation as a primary driver of economic growth. Integral to innovation is the creation of new machines, new methods of production, and new products by inventors. Throughout human history, new technologies have transformed societies (Lenski 1966). Economists have long studied links between innovation, technological change, and economic development. The importance of particular inventions depends on the consequences of implementing them. Estimating the magnitude of technological changes, let alone establishing their wider societal impacts, is extremely hard. Inventions are intrinsic parts of technological change with many inventions being patented. Since the pioneering work of Schmookler (1966), patents have been studied[1] to gain more insight into technologically driven economic and social changes (see Griliches (1984, 1990); Trajtenberg (2002)). Patents differing in importance provide indirect evidence regarding large and small sources of the impact of technology on economic change. One potential indicator of a patent's importance is the extent to which it is used (cited) by later patents. This had led to the development of indices of the importance of patents (e.g. Carpenter et al. (1981); Trajtenberg (2002)). We do not dispute the value of this line of work but adopt a complementary approach. Consistent with the objectives of this book, we try to understand the structure of patent networks while delineating temporal changes in them.

5.1 Patents

A patent is an intellectual property right granted by the Government of the United States of America 'to an inventor to exclude others from making,

---

[1] The patent citation dataset constructed by Hall et al. (2001) for patents granted between 1963 and 1999 created an important treasure trove for work in this area.

Understanding Large Temporal Networks and Spatial Networks: Exploration, Pattern Searching, Visualization and Network Evolution, First Edition. Vladimir Batagelj, Patrick Doreian, Anuška Ferligoj and Nataša Kejžar.
© 2014 John Wiley & Sons, Ltd. Published 2014 by John Wiley & Sons, Ltd.

using, offering for sale, or selling the invention throughout the United States'
or importing the invention into the United States for a limited time in exchange
for public disclosure of the invention when the patent is granted. (USPTO,
http://www.uspto.gov/patents/)

Patents fall into several categories.[2] Approximately 90% of the patent documents issued
by the patent office in recent years have been utility patents intended to protect non-obvious,
novel and useful inventions belonging to four categories (more on these below): processes
(methods), machines, articles of manufacture, composition of matter and improvements in
any of these categories. Most patents are for incremental improvements to known technology
although some patents have been revolutionary. Design patents protect ornamental features
of articles of manufacture.[3] 'New and distinct, invented or discovered asexually reproduced
plant including cultivated spores, mutants, hybrids, and newly found seedlings, other than
a tuber propagated plant or a plant found in an uncultivated state' are protected by plant
patents under the 1930 Plant Patent Act. Reissue patents are granted to correct errors in
already issued utility, design, or plant patents. There are items for which patents are never
granted including abstract ideas, physical phenomena, and laws of nature. Also excluded
are literary, dramatic, musical and artistic works – these can be protected by copyrights.
Inventions deemed impossible, not useful, or offensive to public morality cannot be patented.

Lurking in the language for defining patents are seemingly clear words whose meanings
required specification. The simpler ones are:

- new and novel: In the USA, to obtain a patent, the invention must never have been
  made public in *any* way in *any* place.

- original and non-obvious: This is assessed through a comparison of what is in a patent
  application and what is known already. Those making this assessment have a good
  knowledge of the relevant technical domain(s). The invention must be 'not obvious' in
  the eyes of relevant expert patent examiners.

- useful: An invention has to be a practical device capable of doing something.

Patents, as protections of intellectual property rights, have value for those holding them.
Usually, this is relevant for commercial applications of patents. As such, patents have economic value by excluding other producers from applying the contents of patented items.
When others do this anyway, it is a violation of the patent (patent infringement). These
violations result in legal cases filed in the US court system (see below). Competing claims
for different patents, challenges to existing patents and alleged infringements of patents lead
also to court cases. Patents can be challenged by those seeing them as not being legiti-
mately granted, as can rejections of patent applications. When cases are decided in court,
especially the US Supreme Court, some terms in patent languages are changed. When relevant below, specific Supreme Court decisions are cited in our discussion of utility patent
categories.

---

[2] Our primary source here is http://www.uspto.gov/web/offices/ac/ido/oeip/taf/patdesc.htm (accessed 6/6/2013).
[3] All of the *Star Wars* characters have been patented, as have the distinctive appearance of athletic shoes. See
http://inventors.about.com/od/inventing101patents/f/can_be_patented.htm, accessed 6/6/2013.

The exact (but *not always* unambiguous) nature of the four categories of utility patents was specified further in US Supreme Court decisions:

- Method: *Gottschalk v. Benson* (409US63, 1972) defined this by '(a) process [method] is a mode of treatment [series of steps] of certain materials to produce a given result. It is an act, or a series of acts, performed upon the subject-matter to be transformed and reduced to a different state or thing (items in square parentheses inserted)'.[4]

- Machine: A machine is a concrete object with combinations of parts. From *Corning v. Burden*, 56US252, 1854), the specification of a machine 'includes every mechanical device or combination of mechanical powers and devices to perform some function and produce a certain effect or result'. Implicit in this specification is that the parts interact operationally.

- Article of manufacture: In *Diamond v. Chakrabarty* (447US303, 1980)[5], the Supreme Court relied on a dictionary definition of manufacture (a verb) as 'the production of articles for use from raw or prepared materials by giving to these materials new forms, qualities, properties, or combinations, whether by hand labor or by machinery.' It follows that articles of manufacture resulting from production, in this sense, are tangible items or commodities.[6]

- Composition of matter: 'A composition of matter is an instrument formed by the intermixture of two or more ingredients, and possessing properties which belong to none of these ingredients in their separate state. The intermixture of ingredients in a composition of matter may be produced by mechanical or chemical operations, and its result may be a compound substance resolvable into its constituent elements by mechanical processes, or a new substance which can be destroyed only by chemical analysis (Robinson 1890).' In *Diamond v. Chakrabarty*, the Court added that this term means 'all compositions of two or more substances and all composite articles, whether they be the results of chemical union, or of mechanical mixture, or whether they be gases, fluids, powders or solids'.[7]

---

[4] In this case, the Court ruled that, a proposed way for converting numerical information from binary-coded decimal numbers into pure binary numbers, for use in programming conventional general-purpose digital computers, was merely a series of mathematical calculations or mental steps: it was not patentable as a method.

[5] Chakrabarty, a genetic engineer who worked for General Electric (GE), developed a bacterium designed to break down crude oil. GE's patent application for this was rejected by a patent examiner. At that time, living things were not patentable. This case reached the Supreme Court. It ruled (5-4) a living, man-made micro-organism is patentable subject matter as an article of manufacture or composition of matter within the meaning of the 1952 Patent Act.

[6] This can get quite subtle. *American Fruit Growers, Inc. v. Brogdex Co* (283US1, 1931) ruled 'the addition of borax to the rind of natural fruit does not produce from the raw material an article for use which possesses a new or distinctive form, quality, or property. The added substance only protects the natural article against deterioration by inhibiting development of extraneous spores upon the rind. There is no change in the name, appearance, or general character of the fruit'. It declared the invention of this patent application unpatentable and voided it.

[7] The rationale for dividing utility patents into four categories was driven by a need to place inventions in them, perhaps uniquely. This particular statement by the Court serves to muddy the waters. Items such as fiberglass hulls for boats can be both an article of manufacture and a composition of matter.

This broad classification of utility patents is complemented by a detailed classification of technological content. There are more than 400 three-digit patent classes devised by USPTO with over 120,000 patent subclasses. Hall et al. (2002) (p. 415) state 'This system is continuously being updated, reflecting the rapid changes in the technologies themselves, with new patent classes being added and others being reclassified and discarded.' A footnote discusses granted patents being reclassified. As a practical matter, they developed a two-digit classification with 36 categories before constructing six main categories:

1. Chemical (with six two-digit subcategories: Agriculture; Food; Textiles; Coating; Gas; Organic Compounds; Resins and Miscellaneous Chemical);
2. Computers and Communications (with four two-digit subcategories: Communications; Computer Hardware and Software; Computer Peripherals and Information Storage);
3. Drugs and Medical (with four two-digit subcategories: Drugs; Surgery and Medical Instruments; Biotechnology and Miscellaneous Drugs and Medical);
4. Electrical and Electronic (with seven two-digit subcategories: Electrical Devices; Electrical Lighting; Measuring and Testing; Nuclear and X-rays; Power Systems; Semiconductor Devices and Miscellaneous Electrical and Electronic);
5. Mechanical (with six two-digit subcategories: Materials, Handling and Processing; Metal Working; Motors, Engines and Parts; Optics; Transportation and Miscellaneous Mechanical) and
6. Others (with nine two-digit subcategories: Agriculture, Husbandry and Food; Amusement Devices; Apparel and Textile; Earth Working and Wells; Furniture, House Fixtures; Heating; Pipes and Joints; Receptacles and Miscellaneous Others).

These classifications of patent classes and technological domains (with the detailed subdomains) make it clear, purely in terms of intrinsic content, that patents and patent applications are parts of a complex intellectual arena. It is also one that can become very conflictual, given the potential economic stakes resting on the exploitation of patented inventions. Conflicts over patents arising in the USA frequently end up in the courts. Patent cases have been on the docket of the Supreme Court, as noted above. There is a whole area of Patent Law with specialist lawyers. Inventors seeking patents have to enter a multifaceted domain with features going well beyond the details of the inventions.[8]

Patent applications are submitted to patent offices having authority to evaluate them in the context of prior patents (and knowledge). Patent officers issue final rulings within a patent office regarding the patentability of inventions. The entire process of applying for patents, assessing these patent applications and granting patents (or not) *directly affects* the form of patent citation networks. As described above, patents are sought for *genuinely new* inventions falling within one or more of the technological domains listed above. For all technological areas, their prior knowledge bases are the foundations for evaluating patents. Patent applications must be *very* precise documents stating clearly 1) how potentially patentable items

---

[8] When Googling terms such as 'patents', 'patent applications' and related topics, the results include a plethora of advertisements for lawyers, consultants, and organizations offering their services to inventors. Most often, these services are expensive.

build on prior knowledge, 2) how they are differentiated from this knowledge, 3) the exact nature of their novelty, and 4) the potential usefulness of inventions.[9]

In contrast to scientific citation networks (Chapter 4), for which authors cite whatever earlier scientific papers they select, or the Supreme Court citation network (Chapter 6), where Justices can choose any prior Court decision they wish to cite (while ignoring other decisions with potential value as precedent), patent applicants are severely constrained. They must cite *exactly* the relevant prior patents and *only* those patents (directly relevant prior knowledge). The role of patent officers with special relevant knowledge is crucial. They interact with applicants and can insist on modifications to patent applications if relevant knowledge is omitted or extraneous knowledge is included.[10] The resulting patent citations networks are cleaner and sparser. A strong case can be made for such citation data being more accurate than the other types of citation networks because patent officers are *enforcers of content*, especially citations. Allowing for some inventions not being patented, the resulting network is one where the network boundary has been *well established*. See Section 1.3.1 for a description of the network boundary problem and how this is solved for the patent citation data.

5.2   Supreme Court decisions regarding patents

Before examining the actual citation patterns among patents, we look at some Supreme Court decisions for their role in affecting the patent citation network regarding the upholding or eliminating of patents.

5.2.1   Co-cited decisions

We do this by examining a set of decisions that were co-cited.[11] We focus first on decisions coded by the four areas for utility patents. Figure 5.1 displays this co-citation network with color codes for the four utility areas: method (light sky blue); machine (green yellow); article of manufacture (light purple); and composition of matter (cornflower blue). There are 23 Supreme Court (SC) decisions in this island with nearly all of them falling within the method (10) and articles of manufacture[12] (10) categories. There are two more densely co-cited sets of cases. The one in the lower left is largely made up of decisions involving machine patents while the upper right set has mainly decisions about articles of manufacture.

Such cases reached the Supreme Court in two ways. One route involves claims by patent holders of patent infringement by actors not holding the patent. For these cases, there were

---

[9] Patent applications include statements of the technical problems solved and the operational details for using the invention.

[10] There are no formal constraints on Supreme Court Justices on what they cite, beyond making compelling arguments for (or against) Court decisions. As shown in Section 6.6, ignoring precedent is possible (indeed, it occurs often). For academic journal citation networks, the interactions involve authors, reviewers, and editors. While these process can lead to adding citations (seen as relevant by reviewers) there is little pressure to remove tangential or irrelevant citations.

[11] If C denotes the matrix of the citation network, then the elements $g_{ij} \in G = C'C$, where $i \neq j$, measure the number of times that pairs of decisions are cited together. As described more fully in Chapter 6, high enough g_{ij} serve to help identify elements of G on a co-cited line island of Supreme Court decisions identified in the Supreme Court citation network. See Chapter 6 for further details.

[12] The decision coded as composition of matter could have been coded as method because the Supreme Court ruled it was patentable under both categories.

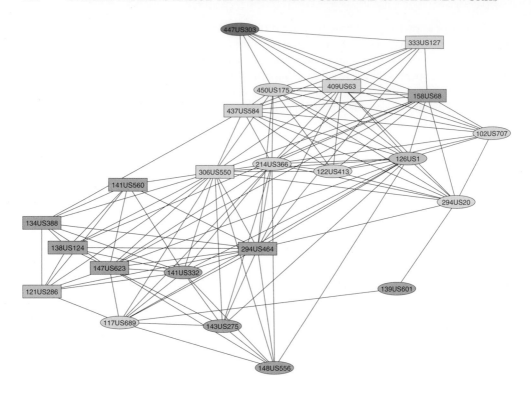

Figure 5.1   Co-cited Supreme Court line island by utility areas.

Color codes are: method (light sky blue); machine (green yellow); article of manufacture (light purple); and composition of matter (cornflower blue). Upheld patents are shown with ellipses while voided patents are in boxes.

two outcomes: 1) the person(s) sued for infringing a patent lost their case and the patent was upheld, or 2) the person(s) accused of infringing the patent won and the patent was voided. The second route took the form of appeals against USPTO decisions rejecting patent applications. Again, there were two outcomes: 1) the SC ruled against USPTO by declaring inventions patentable, or 2) the SC agreed with USPTO that inventions were not patentable. We coded all of these SC decisions into two categories. One combined upheld patents and rulings in favor of an invention being patentable. These are shown with ellipses in Figure 5.1. The other combined decisions voiding patents and decisions ruling an invention as not being patentable. These are shown with boxes in Figure 5.1. The presence of 12 ellipses and 11 boxes suggests no differences overall in the ultimate outcomes regarding patentability.[13] We note that: 1) voided patents are removed from the patent database and cannot be cited subsequently by later patents,[14] and 2) inventions deemed not patentable never receive a patent.

   Figure 5.2 shows the co-cited network with the vertices color coded for their technological areas: chemical (yellow); computers and communications (C&C) (light faded green);

---

[13] We note that six of the ten method patents in this subnetwork were upheld while six of the ten articles of manufacture were voided.

[14] In contrast, the SC decisions voiding patents can be, and often are, cited by later SC decisions.

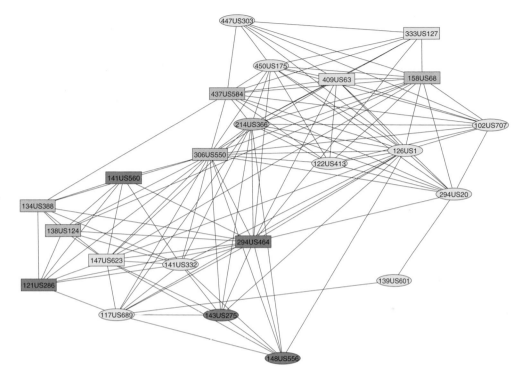

Figure 5.2   Co-cited line island by technological areas.

Color codes are: chemical (yellow); C&C (light faded green); mechanical (light orange) and others (cornflower blue). Upheld patents are shown with ellipses while voided patents are in boxes

mechanical (light orange) and others (cornflower blue).[15] In terms of patents being upheld or voided, there are two seemingly clear features. Seven of the nine patents for chemical inventions were upheld. In contrast, five of the six patents for mechanical were voided.

Half of the lower left denser group of decisions fall within the others area: indeed, all cases in this area are in this group. They were also about inventions in the articles of manufacture category. Two-thirds of the chemical decisions make up half of the upper right group of heavily co-cited decisions although there are three such decisions in the lower left group. Two-thirds of the mechanical decisions are in the upper right group of heavily co-cited decisions. Two of the Computer and Communications decisions are also in the top right group (with the third being the 'lesser' connecting link between the two groups). The linking decision, 294US464, is in the others category. The second linking decision, 306US550, is from the mechanical area. Two of the decisions in Figure 5.1 merit closer attention: they link the two denser groups of decisions. Both voided patents: 1) 294US464 (Patent 1,262,860) concerned a patent in the articles of manufacture category, and 2) 306US550 (Patent 1,930,987) in the method category. There is another 'linking' decision, 139US601 (Patent 270,767), when a patent was upheld.

---

[15] Decisions from the drugs and medical (D&M) area are not in this line island. Neither are there any decisions concerning electical and electronic (E&E) patents.

Briefly, we look at the three linking decisions. The minor one (139US601) dealt with an improvement of telegraph keys in a patent granted to Westinghouse Electric for an article of manufacture. Decided in 1891, it has been co-cited heavily with two decisions dealing with chemical patents for methods. One (117US689) upheld a patent for an improvement in apparatus for a rapid and economical method for producing enameled moldings in 1886. The other (294US20) upheld a patent for an improved egg incubator in 1935. Again, these co-cited links are consistent with the use of general legal patent principles in reaching decisions.

Both 294US464 (decided in 1935) and 306US550 (decided in 1939) voided patents in completely different domains. The first in others dealt with an invention in the machine area while the latter dealt with a method in the mechanical technological area. These two decisions, while heavily co-cited with earlier decisions voiding patents, are the most heavily co-cited with other decisions in this line island. They set precedents for both upholding and voiding subsequent patents.

5.2.2   Citations between co-cited decisions

Figure 5.3 shows the *citations within* the line island of co-cited decisions. Decades are marked on the left with decisions placed approximately in the years when they were made. The color

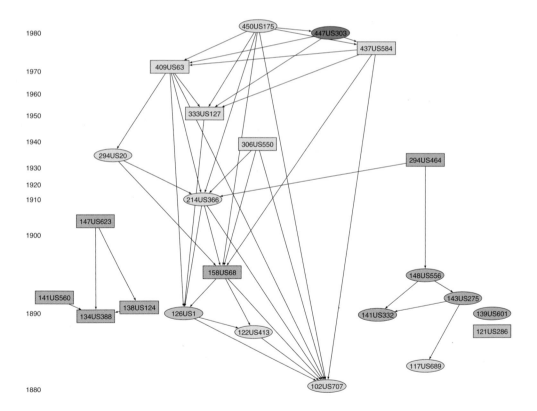

Figure 5.3   Citations within the patent line island by utility areas.

Color codes are: method (light sky blue); machine (green yellow); article of manufacture (light purple) and composition of matter (cornflower blue). Upheld patents are shown with ellipses while voided patents are in boxes.

codes for the utility areas are the same as for Figure 5.1, as is the use of ellipses and boxes. The three earliest decisions all feature patents in the machine area (followed by a method case). All but one of the articles of manufacture decisions come next in time. They are located in different places across the figure as they are roughly contemporaneous. On the left are four cases forming a separate component. Another decision, 158US68, is linked to many machine cases via received and sent citations. On the right are four more decisions with one incoming link from 294US464 to 148US556 which joins them to the main part of the citation network. The later set of SC decisions, with one exception, all feature machine decisions. The two denser patches of co-cited decisions in Figure 5.1 appear to be well separated in time. Decisions concerning articles of manufacture came earlier while those featuring methods came much later.

Both 294US464 (Patent 1,262,860) and 306US550 (Patent 1,930,987) came slightly after the middle of the overall period, consistent with them linking co-cited decisions in Figure 5.1. However, both have zero indegree *in this island*. There are two decisions (lower right) receiving no citations in this line island. Clearly, their co-cited levels came from other SC decisions. We emphasize citations within the Supreme Court citation network differ from citations among patents in the patent citation network. However, decisions made by the Supreme Court have an impact on the patent citation network, albeit an indirect one by voiding patents and redefining patent terms.

When the citation network within this island is drawn for the technological areas, there are no obvious patterns in the sense of the citations being within technological areas. Chemical decisions are scattered across the whole time interval with slightly heavier earlier and later concentrations. The one chemical decision nearest to the middle of the period is not linked with either of the two concentrations. Once mechanical decisions start to appear, in 1890, they are also scattered through time. It appears that *general* legal principles regarding patents are applied in reaching decisions about upholding or voiding patents. The specific technological categories of the inventions appear to have little direct relevance for these SC decisions.

5.3   The 1976–2006 patent data

These citation data are described fully in Appendix A.2. Only directly relevant data are mentioned here. We continue to follow Hall et al.'s labeling scheme for six broad areas of technology: chemical (Chem); computers and communications (C&C); drugs and medical (D&M); electrical and electronic (E&E); mechanical (Mech) and others. Hall et al. argue that this classification must be used with great care. Our intent is to use the patent citation network described in Appendix A.2 to map *broad* trends within and *especially between* these six categories. Given Hall et al.'s (2001) use for the same purpose, this categorization is sufficient for our purposes also. Temporal instabilities of the detailed classifications of specific patents will have little or no effect on these broad classes.

For this chapter the following data items were used:

- patent numbers

- application years (when inventors applied for patents)

- granting years (when patents are granted[16])

---

[16] The USPTO review process takes on average about two years (the variance is about a year). About 95% of all patents are granted within three years of the application year (Hall et al. (2001): Table 1).

- countries (of the first listed inventors on patents)

- broad technological categories (as described in section 5.1)

Additional constructed variables (from Hall et al.'s (2001) 1969–1999 data) include:

- measures of *generality*, g_i: For a patent, i, $g_i = 1 - \sum_j^n s_{ij}^2$, where s_{ij} denotes the percentage of received citations belonging to patent class j out of n patent classes. Patents cited by patents of more fields have higher generality measures. Such patents are thought to have greater impacts on innovation in different fields.

- measures of *heterogeneity*,[17] h_i: This is defined in the same fashion as generality with s_{ij} denoting the percentage of citations from patent i to patents in patent class j.

- mean backwards citation lags: This lag for a patent is the time difference between its grant (or application) year and the citations it receives.[18]

- percentages of self-citation in patents (by patent applicants citing their own prior patents)[19]

- citations (the number of outgoing citations in patents)[20].

5.4 Structural variables through time

Given the size of this citation network (containing 3,210,774 patents), we adopt several strategies for working with 'parts' of it. One reasonable and practical way of obtaining a first impression of this network and, more importantly, how it changes through time, comes from 'shrinking' it according specific variables or data characteristics. (See Section 2.6 for a discussion of extracting subgroups.) The shrunken (easier to manage) network provides clear images of the network through time.

Although Chapters 2 and 3 contain a full discussion of methods used across different chapters, we discuss patent-specific methodological items here. Shrinking networks is discussed further in Section 5.4.1 followed by an account of shrinking patent networks in Section 5.4.2. Structural variables are discussed in Section 5.4.3.

5.4.1 Temporally specific networks

Using *time slices* (time intervals) is a simple way of exploiting time by restricting attention to shorter periods and comparing results across time slices. The initial network is sliced into

---

[17] This is termed 'originality' in the data sources. We prefer to use heterogeneity because the measure takes high values when patents draw on multiple technological areas. A patent can be original even if it does not draw on multiple areas.

[18] For example, Hall et al. (2001) note that patents granted in 2000 will receive, most likely, about half of their overall number of citations within a decade. Within another decade this will increase by 75% implying that they will still be receiving some citations decades later.

[19] The mean percentage of self citation was between 11% and 13.6%. For the 1976–2006 data used here, the corresponding figures are 13.1% and 14.1%.

[20] This was about five citations per patent in 1975 rising to over ten in 1999. This seems to be due to changing USPTO practices (including greater computerization) and increasing numbers of patents being granted over time.

multiple networks each with a specified time interval. These networks are called *temporally specific networks*. To define and observe temporally specific networks we use a *sliding time window* designed to study temporal networks (Doreian 1979–1980). The 'width' of the window is specified by a time slice shorter than the whole period for the temporal network. The window is started at an initial time point and is slid across the network to a terminal time point. As it slides across a network, it defines *overlapping* time slices. Having successive time slices share common data is useful for mapping incremental changes in network structure over time. For our purposes here, this is more useful than comparing the structures of the initial and final periods: while doing this, it shows the *evolution* between the initial and final periods.

For using sliding windows, three decisions are needed about: 1) when to start; 2) the width of the window; and 3) when to end. The choices rest on the data across which the window slides. Every dataset comes with specific features driving these decisions. Our data are limited to the years between 1976 and 2006. The final year for these data (2006) is unproblematic. Making the first two decisions required closer attention to the data.

Identifying the starting year

The starting year for these data complicates the first decision. Patents issued before 1976 are not in the dataset. Patents granted in 1976, or shortly thereafter, will have cited patents before 1976. As a result, if the constructed citation network was taken at face value, it would show no outgoing citations to older patents from before 1976. In short, the citation data are censored on the left. To deal with this problem, the outdegree (the number of citations from a patent) for each grant year was inspected. The results are shown in Figure 5.4 with plots of the percentages of low outdegrees for every grant year.

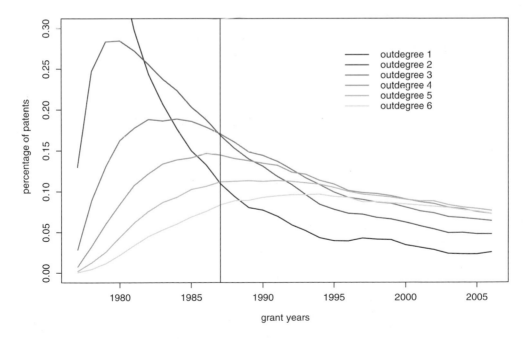

Figure 5.4   Percentage of low outdegrees (citations from a patent) by granting years.

Hall et al. (2001) (Figure 6.7) show the average number of citations made in the 1980s was about 6 (rising to 10 in 1999). Figure 5.4 shows plots of the percentage of outdegrees of 1–6 across the period covered by our data. The lines were obtained simply by connecting the values for different years.[21] For patents granted in 1976, only outdegrees of 1–4 are present. The trajectory for patent outdegree of 1 starts well above the 30% shown in the figure. It declines dramatically across all years. The trajectories for outdegrees 2–4 climb and peak well before 1990 and decline thereafter. The trajectories for outdegrees of 5 and 6 climb for a longer period but they also decline subsequently. Towards the end of the period covered, all trajectories are dropping. These trajectories suggest that 1987, marked by the vertical line in Figure 5.4, is an appropriate initial time point for a sliding window.

Defining the width of the sliding window

Selecting time intervals for the sliding window depends not only on the length of the overall time period of the citation network. It depends also on the temporal nature of the data. For these patent data, information on the time difference between granting years of patents and the citations they made is critical. This difference is the *backward citation lag*.[22] Its distribution (for all the patents) is shown in Figure 5.5. Only references to older patents were taken into account (with anomalies and possible data errors excluded). The median backward citation lag is 7 years with the mean lag being 8.46: 8 years was taken as an appropriate time interval for the sliding window for this network.[23]

5.4.2 Shrinking specific patent citation networks

Using a temporal network with time-indexed vertices and lines, the relevant Pajek commands for extracting time slices, shrinking an extracted network, and repeating the command for each network in a time slice[24] are:

```
Network/Transform/Generate in Time/Interval [from to]
Operations/Network+Partition/Shrink Network/ Partition []
Macro/Repeat Last Command (F10)
```

Using 1987 as the start year, a width of 8 years for the sliding window, and 2006 as the final year, allowed the construction of 13 temporally specific networks for each eight year period using the width of the sliding window. These are the networks we shrank by using the six broad technological categories listed in Section 5.1. Each patent belongs to one of these six categories. The vertices representing patents *in* a category were joined to form a single vertex representing the technological category. This was done for every category. For each pair of such technologically defined vertices, all of the citation ties from vertices (patents) in one category to patents in the second category were joined to form a single-valued arc

---

[21] Regression methods were used. The residuals for all lines were small and no serious leverage points (using Cook's distance) were identified.

[22] Hall et al. (2001) (p. 18) note: 'The backward citation lag compounds of the time differences between the grant year of the citing patent and that of the cited patents.'

[23] Kejžar (2005) suggests citations younger than 8 years (about 64%) represent references to *concurrent research and development* while the rest mostly belong to older *well-known methods*.

[24] In the pop-up window, neither the network nor the partition is fixed across all repetitions.

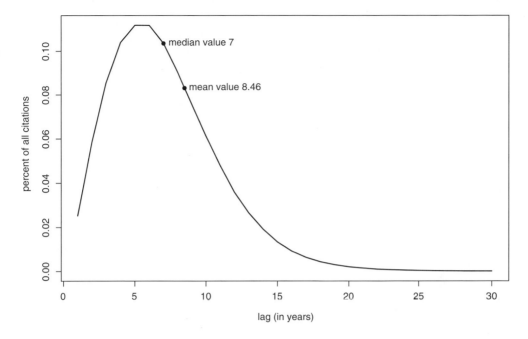

Figure 5.5 Backward citation lag in years.

from the first technological vertex to the second. This was done for all pairs of technological vertices. The result, for each time slice, was a network with six[25] vertices and reciprocal valued arcs, of different strengths, between each pair of technological vertices.

5.4.3 Structural properties

For each temporally specific network of a time slice and the network with technologically defined vertices created from it, a wide range of measures can be computed. These include transitivity, degree centrality,[26] hub and authority scores and page ranks. However, care is needed in distinguishing the *whole network*, each *temporally specific network* with techno- logical vertices, and each *technologically defined network*. The above listed measures make most direct sense for the whole network and the temporally specific networks when the technological network has a small number of vertices. Some measures are specific to the network for which they are computed. For example, the *vertex size* of each shrunken vertex (one for each characteristic) is the number of vertices (patents) that went into it.[27] Degrees in the original and temporally specific networks are the outdegree and indegree values. Degree scores in the technologically defined network are, in essence, the sums of the degree scores of the vertices into it. Computation of the other centrality measures in the technologically

---

[25] If the two-digit technological codes were used, there would be 36 vertices, and for the three-digit codes there would be more than 400.

[26] There is an unfortunate tendency to compute a wide range of centrality measures without considering which is most appropriate for a specific network. In this context, neither closeness nor betweenness centrality has value.

[27] Trivially, in the whole network this value would be 1 and therefore uninteresting.

defined (shrunken) networks are performed for the valued network and cannot be obtained directly from measures in the corresponding larger network. Hub and authority scores can be computed for all three types of networks. Other measures – for example, shortest path length, network clustering coefficient,s and diameter – can be computed only for the whole network and temporally specific networks. We consider some of these properties for the networks defined by the six main technological categories in Section 5.5.

5.5   Some patterns of technological development

Using the sliding window approach outlined in Section 5.4, 13 temporally specific networks were constructed. The earliest one includes patents from 1987–1994. The latest has patents from 1999–2006. The sliding window was moved one year from one temporally specific network to create the next temporally specific network. Using the structural property of broad technological areas meant that each network has six vertices. The lines between them represent the citations between technological categories. The newly constructed variable, vertex size, was saved as a vertex property. Four of these technologically defined networks are shown in Figure 5.7.

However, before visually examining details of these short networks, it is useful to consider the growth in the number of patents by technological categories over time. These are shown in Figure 5.6 for 1987–1999 (summed over the eight-year time intervals defined in Section 5.4.1). Computers and communications (C&C) had the second lowest number of patents in 1987. However, it had the steepest rise in the number of its patents, finishing with the highest number of patents in 1999. Electrical and electronics (E&E) started at the third lowest level in 1987, had the second largest increase in patents and finished with the second

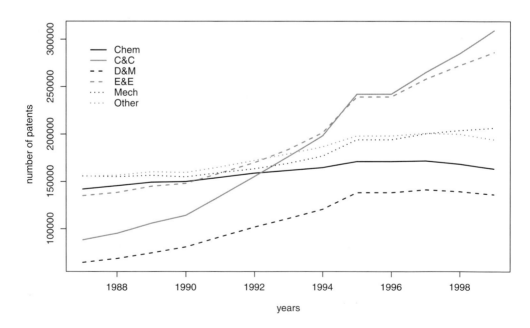

Figure 5.6   Growth in the number of patents by categories from 1987 to 1999.

highest number of patents. Together, these two categories dominated both the growth of patents issued during this period and the absolute levels of granted patents towards the end of it. Undoubtably, this was driven by the development of the computer industry, its rapid emergence, and its derivatives.

While drugs and medical (D&M) started at the lowest volume of patents and remained there, it had the third steepest rise through 1995 before leveling off. The sharp early rise was likely due, in part, to an increasing use of medications to handle a variety of problems, especially psychological problems. Another reason could be the aging population of the USA; disproportionate amounts of medical costs in the USA are incurred in treating the aged in efforts to improve the quality and length of their lives. See, for example, Berk and Monheit (1992, 2001); Conwell and Cohen (2005); Yu and Ezzati-Rice (2005). These two goals need not be compatible. The treatments include greater uses of medications and surgery. There was also the rise of sports medicine. Both mechanical and others had the highest levels of patents in 1987 but grew only slightly before finishing third and fourth in total patent volume in 1999. Chemical (Chem), one of the drivers of technological change following WWII, barely grew through 1995 before leveling off and then declining slightly.

Figure 5.7 shows four networks with citation volumes between technological areas, each defined by the eight-year width of the sliding window. The first covers 1987–1994, the second 1991–1998, the third 1995–2002, and the final one 1999–2006. The changes in vertex sizes across the four panels in Figure 5.7 are fully consistent with the trends shown in Figure 5.6. The number of granted patents increased throughout this period. This accelerated for C&C, E&E, and D&M (except at the end). More interesting here are changes in the flows between technological areas, as shown by the thickness of the arcs. They represent the total number of citations from one technological category to another. Of course, citation volumes are highest within technological areas, sometimes by several orders of magnitude. These could have been included as loops. However, given our focus on citation flows *between* technological areas, loops were removed from Figure 5.7.

The most noticeable features of Figure 5.7 are the reciprocal arcs between C&C and E&E. They were the largest flows in 1987–1994 and roughly equal in volume. Both grow in every time slice. These changes were particularly dramatic in the last two periods. After 1987–1994, the citation volume from C&C to E&E exceeded the reverse flow. For 1999–2006 they are, by far the dominant flows as they were for 1995–2002, albeit to a lesser extent. Although most of the citation flows between technological areas increase across the four periods, further evidence of the 'patent citation dominance' of C&C and E&E comes from considering the flows between them and the other technological areas.

While mechanical remained stable with regard to patent citation volumes, its reciprocal flows with both C&C and E&E increased across all periods. In 1999–2006, these were the second largest reciprocated flows. Indeed, the flows between these three technological areas grew to be the largest in 1999–2006. The citation volumes between mechanical and others were the second largest in 1987–1994 and grew thereafter. It appears that patents in mechanical drew more heavily on patents in others in the last two periods than the reverse flow. The reciprocal flows between chemical and others grew. Also of interest is the growth in the reciprocal flows between chemical and D&M through the first three periods before diminishing. In terms of patents, this suggests that the initial growth for D&M shown in Figure 5.6 may be driven more by the increased use of medications.

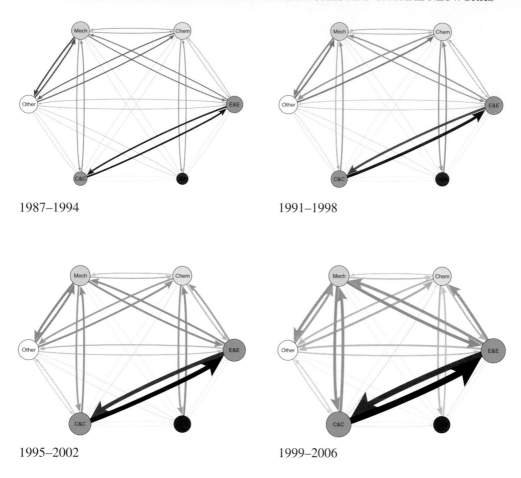

1987–1994

1991–1998

1995–2002

1999–2006

Figure 5.7   Shrunken networks in the four year time windows for citation flows.
Loops were removed. Given the large volumes and vertex measures, line values were multiplied by 0.0002 and vertex values by 0.002

5.5.1   Structural properties of temporally specific networks

Hub and authority scores (Kleinberg, 1998) are particularly suited for studying the patent networks considered here. Two weights were computed for each vertex, one for hubs and one for authorities. A good *hub* is a vertex in a network with many good authorities pointing to it and a good *authority* is one pointing to many good hubs. Let $\mathcal{N} = (\mathcal{V}, \mathcal{L}, w)$ denote a directed network with weights on lines. $\mathcal{V}$ represents a set of vertices, $\mathcal{L}$ a set of lines, and $w : \mathcal{L} \to \mathbb{R}$. For every vertex $v \in \mathcal{V}$, two weights, h_v and a_v, were constructed. The hub score, h_v, for v is given by

$$h_v = \sum_{u \,:\, (u,v) \in L} a_u.$$

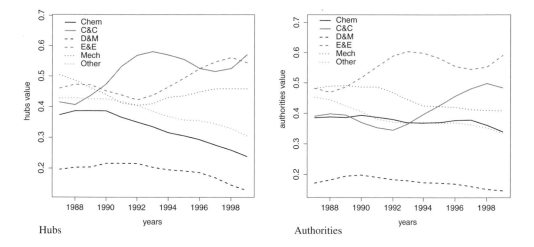

Figure 5.8   Hubs and authorities weights for categories over time.

The authority score, a_v, for v is obtained from

$$a_v = \sum_{u:(v,u)\in L} h_u.$$

An iterative method for establishing h_v and a_v computes the principal eigenvectors of matrices $\mathbf{W}^I\mathbf{W}$ and $\mathbf{W}\mathbf{W}^T$ where $\mathbf{W}$ is the adjacency matrix of the network. (See Section 2.11 for more details on hub and authority scores.)

Figure 5.8 shows the temporal distribution of the hub and authority scores for the six technological areas. Hub scores are on the left with authority scores on the right. The order of hub scores at the first time period is mechanical, E&E, others, C&C, chemistry, and D&M. Thereafter, the temporal trajectories differ dramatically. Both C&C and E&E have reversed fluctuating trajectories. However, both have the highest hub scores among the technological categories in 1999. Hub scores for mechanical drop until 1992 before climbing to finish with the third highest value in 1999. The other three technological areas have hub scores declining throughout this period and finish at the bottom three levels. D&M ranks the lowest throughout, consistent with being less well linked to the other technological areas.

Viewing the right panel of 5.8 shows only C&C and E&E as having higher authority scores in 1999 compared to 1987. In the main, all authority scores for the other technological areas decline. The fluctuations for C&C and E&E are reversed in this panel. Further, they are reversed compared to the changes in their hub scores. These plots appear to supplement the results shown in Figure 5.7. Overall, the important hubs are mechanical, C&C, and E&E. Both C&C and E&E became especially influential over the last quarter of this interval. The other three technological areas lost some of their importance. The picture is similar for authority scores. However, E&E was by far in the best position in 1999, while C&C gained advantage towards the end of this period compared to mechanical, chemical and others. Kejžar (2005) suggests that authorities represent technological categories playing an important role in *setting the base* for the other areas. Large hub values correspond to categories *drawing upon many other* technological areas. This clarifies the initial place of E&E and the late rise of C&C. These results help to account for the huge boom for C&C in the 1990s due, most likely, to the development of the Internet. D&M depends primarily on chemical in all panels of Figure 5.7, especially for 1999–2006. As a result, both its hub and authority scores are small throughout this period.

The flows presented in Figure 5.7 used the actual patent citation flow data. However, Figure 5.6 shows the six technological categories differing in terms of the numbers of patents they contain. This raises a simple question: how much do patents from their technological areas cite patents outside their areas *relative to* their sizes? To respond to this query we used *relative citation* frequencies. These line weights $w_{l(rel)}$ were calculated from

$$w_{l(rel)} = \frac{\#\text{citations on } l}{\#\text{patents of } u} = \frac{w_l}{p_u}.$$

For a line $l \equiv (u, v) \in \mathcal{L}$ where $u, v \in \mathcal{V}$, p_u denotes the property of a vertex u. To explore the role of these relative citation weights, we did not simply repeat the analysis shown in Figure 5.7. From Figure 5.6, the start of a steep increase in the number of patents for categories begins in 1996. It made more sense to focus attention on the period after 1996. We note also, following an inspection of networks prior to 1996, that the difference between using simple and relative weights was negligible. Paying close attention to the period 1996–2006 provides an opportunity for a more detailed use of the sliding window technique.

Figure 5.9 shows the between-technological areas networks for 1996–2003, 1997–2004, 1998–2005, and 1999–2006. As before, the sliding window has a width of eight years: each pair of successive networks has seven years in common. This permits more short-term comparisons than were made in Figure 5.7. In the following, we make two comparisons. One features the four networks shown in Figure 5.9 while the other focuses on the 1999-2006 panels of Figure 5.7 and Figure 5.9.

The dominant features in Figure 5.9 are the reciprocated relative flows ($w_{l(rel)}$) between C&C and E&E, as was the case for the actual flows in Figure 5.7, but with attention on a narrower time frame. Little changes from 1996–2003 to 1997–2004. For both (overlapping) periods the relative patent citation flow from C&C to E&E is higher than the reciprocal flow. This is reversed for 1998–2005; the change was dramatic. The sizes of the relative flows between C&C and E&E for 1996–2006 reverts to the pattern of 1997–2004, suggesting that something important occurred between 1997–2004 and 1999–2006. This impression is reinforced by the heavier citation flow from E&E to mechanical for 1998–2005, given that the flows between these two technological areas were close to parity for all other periods.

The third highest relative citation flow for both 1996–2003 and 1997–2004 is from D&M to chemical. The reciprocal flow is far weaker. However, the citation flow from D&M to chemical drops dramatically in 1998–2005. This citation flow strengthened for 1999–2006 but remained well below the levels of the first two networks in Figure 5.9. The reciprocal flow from chemical to D&M changed little across the four periods. Two other changes in the pattern of the relative citation flows in 1998–2005 are noteworthy. The relative flow from E&E to chemical is largest in this period and is the fourth highest. While considerably smaller, the flow from E&E to others is also highest in this period.

Changes in the structure of patent citation networks can occur for several reasons including: 1) shifts in the volumes of patents across technological areas; 2) unusually large increases in the number of citations from patents in one technological area to patents in another area in a time interval; and 3) changes in USPTO procedures for processing patents. Looking at the total citation flows, as in Figure 5.7, focuses attention on citation volumes alone. When

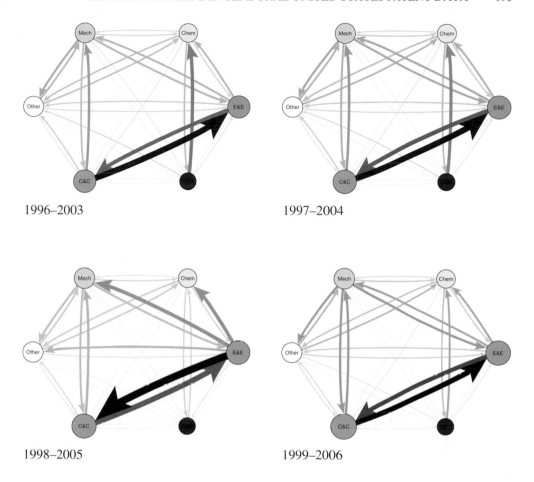

1996–2003   1997–2004

1998–2005   1999–2006

Figure 5.9   Shrunken networks in the four time windows for relative citation flows.
Loops were removed. Given the sizes of the flows, line values were multiplied by 0.05 for a more readable figure.

relative citation flows are used, some control of differential numbers of patents by technological areas is imposed. The overall features discerned from Figure 5.7 are not contradicted by
the results in Figure 5.9. However, there were subtle shifts in emphases when relative citation
weights were used. We suggest using both types of patent citation flows for assessing this
kind of temporal change to obtain a more nuanced assessment of structural changes.

5.6 Important subnetworks

For another approach to discerning the structure of a network, examining some of its *important
parts* is useful. Such substructures give alternative impressions of the structure of the entire
network and can be combined under the divide and conquer strategy outlined in Chapter 2.

Patent data have two time stamps for each patent. One is the year when the application was
made. The second is the year when it was granted. Necessarily, the application year is closer
to when an invention was made. As such, it is one marker for defining the age of a patent.
However, patents (and the patented inventions) become public *only* when they are granted.

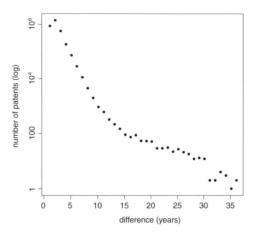

Figure 5.10   The distribution of difference between application and grant years for patents.

A patent's grant year is the second marker for determining an invention's age.[28] As patents cannot be cited until they are in the public realm, granting years are more important for patent citation analyses. It is useful to examine differences between application and granting years for patents. Hall et al. (2001) determined that the average difference was about two years (with a standard deviation of about one year). For the 1976–2006 dataset used here, the average difference is 2.14 years (with a standard deviation of 1.16 years). The distribution of the application–grant year difference is plotted in Figure 5.10. As the y-axis uses a logarithmic scale, there were only several thousands of patents having year differences larger than 10. Even so, there is considerable variation in the differences between granting and application years due solely to variations in USPTO review process. One contributing reason for these differences is the number of citations in a patent because each has to be checked by a patent examiner.

   To diminish impacts on our results of some patents being granted more quickly than others, we restricted attention to patents granted in 1976–2004. The resulting network still contains 2,836,627 patents. It is weakly connected with neither loops nor non-trivial strong components.

5.6.1   Line islands

Line islands, introduced in Section 2.9, were identified in the network $\mathcal{N} = (\mathcal{V}, \mathcal{L}, w)$ where the weights, w, are the normalized SPC counts (Section 3.2). These weights capture the *relative importance* of each citation arc in the network (Kejžar 2005). Obtaining line islands is done by specifying the smallest island size, k, and the largest island size, K, where $k \geq 2$ and $K \leq n$. In an exploratory fashion, K was varied from 500 to 5500 in increments of 500. The object was to identify 1) line islands with large *minimal* internal line weights, and 2) line islands with the *largest* number of vertices. Each line island has one or more *themes*. If a line island is primarily within one technological area, this area identifies its theme. Alternatively,

---

[28] Recall the important requirement of there being no prior public knowledge when considering a patent application.

Table 5.1 Sizes of the largest islands for selected $[k, K]$ values.

[k,K]	A	B	C	D	E	F	G	H	I	J
[2, 500]	500	500	499	271	188	167	79	71	65	54
[2, 1000]	**799**	**761**	730	271	188	167	79	71	65	54
[2, 1500]	<u>799</u>	<u>761</u>	730	271	188	167	79	71	65	54
[2, 2000]	**2000**		730	271	188	167	79	71	65	54
[2, 2500]	**2498**		730	271	188	167	79	71	65	54
[2, 3000]	**2943**		730	271	<u>188</u>	167	79	71	65	54
[2, 3500]	**3500**		730	271		167	79	71	65	54
...	...									
[2, 5000]	**4560**		<u>730</u>	271		167	79	71	65	54
[2, 5500]	**5495**			271		167	79	71	65	54
...	...									

Changes are in bold. Merges of islands are underlined.

if patents in an island come from different technological areas, the island has multiple themes suggesting that it is *diffused* and that many *knowledge spillovers* (between technological areas) occurred.

Table 5.1 shows the sizes of the largest ten line islands obtained by using the values of k and K listed in the left column. The line islands are labeled A though J. For the pair ($k = 2$, $K = 500$), ten islands were identified. When K was raised to 1000, islands A, B, and C were all larger (with additional patents joining those in the islands for $K = 500$) while the remaining islands were unchanged in size.[29] Nothing changed for $K = 1500$. When $K = 2000$, island B merged into island A. When K was raised (in increments of 500) to 3500, island A continued to expand. For $K = 3500$, island E merged into island A. Finally, for $K = 5000$, island C merged into island A. The result, as shown in the bottom row of Table 5.1, was seven identified line islands. One island is very large. The rest are less than twenty times smaller. Each can be studied for their themes and potential knowledge spillovers.

In selecting islands to study, choices are made. However, they need not be made blindly. The description for Table 5.1 has an agglomorative spirit in the cluster analytic sense. An alternative language using the term *cut size* is useful. Considering Table 5.1, for a cut size between 5000 and 5500, island C is cut from island A with A becoming smaller. For a cut size between 3000 and 3500, island E is cut from island A. Further, for a cut size between 1500 and 2000, island B is cut from A. Using lower cut sizes leads to identifying more line islands. This can be studied to inform choices of which line islands are selected for further attention. We use the term *maximum cut size* for the largest cut size within the kinds of intervals used in Table 5.1.

The number of identified islands is plotted against maximum cut sizes in the left panel of Figure 5.11. As expected, the number of islands drops as the maximum cut sizes increase: islands are merged. Most of the merges occur for small islands with maximum cut values well below those shown in Table 5.1. The right panel Figure 5.11 shows the corresponding plot of line island weights against maximum cut values. Not surprisingly, these weights decline

---

[29] Except for the merging of islands, these islands were unchanged as K increased.

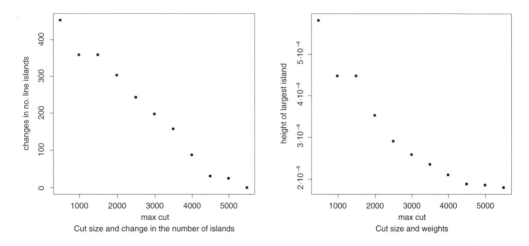

Figure 5.11   Plots of the changes in number of line islands and weights against maximum cut sizes.

For a max cut size [2, 5500] the number of changes is 0. The distribution of weights is very right skewed with a median value of $6 \cdot 10^{-7}$. The values have been normalized with the small values due to the large size of this networks

as the maximum cut values increase: smaller line weights are a part of having larger line islands.[30]

We focus on island A, the largest island. To examine how this island changes when other line islands are merged with it, percentages of technological areas in this island were computed as K was raised. These are shown in Table 5.2. When $(k, K) = (2, 500)$, island A largely featured patents from C&C. When K increases to 1500, the percentage of C&C patents in the island increases slightly. For $(k, K) = (2, 2000)$, when island B is merged into this island, the percentage of C&C patents drops sharply to 64.7% while the proportion of chemical patents increases to 20.3%. Clearly, the theme of island B was predominantly chemical. Even so, we note that the percentage of the other technological areas in island A changed also, albeit by small amounts.

As K increases, the percentages for the technological areas change little in this line island. When $(k, K) = (2, 3500)$ is reached, the percentage of patents from mechanical jumps. Island E merged with island A when $(k, K) = (2, 3000)$, suggesting it had mechanical as a dominant theme. Similarly, the percentage of patents from E&E jumps for $(k, K) = (2, 5500)$ shortly after island C merged with island A. Island C appears to have primarily E&E as its theme. Overall, smaller islands are more homogeneous. As the maximum cut sizes are increased, islands become more heterogeneous. While this is not surprising, it permits an examination of potential knowledge spillovers between technological areas.

5.6.2   Line islands with patents tagged by keywords

Line islands are discerned simply by considering the weights of the arcs. In Section 5.6.1, this idea was extended to include the technological areas into which patents have been

---

[30] The largest normalized weights for line islands range from $6 \cdot 10^{-4}$ to $3 \cdot 10^{-3}$.

Table 5.2 Percentages of patents in different categories for the largest island (island A) for selected $[k, K]$ values.

[k,K]	Chem	C&C	D&M	E&E	Mech	Other
[2, 500]		**90.80**		3.00	4.80	1.40
[2, 1000]		**92.37**		2.75	3.63	1.25
[2, 1500]		**92.37**		2.75	3.63	1.25
[2, 2000]	20.30	**64.70**	0.20	6.70	5.40	2.70
[2, 2500]	17.77	**67.49**	0.16	6.69	5.08	2.80
[2, 3000]	15.97	**68.23**	0.17	6.66	5.91	3.02
[2, 3500]	14.23	**65.06**	0.34	6.97	10.74	2.63
. . .	. . .					
[2, 5000]	12.04	**65.77**	0.29	7.61	11.71	2.57
[2, 5500]	10.35	**57.05**	0.73	19.60	9.99	2.26
. . .						

About 0.03% of patents have no category. They were omitted. The largest percentage in each line island in bold. The largest changes are underlined. The full names of the technological areas are in Appendix A and in Section 5.1

located. However, these six broad technological areas are heterogeneous regarding the details of the subclasses for patents. An element of greater refinement regarding parts of identified line islands is available through using additional information in the patents. Trivially, patents have titles. Examining patent titles permits the construction of patent descriptions in terms of *keywords* extracted from titles. These titles were obtained from the searchable USPTO patent website. The words from these titles were delimited by characters differing from alphanumeric characters and lemmatized using the MontyLingua Python library with the stopwords removed. In this way, 132,829 keywords were identified and assessed for the 1976–2004 network. This was supplemented by using also the titles in patents.

We illustrate this idea with a seemingly trivial example. Before examining keywords, it is possible to work with patent titles. Figure 5.12 shows a tiny line island having the largest line weights subject to the constraint of having at least three vertices with a maximum cut value of 5500. While patent titles provide a sense of island content, a far more useful understanding comes from using their keywords. The titles in Figure 5.12 were used to obtain patent descriptions with *keywords*. The resulting two-mode network is shown in Figure 5.13. Reading clockwise from the top left: Patent 6567850 has eight keywords; Patent 6615234 has five; Patent 6806976 has eight; and Patent 6748471 has ten. Of greater interest is the *sharing* of keywords by patents. For example, all four patents have the keyword 'method', the only keyword shared by Patent 6615234 with other patents. Patents 6567850 and 6748471 share the keywords 'method' and 'request' while Patents 6748471 and 6806976 share four keywords.

A slightly larger example is shown in Figure 5.14 obtained by requiring at least ten vertices, again with a maximum cut size of 5500. Shown on the left is the line island of patents with their titles. Arcs between patents have different weights. The right panel of Figure 5.14 shows the corresponding two-mode network of patents and keywords. Lines

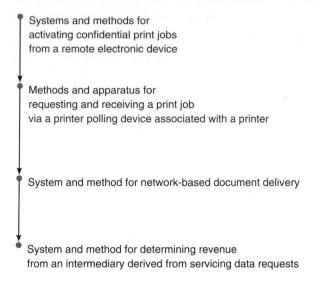

Figure 5.12   A line island with heavy weights and patent titles.

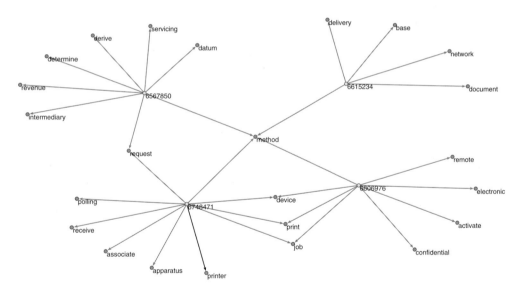

Figure 5.13   The line island from Figure 5.12 with patents and keywords.

crossing indicate patents sharing keywords.[31] The theme of this line island is computers and communications. Most of the patents in this island are dated between 1980 and 1984, a time of rapid growth of low-cost personal computers reaching the market.

---

[31] The darker lines from Patent 4244032 are because it has two words twice in its title. Such duplicates can be removed if needed.

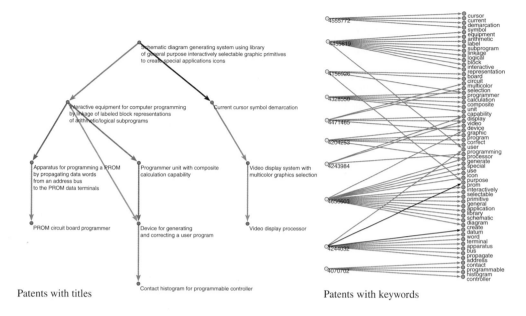

Figure 5.14 A line island ten vertices and heavy weights.

Figure 5.15 represents the patent citation for line island D (from Table 5.1). This is the second largest at the line island for $(k, K) = (2, 5500)$) with 271 patents. There are two more densely connected subnetworks in this island with a small number of arcs linking them. However, the strength of connections between the subnetworks is higher than the arcs from the island to the rest of the patent network.

One issue arises when using the line islands algorithm: usually, many islands are identified, especially for large networks. Once such islands have been established for patent networks, one important task is identifying their themes. This was illustrated with the small examples shown in Figure 5.13 and on the right of Figure 5.14 where the number of keywords was small. However, doing this can be tedious when the islands are large.

We examine briefly island F from Table 5.1 to demonstrate a way of identifying island themes in a relatively quick way. This island is the third largest with 167 patents. It does not merge with island A when $K = 5500$. Given a two-mode network of patents and keyword (patents × keywords), a one-mode keyword network (keywords × keywords) can be constructed. Two keywords are directly linked if there exists one (or more) patents using both in their titles. The keyword network is valued with the values counting the number of times *pairs* of keywords are used together in patent titles. For island F, the resulting network is shown in Figure 5.16.

Examining keywords in Figure 5.16 reveals a mixture of keywords for technologically specific areas as well as more generic keywords. The minimum arc value for this network was set at three (jointly used in at least three patent titles). There are two keywords with the highest degree values. Catheter is a specific term while method is generic: both are used together very frequently. Keywords directly linked with catheter with high counts include angioplasty, dilation, balloon, and vascular. This suggests that the primary theme of island F is D&M. More specifically, this theme is about medical solutions to cardiovascular problems. Some

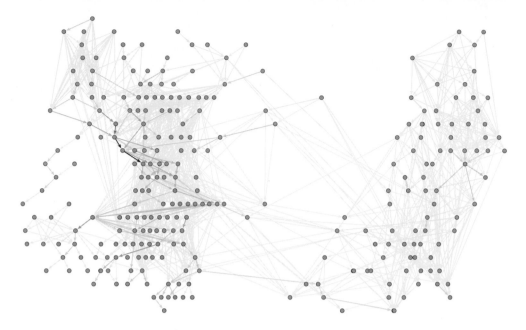

Figure 5.15   The second largest patent line island.

The maximum cut size is 5500.

generic keywords directly linked to catheter include treatment, sleeve, coaxial, diameter, wire, perfusion, and steerable. All are directly related to methods used when dealing with cardiovascular problems. Assembly, another generic keyword, is directly linked to dilate and vascular. Both fit with dealing with heart problems.

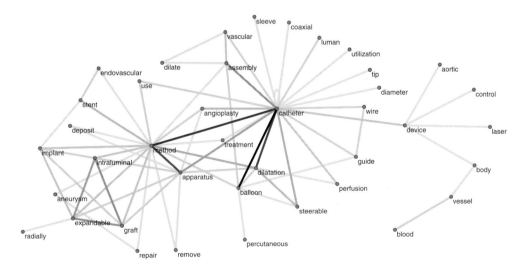

Figure 5.16   Keyword network of line weight three or higher for the third largest line island ($[k, K] = [2,5500]$).

Table 5.3 Size of the largest vertex islands for values of $[k, K]$ boundaries.

[k,K]	a	b	c	d	e	f	g	h	i	j
[2, 500]	**495**	483	210	154	114	101	93	**85**	84	78
[2, 1000]	<u>979</u>	948	210	154	114	101	93		84	78
[2, 1500]	1079	948	210	154	114	101	93		84	78
[2, 2000]	**1079**	**948**	210	154	114	101	93		84	78
[2, 2500]	<u>2438</u>		210	154	114	101	93		84	78
[2, 3000]	**2928**		210	154	114	**101**	93		84	78
[2, 3500]	<u>3435</u>		210	154	114		93		84	78
. . .	. . .									
[2, 5000]	4987		210	154	114		93		84	78
[2, 5500]	5485		210	154	114		93		84	78
. . .	. . .									

Changes are marked in bold, merging islands are underlined.

There are many generic keywords in Figure 5.16. At face value, they could be viewed as marking different themes. Directly linked to both catheter and method is apparatus (as is angioplasty and treatment). This makes it clear that method is less generic than it appears because the method is about treating heart problems. Catheter is directly linked to device. Trivially, it is a device. But device is linked directly to aortic and body. In turn, body is linked to blood through vessel. The generic terms are, in this island, contextually specific. Given that island F never merged with island A, it is highly unlikely that the theme of treating cardiovascular problems using catheters and related techniques appears[32] in island A.

5.6.3 Vertex islands

Another way of extracting subnetworks is to identify vertex islands (see Section 2.9). The network from which vertex islands were extracted is denoted by $\mathcal{N} = (\mathcal{V}, \mathcal{L}, p)$, where the vertex property, p, measures indegrees (citations received from other patents). Vertex islands with large minimal properties correspond to the subnetworks whose vertices have important citation information. As was the case for line islands, maximum cut values for vertex islands can be varied. Table 5.3 shows sizes of the largest vertex islands for different $[k, K]$ values. Again changes are in bold with merging of islands underlined. (We use lowercase, Roman letters to label these islands to distinguish them from the islands in Table 5.1.) As K increases, vertex islands merge: the number of vertex islands gets smaller (similarly to Figure 5.11 left for the line islands).

Figure 5.17 shows the vertex island, island e, the fourth largest island in the bottom row of Table 5.3 with 114 vertices (patents). Vertex sizes represent the indegrees from the original network. Although the underlying logic of line and vertex islands are the same with regard to identifying subnetworks, there is an important difference between them. With line islands, attention is on the values of arcs in the island, as in Figure 5.15, while attention is on vertex sizes in vertex islands, as in Figure 5.17. Each provides useful information.

---

[32] From Table 5.2, a little over 0.7% of patents in island A are from D&M. While some D&M patents are in this island, it will not be a theme of the island.

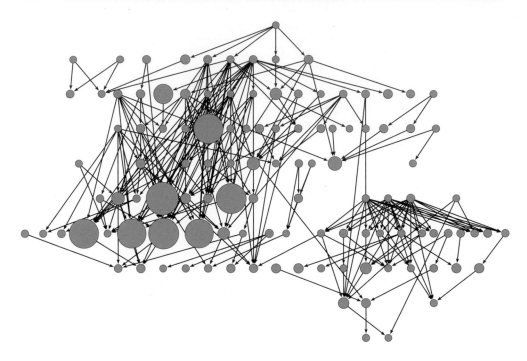

Figure 5.17   Vertex island e from Table 5.3.
Vertex sizes represents indegrees from the original network.

However, when using indegrees as the property for producing vertex islands in patent citation networks, additional care is needed. In island e, the indegrees are large, varying between 80 and 1678.[33] We know that the means and medians for the number of citations differ across scientific disciplines (Hall et al., 2001). Among technological areas, the D&M category has by far the lowest average number of citations. As pointed in the subsection 5.4.1, citation volumes grow with the time of patent granting years. For vertex islands, however, vertex properties are important, with line values secondary. Considering substantive issues and potential methodological problems, for patent citation networks, line islands seem much more useful. We do not consider vertex islands further.

5.7   Citation patterns

Every vertex in a patent citation network has its publication year. Temporal *received* citation distributions can be constructed for each vertex (patent). As noted in Section 5.6.3, the number of citations received by a patent carries information about its importance, its usefulness, and its value for other inventions for which subsequent patents are granted. These citation counts can change as new patents are granted and added to the patent citation network. Examining the number of citations received by patents *through* time provides additional information about patent contents. Some patents are cited more when they are younger compared to other patents. Others are cited when they are older. The received citations for some patents grow

---

[33] This is the largest span for the islands shown in Table 5.3.

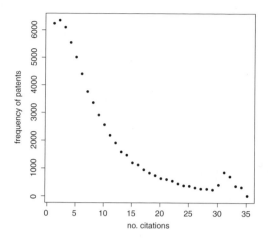

Figure 5.18 The indegree citation distribution for patents from 1976 with at least one citation.

over time while those received by others decline. Of course, many other temporal patterns of received citations are possible. An interesting feature of all patents is *when* they are used after being noticed. Research is a very dynamic activity. As research foci in technological areas change over time, information about the distribution of received citations provide clues about the development of broad technological research areas.

Each patent has its own (changing) received citation pattern. To obtain a general sense of these patterns, it is useful to cluster them into a small set of clusters, each with a distinctive pattern. One exploratory approach is to cluster symbolic data objects as described in Section 3.10. The dataset we use here contains all patents granted in 1976: 65,054 patents received at least one citation through 2006 after each patent had existed for 31 years. Figure 5.18 shows the distribution of these patents and the number of citations they received. This distribution is almost scale-free, a typical characteristic of many distributions of received citations in citation networks (Newman 2003, 2005). As a result, our dataset disproportionately has more units with low citation levels.[34] However, seldom-cited patents are unlikely to be important: their patented inventions have little impact on subsequent inventions for which patents were granted. Attempting to get a better sense of changes in technological areas requires attention to those patents cited *more often*. Seldom cited patents were therefore excluded. Somewhat arbitrarily, we restricted our analysis to patents cited at least 20 times.

Given patent granting years, after which patents can be cited, a temporal citation frequency distribution can be created:

$$X_p = [f_1, f_2, \ldots, f_q]$$

where f_i represents the number of citations for patent p in i-th year. Here, q is the number of years for which citations are observed. For this analysis, $q = 31$. When clustering symbolic data, such distributions need not be converted (e.g. normalized with respect to the number

---

[34] This holds also for patents granted in other years.

of citations). The count of citations received has a meaningful metric of direct interest when studying received citation distributions. To cluster these distributions we combined the adapted leader method with the hierarchical clustering approach.

5.7.1   Patents from 1976, cited through to 2006

Restricting attention to patents with at least 20 citations over 31 years reduced the number of units to 6774 patents. In the first step, 40 clusters were obtained with the adapted leaders algorithm (using $\delta_3 = (p_j - t_j)^2/t_j$ as the error measure (dissimilarity) where p_j denotes the relative size of the variable, and its component j, and t_j represents the cluster representative (leader) for variable's component j). Since the adapted leaders algorithm is based on local optimization, the procedure was repeated 100 times: the best result– the clustering with the smallest error (having the lowest values of $P(\mathbf{C})$)– was retained. Figures 5.19 and 5.20 show

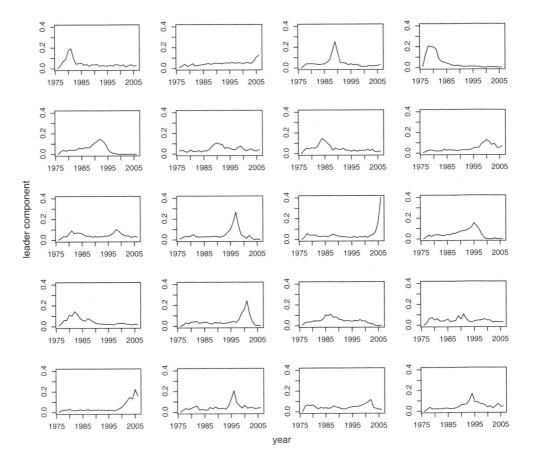

Figure 5.19   Leaders for the first 20 clusters using leaders algorithm with δ_3 as the error measure.

The units are all patents from year 1976 receiving at least 20 citations over 31 years. Lines are used instead of dots for greater visual clarity

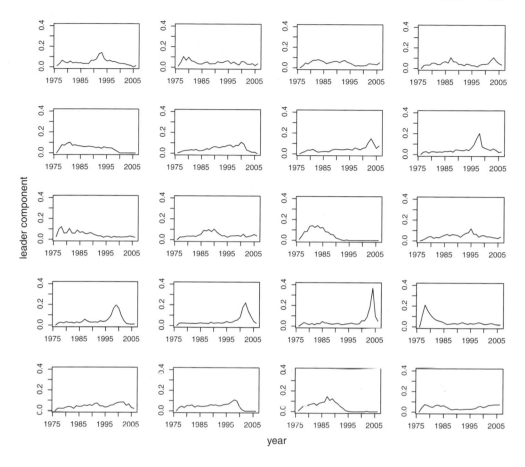

Figure 5.20   Leaders for the second 20 clusters using leaders algorithm with δ_3 as the error measure.

the profiles of these 40 leaders. Each box in these figures shows a profile for temporal patterns for received citations. These 40 profiles can be interpreted separately. Patents in the top left box of Figure 5.19 had an increase in citations shortly after their publication years. Thereafter, their citations dropped and remained low. In the second box on the top row of Figure 5.19, these patents had slow steady increases in citations before a sharper increase started around 2003. Moving to the next box, a steady level of citations to these patents was interrupted by a spike in the late 1980s. In the final box in the top row of this figure, the patents had a sharp jump in received citations shortly after they were granted. This level of citations was sustained for a few years before dropping sharply. The third box in the third row of Figure 5.19 has a profile for patents having a modest level of citations before the citations skyrocketed in the early 2000s. While interpreting every profile is possible, this gets to be tedious. Also, there are some commonalities across subsets of profiles. It is more efficient to cluster these profiles in a second step. More importantly, interpreting these clusters is straightforward and provides a more general (and easily digestible) summary of the patterns in the patent citation profiles.

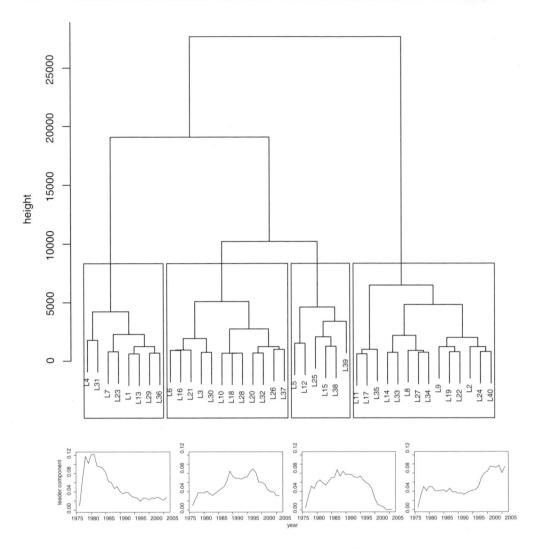

Figure 5.21   Dendrogram for hierarchical clustering on the 40 leaders with δ_3 as the error measure, 1976–2006.

The four summary leader patterns are shown below the corresponding red boxes of the dendrogram.

The 40 profiles in Figures 5.19 and 5.20 were, in the second step, clustered by using the *adapted hierarchical algorithm* (using the δ_3 error measure). The dendrogram is presented in Figure 5.21. The partition into four clusters, marked with red boxes, is very clear. For each of these clusters, the summarized profiles of the leaders are shown below the dendrogram. These profiles differ: these differences merit closer attention.

The first (leftmost) pattern is for patents being very popular soon after they were granted: they had an early impact. Thereafter, interest in them decayed slowly over time. Even so, they were still being cited at a modest level 31 years after they were granted. The second pattern differs dramatically. Patents with this profile did not have a dramatic early impact. Yet the

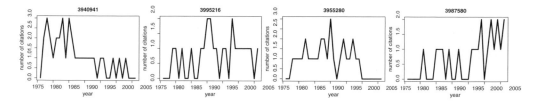

Figure 5.22   Example profiles of patents in each identified 1976 patent cluster.

growth of interest in them was maintained over a longer period before dropping. However, despite this drop, interest in them in 2006 was above the interest in patents of the first cluster.

Patents in the third cluster had an early rise in interest, reaching a peak between the peaks in early interest for the first two clusters. The salient feature of this profile is that interest in these patents was sustained at a high level for longer than the patents in the second cluster. In contrast to the patents in the second cluster, starting shortly before 2000, interest in these patents plummeted to near zero levels. Patents in the fourth cluster enjoyed the same rise in initial interest. But interest stayed at a lower level, even dropping slightly, before taking off to finish at a high level of interest. This takeoff coincided with the drop of interest in the patents of the second cluster.

Figure 5.22 shows the actual temporal profiles of four patents with one from each 1976 patent cluster in the same order of the clusters shown in Figure 5.21. The profiles are more jagged than the smoothed profiles shown in Figure 5.21. The first patent, number 3,940,941, was issued for the invention of new anchor bolts and a new method for installing them in order to strengthen the roofs of underground mines. It belongs to the broad method category and the two-digit mechanical area. Issued to a French company, it had high citation volumes shortly after it was granted with the citations to it diminishing thereafter. The second patent, number 3,995,216, shown in Figure 5.22, was granted to IBM for an apparatus designed to measure the number of surface states at or near the insulator–semiconductor interface in a metal–insulator semiconductor. It belongs to the method and computers and communication categories. It had two separated spikes in received citations during the time period considered.

The third profile is patent number 3,955,280 issued to an Israeli inventor for a new type of dental implant capable of absorbing the shocks associated with chewing. It belongs to the articles of manufacture and drugs and medical categories. This invention had a single spike of received citations coupled to citations around this peak. The final patent in this figure, number 3,987,580, was granted for a separably connective toy with separate geometric flexible parts. This patent belongs to the articles of manufacture the others (as an amusement device) categories. It had a moderate citation level before spiking at the end of this time period.

5.7.1.1   Utilizing supplementary variables for 1976 patents

The clusters obtained in Section 5.7.1 can be further characterized by using supplementary materials in the form of patent variables. We examined differences between the variable distribution of the whole dataset and the variable distributions within each cluster. We present results for using 1) the two-sample Kolmogorov–Smirnov test (Bickel and Doksum, 1977) for numerical variables, and 2) Pearson's χ^2 test of independence for the categorical variables. We note some potential drawbacks. Using the Pearson's χ^2 test with more than

Table 5.4 *p*-values for Kolmogorov–Smirnov statistical tests for the four final 1976 clusters.

variable\cluster	1	2	3	4
Number of citations	< 0.001*	< 0.001*	< 0.001*	< 0.001*
Generality	0.027	0.149	< 0.001*	0.433

Statistically significant results are marked by * (with $\alpha = 0.005$).

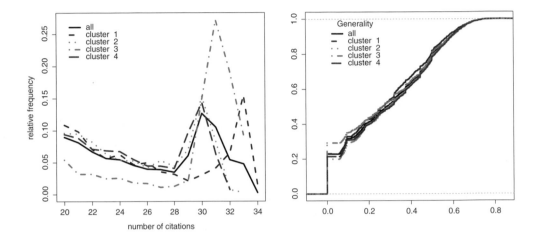

Figure 5.23   Citations received and generality for the four clusters of Figure 5.21.

Left panel: citation distributions for the four clusters and overall (black solid line), Right panel: empirical cumulative distribution functions for generality in all four cluster and overall (black solid line).

two categories has the risk of leading often to significant differences. Having large clusters can lead also to significant results. With the Kolmogorov–Smirnov test, small deviations between two variables can be deemed significant. Caution is merited in interpreting these results to avoid over-interpreting them.[35]

Variables available in the patent dataset (Section 5.3) include the number of citations received and measures of generality. Unfortunately, for patents granted in 1976, information about citations made, measures of heterogeneity, and the percentage of self-citation did not exist.[36] Table 5.4 shows the *p*-values for eight statistical tests (four for the number of citations and four for generality). For judging significance, $\alpha = 0.005$. All four clusters differ significantly in the overall number of received citations (Table 5.4). The left panel of Figure 5.23 show similarities in the sense of having peak citation levels of 30 or more. Cluster 3 has the highest relative frequency at 31 citations, Cluster 1 has it at 33, with the other two clusters having their peaks at 30. However, interpreting these profiles is better done

---

[35] Also, when making multiple statistical tests with the same dataset, the chances of getting at least one statistically significant result increase. Since the issue is not a standard multiple comparison test, one possibility to use is the idea of the Bonferroni correction and use much smaller α values than $\alpha = 0.05$ or $\alpha = 0.01$. We use $\alpha = 0.005$.

[36] In Section 5.7.2, we do include these additional variables for patents granted in 1987.

in conjunction with the profiles in Figure 5.21. Cluster 1 has patents with the largest received number of citations which came mainly in the first half of 1976–2006. Cluster 3 also includes many more well-cited patents distributed more evenly over time. Given the results in Table 5.4, this is the only cluster where generality needs to be considered. According to the right panel of Figure 5.23 it was less general: it has the only line slightly above the overall line.[37] This may be related to the sharp drop of citations over the last eight years.

5.7.2 Patents from 1987, cited through to 2006

It is possible that received citation profiles change over time with results depending on the time interval chosen. We pursue this idea here by using a dataset defined by patents granted in 1987. While this reduces the number of years over which received citation distributions are defined, there is an additional reason for performing this analysis. We were able to include patent heterogeneity and self-citations as two more supplementary variables became available. A total of 12,291 patents granted in 1987 were cited at least 20 times to the end of 2006. The time interval for citations is 20 years. The methodology used is exactly the same as in subsection 5.7.1. After establishing 40 leader profiles, these were clustered into four clusters. The resulting dendrogram and the corresponding four leader profiles are in Figure 5.24. The first cluster, on the left of Figure 5.24, features patents in which interest increased continuously until 2004 before dropping. Interest still remained high in 2006. Patents in the second cluster had a sharp increase in interest until 1989 when interest in them reached a peak. Thereafter, there was a decline to a modest level of interest in 2006. Interest in the patents in the third cluster rose steadily through 1998. This was followed by a sharper decline in interest than for the patents in the second cluster. Interest in the patents in the fourth cluster also rose sharply, reaching a peak shortly after 1989. This interest was maintained at a (slightly lower) high level through 1995. Interest in these patents then dropped to near zero. In the 2000s, this interest dropped to zero.

Figure 5.25 shows the actual temporal profiles of four patents with one from each 1987 patent cluster as in the same order of the clusters shown as in Figure 5.24. The first patent, number 4,663,769, was issued to Motorola Inc. for dealing with asynchronous communication with a clock acquisition indicator circuit for non-return-to-zero data. Belonging to the method and computers and communication categories, its spike in interest occurred at the end of the time period considered. The second patent, number 4,692,352, was awarded to a Japanese corporation, Matsushita Electrical Industrial, for a method and apparatus for drawing a thick film circuit. Belonging to the method and the electrical and electronic categories, it had the reverse profile of an early peak in citation followed by a decline in citations.

The third patent in Figure 5.25, number 4,673,695, was granted to the USA Government for the invention of low density, microporous polymer foams created by a solution of a polymer and a solvent through a rapid cooling process. This patent belongs to the composition of matter and chemical categories with a profile having a steady growth in citations to a peak followed by a sharp decline. The final profile in this figure is for patent number 4,650,479 awarded the MMM Company for a sorbet sheet product useful for disposable diapers, incontinence devices, and sanitary napkins. Belonging to the composition of matter

---

[37] From the result of the statistical test alone, the direction of the deviation was not known.

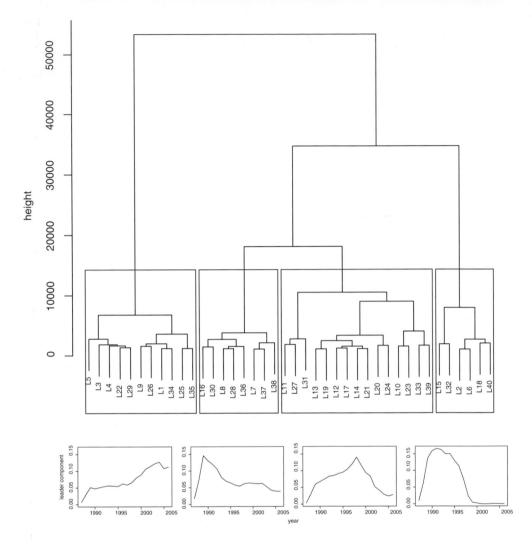

Figure 5.24   Dendrogram for hierarchical clustering on the 40 leaders with δ_3 as the error measure, 1987–2006.

The four summary leader patterns are shown below the corresponding red boxes of the dendrogram

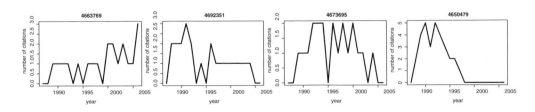

Figure 5.25   Example profiles of patents in each identified 1987 patent cluster.

Table 5.5 *p*-values for Kolmogorov–Smirnov statistical tests for the four final 1987 clusters.

variable\cluster	1	2	3	4
Number of citations	$< 0.001^*$	$< 0.001^*$	$< 0.001^*$	$< 0.001^*$
Generality	0.001^*	0.006	0.099	$< 0.001^*$
Heterogeneity	0.332	0.551	0.999	0.400
Self-citation	0.785	0.989	0.912	0.999

Statistically significant results are marked by * (with $\alpha = 0.005$).

and drugs and medical (miscellaneous) categories, it had an immediate impact in terms of citations followed by a decline to zero.

5.7.2.1 Utilizing supplementary variables for 1987 patents

The results of considering the supplementary variables for these 1987 data are shown in Table 5.5. As for the 1976 data, all differences between the clusters for the number of citations and the overall distribution are significant. The plots of citation numbers for the four clusters are shown on the left of Figure 5.26. Regarding generality (being cited by patents in different technological areas), two clusters – the first and fourth clusters – are significantly different from the overall citation level. From the right panel of Figure 5.26, the first cluster has higher generality while the fourth cluster has lower generality.[38] In this regard, the fourth cluster is like cluster 3 of the 1976 data. Another difference between the results for the 1976 and 1987 patents is a contrast between cluster 1 from 1976 and the second cluster for 1987. Both clusters had early peaks in their interest value. However, cluster 1 in the 1976 data had a higher citation frequency while the second cluster of the 1987 had a lower such frequency. Finally, the supplementary variables heterogeneity and self-citation levels by patent authors to their earlier patents do *not* differ across the four clusters.

5.8 Comparing citation patterns for two time intervals

We consider an alternative way of comparing citation frequencies for two different time intervals provided they have the same length. The two time intervals used here are 1976–1995 and 1987–2006.[39] Again, attention was restricted to patents receiving at least 20 citations. There were 6774 patents for the first interval and 12,291 patents for the second. We use a general linear model (with possible random effects). The variables used for the comparison are the numerical citation frequency, f (years from the grant year, t), and a binary variable, i, for the two time intervals. Frequency is the response (dependent) variable with the other two used as predictors. A unit of observation is one year of citations for one patent. Operationally, each patent was split into 20 units, one for each year of the interval. Before stating and estimating a model, it is useful to examine the mean response profiles to discern trends of the curves.

---

[38] For greater visual clarity, the empirical cumulative distribution functions are plotted for only the first and fourth clusters. The distribution of generality measures for the other two clusters do not differ from the overall distribution.

[39] This provides some continuity to the results shown in the previous two sections by using start points of 1976 and 1987.

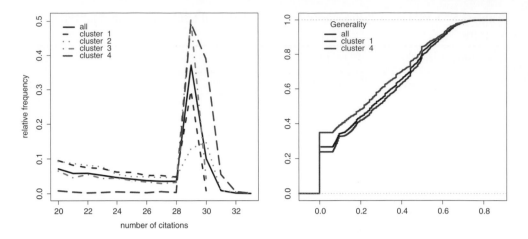

Figure 5.26 Plots of citation levels and cumulative generality levels.

The start year is 1987. Left panel: citation distributions for the four clusters and overall (black solid line). Right panel: empirical cumulative distribution functions (ecdf) for generality for only statistically significant clusters and overall (black solid line).

These are shown in Figure 5.27. Examining these profiles suggests that there are two break points in the number of citations: one occurs about four year after the patents were granted, with the second occurring 15 years after the granting year. Accordingly, the following model for the mean response profiles was specified where μ_f is the yearly frequency mean:

$$\mu_f = \begin{cases} \beta_i + \beta_1 \cdot t & t \le 4 \\ \beta_i + 4 \cdot \beta_1 + \beta_2 \cdot (t - 4) & 4 < t \le 15 \\ \beta_i + 4 \cdot \beta_1 + 11 \cdot \beta_2 + \beta_3 \cdot (t - 15) & t > 15 \end{cases}$$

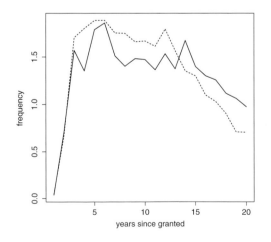

Figure 5.27 Observed mean response profiles. Solid line – 1976–1995, dashed line – 1987–2006 .

Table 5.6   Estimated model parameters for comparing two periods.

	Estimate	Std.Error	t-value	Estimate	Std.Error	t-value
(intercept)	−0.504	0.017	−29.4 **	−0.311	0.038	−8.2 **
i_{1987}	0.033	0.008	3.9 **	−0.200	0.042	−4.8 **
t_1	0.603	0.004	126.2 **	0.496	0.012	42.8 **
t_2	−0.052	0.001	−50.1 **	−0.024	0.002	−9.8 **
t_3	−0.138	0.003	−53.0 **	−0.087	0.006	−13.8 **
$i_{1987} : t_1$				0.129	0.013	10.1**
$i_{1987} : t_2$				−0.033	0.003	−11.9**
$i_{1987} : t_3$				−0.062	0.007	−8.9 **
adj. R^2	0.070			0.072		

Left: no interaction terms, Right: the interaction terms are time × time interval.

Two models were estimated, one without interaction terms and one with them. The results are shown in Table 5.6. The model without interaction terms is statistically significant with all β coefficients also significant. The variance explained is low (about 7%). For our purposes here – comparing the patterns in two non-overlapping time intervals – these two variables suffice. The model suggests that: 1) there is a steep rise of citations for the first 4 years (β_1 is positive) to a peak; 2) there is a (slightly declining) plateau due to small negative estimates for β_2 for 4–15 years since the patents were granted, and 3) the number of citations start to decline more steeply due to the larger negative estimate for β_3.

Considering also interactions between the variables time and time interval (1976 vs. 1987) and comparing it to a simpler (no interaction) model, the new more complex model

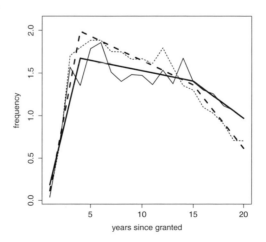

Figure 5.28   Observed mean response profiles and predictions from the regression model with interactions.

The solid line is for 1976–1995 and the dashed line is for 1987–2006. The predictions from the regression model are shown with thick lines.

has a significantly better fit (ANOVA, p-value < 0.001), although the increase in explained variance is minimal. The interactions between time (years from grant year) and time interval are all statistically significant. It suggests that the number of citations rises more steeply and declines more rapidly for the later time period (1987–2006). The conclusions can be seen more clearly with the predicted mean response from the model presented in Figure 5.28.

5.9 Summary and conclusions

Inventions drive technological, economic, and social change. Over time, protections were created to shield inventors from others appropriating their inventions unfairly by using patents as protections of intellectual property rights. When applications are made for patents, these applications are required to cite relevant prior patents.

There are two broad sets of categories into which patents are placed. The first provides definitions for utility patents with four categories: methods (processes); machines; articles of manufacture and composition of matter. The second is a classification of patents into technological areas. We used a widely accepted six-category system to study broad temporal changes: chemical; computers and communications; drugs and medical; electrical and electronic; mechanical, and others.

We considered the role of the US Supreme Court on the patent system. Most often, patents get on the Supreme Court docket via cases involving patent infringement. We looked at some decisions to get a sense of how the Court reaches decisions about patents, by considering a line island of heavily co-cited Court decisions about patents covering almost a century (1888–1981). The large majority of these decisions featured machines or articles of manufacture. The heavily co-cited line island has two clear different regions linked by a trio of decisions. One region is dominated by cases on articles of manufacture (earlier in time) with the other dominated by machines (later in time). The Court uses general legal principles in reaching decisions about patents: neither the technological area nor the categories for utility patents appear to have a direct influence. There is some evidence that chemical patents were more likely to be upheld while mechanical patents were more likely to be voided.

Growth in the number of patents granted in the six broad technological areas differed greatly from 1987 through 2006. By far the greatest *numbers* of patents in 2006 were for patents featuring computers and communication and patents in the electrical and electronic domains. The most dramatic *increases* featured patents for computers and communication, patents in the electrical and electronic area, plus patents for drugs and medicine. Flows of citations between technological areas changed also. The most dramatic increases were for reciprocal citation flows between computers and communication patents and electrical and electronic patents, especially during 1995–2006. Also increasing were reciprocal citation flows between patents in these two categories and mechanical patents. Reciprocal flows between chemical patents and patents for drugs and medicine were noteworthy for 1995–2002. All these changes were coupled to wider social change in the USA.

Coupling two-mode data for patents and keywords permits the identification of the technological contents of an island regarding its theme(s). Examples of doing this were provided for a line island for computers and communications patents and another for patents in drugs and medical. The use of keywords for the drugs and medical line island narrowed its focus (theme) to treatment of heart problems by considering the one-mode keyword network for the island. This approach – using one-mode patent citation networks, two-mode patent and keyword networks and one-mode keyword networks – can be extended to all of the identified line islands.

Important inventions in patents are cited more often. Over time, the number of citations received by these patents increases. Every patent has a received citation temporal profile for each year following the year it was granted. These profiles were clustered. The final result was a clustering of patents into four clusters, each with a very different temporal profile. One set of patents were cited heavily shortly after they were granted with the subsequent citations dropping. Their greatest impact was short-term. The second set had a slower initial rate of citation, reached a sustained level of citation (with two mini-peaks), and then had declining citation levels. Their greatest impact was more mid term. The third set had citations growing to a sustained high level before dropping sharply. By the end of the period, they were invisible. They had a sustained mid term impact. The final set of patents had their greatest impact at the end of the period. Clearly, the impact of patents have varying timescales.

This line of analysis was repeated for patents issued in 1987 (and cited through 2006). There were great similarities but with finer-grained details meriting greater attention. For both sets of patents (defined by two granting years), the clusters differed in terms of the numbers of citations received overall. Only one cluster of the 1976 patents differed in terms of generality. For the 1987 patents, two clusters differed, with one having less generality and the other more. There were no differences in terms of heterogeneity and level of self-citation for the 1987 patents. One final round of analysis showed differences between pairs of periods (having the same duration) in terms of break points in the citation trajectories.

Some lines of work, based on the results of this chapter, merit deeper exploration. They include: 1) a more detailed examination of the trends in citations between technological areas, as shown in Figures 5.6 and 5.7, by examining all successive pairs of overlapping temporally specific networks; 2) obtaining a tighter link between Supreme Court deliberations and patent citations; 3) examining more line islands to establish their themes through the use of keywords, and linking them to wider societal changes; 4) examining links between patents in different line islands because additional relevant citation information can be located there to examine changes within and between technological areas; 5) examining patterns in temporal received citations for the patents granted after 1976 and before 1987 to get a potentially wider set of general profiles, and 6) obtaining a better understanding of the temporal nature (timescales) of the utility of patents.

6

The US Supreme Court citation network

Fowler and Jeon (2008) compiled a large citation network of Supreme Court decisions citing earlier decisions. The citations came from 30,288 majority opinions written between 1789 (the year of the first Court decision) and 2002.[1] This was a Herculean task. Their primary intent was tracing the evolution of *stare decisis* (Justices accepting a rule to follow prior legal precedents). Using this network to compute indegree, outdegree, eigenvector centrality measures, and hub and authority scores, they also characterized each decision by these measures. We use their data for a different, but complementary, purpose – locating *sets of decisions* in their historical, legal, and judicial contexts. Rather than examine citations to and from *single* decisions, we study *co-cited* decisions (see Section 6.2). We then couple some of these decisions to laws enacted by the federal and state governments.

This chapter is structured as follows. A brief introduction to the Supreme Court is in Section 6.1. Summaries of co-citation in citation networks and line islands (see Section 2.9 for details) are in Section 6.2 together with some preliminary results for identified line islands. An examination of cases in an island focused on Native Americans follows in Section 6.3. Cases featuring 'threats to social order' in another island are in Section 6.4. They are coherent examples of using line islands to understand this long temporal network spanning more than two centuries. Section 6.6 discusses an infamous Supreme Court decision (*Dred Scott*) for two reasons. First, it is not an 'important' decision for Fowler and Jeon (2008) despite appearing in many discussions of race, citizenship in the USA, and the role of the Supreme Court. Second, it points to some limitations in using *only* majority opinions to construct citation networks for Supreme Court decisions. We provide some implied consequences and suggest lines of future work in Section 6.7.

---

[1] Some early decisions cited decisions of prior courts dating back to 1754.

Understanding Large Temporal Networks and Spatial Networks: Exploration, Pattern Searching, Visualization and Network Evolution, First Edition. Vladimir Batagelj, Patrick Doreian, Anuška Ferligoj and Nataša Kejžar.
© 2014 John Wiley & Sons, Ltd. Published 2014 by John Wiley & Sons, Ltd.

6.1 Introduction

The United States (US) Constitution, signed on 17 September 1789 after many months of debate, went into effect on 4 March 1789 following its ratification. The opening sentence reads: 'We the people of the United States, in order to form a more perfect union, establish justice, insure domestic tranquility, provide for the common defense, promote the general welfare, and secure the blessings of liberty to ourselves and our posterity, do ordain and establish this Constitution for the United States of America'. Containing seven Articles, it became the supreme law of the newly formed country. The words of the initial Constitution, together with those in its subsequent and included amendments, are important for interpreting judicial decisions.

Ten constitutional amendments were proposed by Congress in September 1789. The first established a US Congress consisting of a Senate and a House of Representatives. The second stipulated that executive power be vested in the President while the third located US judicial power in the Supreme Court. Following ratification by three-fourths of the states in December 1791, these ten amendments became the Bill of Rights. Starting in 1798, 17 additional amendments were adopted and incorporated into the Constitution. The last Amendment was ratified in 1992.

Details of the 'federal court system' remained unspecified until Congress passed the Judicial Act later in 1789. The Supreme Court sits on top of the judicial system. Figure 6.1 shows the current structure. The Supreme Court has exclusive original jurisdiction (the right to hear cases first) for all civil cases between states and between a state and the USA. It also has appellate jurisdiction over (the right to review) all decisions made by lower courts. Cases reach the Supreme Court over the links shown in Figure 6.1. The upwards flows are depicted by arrows. In addition, the Supreme Court can signal lower courts to intentionally send them

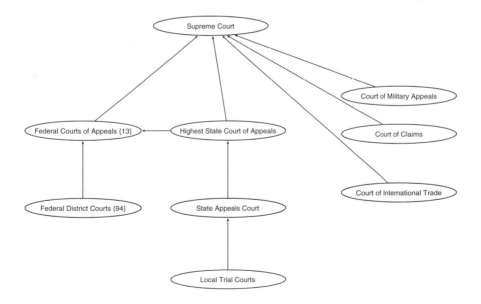

Figure 6.1 The structure of the US court system.

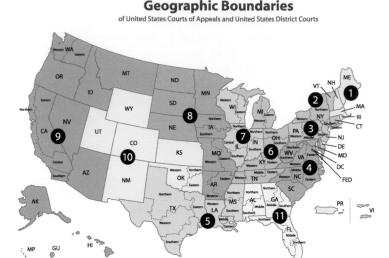

Figure 6.2   An outline map of eleven circuit courts and their district courts. *Source:* U.S. Government.

specific cases or return cases to lower courts to have them reconsidered under Supreme Court directions.

There are 94 US Federal District Courts whose decisions can be appealed to 13 US Federal Courts of Appeals. Decisions of these courts can be appealed to the Supreme Court. States also have many local trial courts. Their decisions can be appealed to State Appeals Courts. These decisions can be appealed to the highest State Court of Appeals, most often the State Supreme Court. Decisions of the top state courts, on appeal, reach the Supreme Court directly or via specific US Federal Court of Appeals with jurisdiction. Figure 6.2 shows 11 circuit courts of appeals.[2] Cases can reach the Supreme Court directly from the three courts on the right of Figure 6.1.

Cases reach the Supreme Court over many routes. Their origins can be anywhere in the USA. While general issues have been raised in geographically dispersed places, some regions raised issues unique to their areas. This alone creates complexity. Court decisions can be classified according to the parts of the US constitution on which they rest. Vile's (2010) classification used 15 broad categories based on components of the constitution, some with multiple parts. Cases can be categorized also in terms of their immediate *substantive* issues. The codebook constructed by Benesh and Spaeth (2003) has over 260 categories for issues.[3] Moreover, cases often involve combinations of substantive issues with constitutional principles applied to them. Characterizing all cases within a single organizing system is extremely complex. As a result, there is no universally accepted complete classification of all Supreme Court decisions.

---

[2] The DC Circuit and the Federal Circuit, both in Washington, DC, are the other two courts in the the Federal Courts of Appeals system. Source of Figure 6.2: http://www.uscourts.gov/uscourts/images/CircuitMap.pdf.

[3] This includes almost 60 categories for criminal law, more than 40 for civil rights, over 50 concerned with judicial power, and more than 30 regarding economic activity.

Convential wisdom states that Supreme Court Justices base their opinions and decisions on the Constitution's text. This document's ambiguity greatly complicates this task. Ambiguity was built into US legal documents at the start. For example, the Declaration of Independence (1776) included 'We hold these truths to be self-evident, that all men are created equal, that they are endowed by their Creator with certain unalienable rights that among these are life, liberty and the pursuit of happiness'. Yet the term 'men' did not include women, slaves, or men without means. Moreover, the terms 'liberty' and 'pursuit of happiness' were never defined – a persistent source of disagreement in US courts.

Deep conflicts over federalism versus state sovereignty and slavery were present during the Philadelphia convention that created the US Constitution. They could never have been resolved to the satisfaction of all contending parties. The document's ambiguity allowed enough signatures so that it could be adopted.

Irons (2006) (xv) observed 'Just over a hundred people have served on the Supreme Court in just over two hundred years. All but two have been white, all but two have been men, and all but seven have been Christian. Many of the landmark cases these Justices have decided were brought by blacks, women, and religious and political dissenters. In a very real sense, the history of the Supreme Court reflects the appeals of powerless "outsiders" to the powerful "insiders" who have shaped the Constitution's meaning over the past two centuries'. This insight is crucial for understanding the co-cited decisions considered in Section 6.2.

These complexities (studied in specific cases) justify Fowler and Jeon's (2008) strategy of examining citations to and from single decisions for grasping an understanding of the Court. We extend this by using network tools to identify sets of decisions. When identifying cases, we use the names of the parties in the cases reaching the Supreme Court and their official identification labels, xxxUSyyy, where xxx denotes the volume number containing a decision and yyy is the beginning page number of the decision's text within the volume. Together, these numbers uniquely identify each Supreme Court decision. For example, the Dred Scott case considered in Section 6.6 is identified as 60US393.

6.2   Co-cited islands of Supreme Court decisions

In order to move beyond looking at *single* decisions to focus on *sets* of decisions, some method is needed for identifying these decisions in a general fashion. Specific Court decisions cite earlier decisions for many reasons, including constitutional principles, substantive content, and establishing context. Methodologically, it would be very inefficient, given the number of decisions in the database, to try coding the many features of cases and decisions. Our proposal for identifying sets of decisions is to discern the extent to which earlier decisions are co-cited by later decisions. Earlier decisions *co-cited* by later decisions are far more likely to share common features, making them more similar than any arbitrary pair of cases. When sufficient numbers of decisions are co-cited they are likely to form coherent parts within the citation network. Examining co-cited decisions helps us to understand key parts of this Supreme Court network, and establishing line islands is a very effective way of doing this.

The formal foundation for considering co-cited decisions is easy to state. Let C denote the matrix of the citation network. Then $G = C'C$ captures decisions being cited together. Let g_{ij} denote the elements of G. The diagonal element, g_{ii}, is the number of times the ith decision is cited. These elements were used by Fowler and Jeon (2008) in identifying their 'important' and noteworthy decisions. The off-diagonal element, $g_{ij}(= g_{ji})$, is the number of times the ith and j^{th} decisions are *jointly* cited by later decisions. The co-cited network

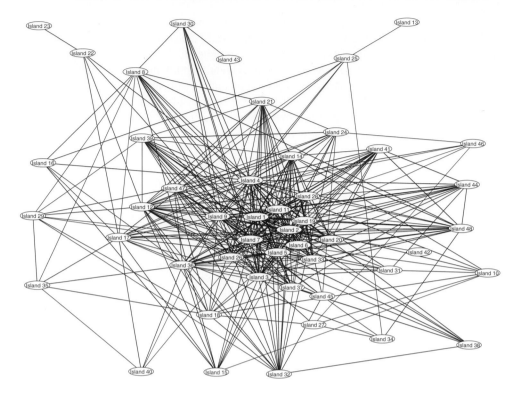

Figure 6.3   The network of 48 co-cited islands.

is symmetric with valued edges. If two decisions are jointly cited frequently they are likely to have common features meriting further attention. This is the basic premise for what follows.

From Chapter 3, informally, network *line islands* have vertices linked by edges with larger values than edges between them and vertices outside the island. In Pajek, islands are identified by specifying the minimum and maximum number of vertices in islands. Setting these values to 10 and 100, respectively, identified 48 islands with a total of 988 decisions.[4] Despite the use of 'islands' as a term, line islands can be linked with other line islands. Vertices in a line island, while more heavily linked internally, can be co-cited frequently with decisions in other line islands, albeit less heavily. The network of these 48 islands is shown in Figure 6.3 where lines represent high enough valued co-citation links.

The numbering of the islands follows the number of decisions they contain. Island 1 contains the most decisions with island sizes decreasing (not strictly monotonically) as the island numbering increases. Given the presence of 48 linked line islands, our strategy of examining single islands barely scratches the surface and points to an extended future agenda. Even so, the contents of islands are revealing for understanding the Supreme Court citation network.

---

[4] Neither the number of identified islands nor their composition changed when the maximum island size was increased.

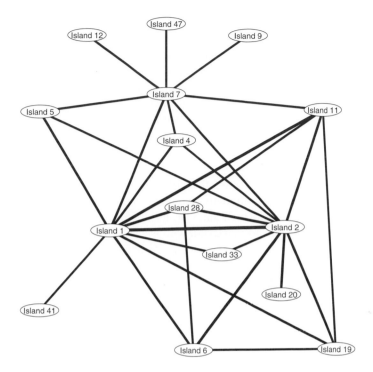

Figure 6.4   A reduced network with 15 co-cited line islands.

Obviously, this network has a core of line islands with other islands being peripheral.[5] To look more closely at the most heavily co-cited line islands, the network shown in Figure 6.3 can be pruned. Removing links below a (high) threshold and eliminating the resulting islands creates smaller line island networks. Part of the core of Figure 6.3 is shown in Figure 6.4. Clearly, we do not have the space to study every identified line island in detail. So we focused on some islands to 1) examine closely their decisions for substantive content and legal principles; 2) extract the citation network for the decisions *within* line islands; 3) look at citations between these decisions, and 4) interpret their patterns and couple them to their judicial and historical contexts. We selected a peripheral line island and a core line island from Figure 6.4 for further exploration.

There is another contrast between what we do and what Fowler and Jeon (2008) did. They used their citation network to mark 'important' decisions through counts (and weighted counts) of received and outgoing citations. They also operationalized a different conception of the importance of decisions with two dummy variables: these appear in Hall (1999), also known as the Oxford List, and appearance in the Legal Information Institute's (2005) list. While authoritative, these are restricted lists, the criteria for inclusion not being entirely clear. We expanded these lists to include the decisions considered in Irons (2006); Powe (2009);

---

[5] While being peripheral is often thought to imply being 'less important' in a network, the peripheral islands are likely to identify also linked constitutional principles and substantive concerns. Being peripheral need not imply being less important.

Vile (2010); and in Wikipedia's 'landmark' decisions.[6] We interpret decisions in contexts defined by line islands with special attention to important decisions. Co-cited decisions can be coupled usefully to importance to expand such decisions to a broader judicial context. Important decisions are defined in writing grand histories of the Supreme Court. While islands without such decisions are likely to be overlooked, they also have coherence regarding more prosaic decisions. Having an expanded list of important decisions is helpful for assessing specific decisions. As there are too many islands to discuss in this chapter, we consider two islands to illustrate what can be done using co-cited islands to help understand this Supreme Court network. The first is a middle-sized island permitting an easier visualization before we consider the largest island in Section 6.4.

Irons argued that the people suffering the most from the Puritans and other early colonial settlers were Native Americans.[7] 'Indians were the losers in virtually every battle with the colonialists who forced them from their lands (Irons (2006): 15)'. This was the beginning of a persistent pattern. Given this history, we start with island 9 (the ninth largest) in Section 6.3. All 27 decisions involve Native Americans with 26% having been deemed important.

6.3 A Native American line island

While it could be claimed that it is enough to note that all decisions in this island involve Native Americans, we do not think this is sufficient. Having identified these decisions, we focus on the citation structure between them and their content. The citation network of the Court decisions in this co-cited line island are shown in Figure 6.5. Arcs are citations from later to earlier decisions. Decades are marked on the left to provide a sense of when decisions[8] were made. These decisions cover more than 150 years and while they show many clear consistencies over this time span there are some departures from this consistency. Markedly different themes emerge when the content of decisions is considered.

6.3.1 Forced removal of Native American populations

All European colonial powers practiced the *discovery doctrine*: when they discovered new (to them) lands they claimed these lands as their own. The USA extended, and practiced, this internally to move Native American peoples off their traditional lands. The so-called 'Indian problem' was anything getting in the way of whites (mainly) appropriating lands and taking advantage of Native American populations. The term 'Indian matters' covered a wide variety of topics ranging from large to small. The former are easy to characterize: outright theft of Native American land either by treaty or legal constructions justifying the theft. The most infamous set of the consequences resulting from such land thefts became known as 'The Trail of Tears' after the Cherokee and Choctaw nations were moved forcibly from their lands to

---

[6] Not surprisingly, some decisions appear on multiple lists.

[7] These peoples were referred to as 'Indians' even in Supreme Court opinions. We use the term Native Americans instead but exceptions occur when we quote from other sources including the names and text of Supreme Court opinions. Also, we use 'nations' rather than 'tribes' to refer to Native American sovereign groups. We include parts of 'tribes' within this designation.

[8] Two decisions (411US145 and 411US164) are linked by an edge. This is a rare instance of decisions citing each other. They were decided 22 days apart by the same court . It is a rare example of local configurations inconsistent with the predominately acyclic nature of a citation network. See Chapter 3 for the methods used for dealing with this problem.

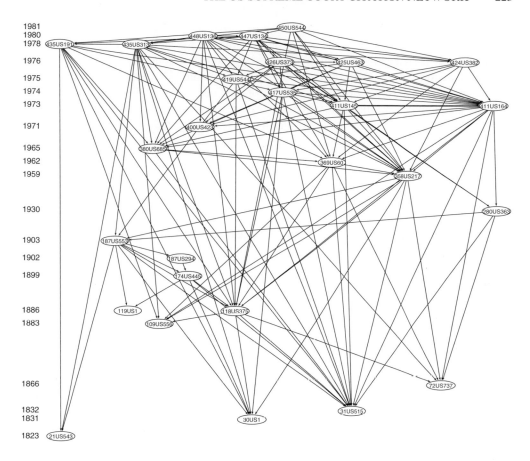

Figure 6.5   Citation network within the Native American island.

a part of what is now Oklahoma.[9] The Supreme Court decision of 1831, *Cherokee Nation v. Georgia* (30US1), stated it had no jurisdiction and permitted this eviction.[10] It is the second earliest decision in Figure 6.5.

The Indian Removal Act of 1830 was a salient piece of legislations with multiple interpretations depending on the interests in play at the time. On the one hand, Georgian politicians and decision makers wanted to engage in ethnic cleansing: white settlers wanted these lands and campaigned for the removal of the Cherokee. At face value, they were successful in the passage of the Indian Removal Act which was upheld by the Court in 1831 and informed *Cherokee Nation v. Georgia*. A case can be made that while Congress and the Supreme Court authorized Andrew Jackson, as US president, to 'negotiate' with Native American leaders in the Southern United States for the removal of their nations to federal territories west of

---

[9] Over 25% of the people forcibly removed died during the Trail of Tears. For them, it was, literally, a death march.

[10] Chief Justice John Marshall's opinion included 'the framers of our Constitution had not the Indian tribes in view when they opened the courts of the union to controversies between a State or the citizens thereof, and foreign states.'

the Mississippi River[11] it was also an attempt to curtail states' rights. Yet by asserting that it had no original jurisdiction in the matter because the Cherokee was a 'domestic dependent nation' – a term created in Marshall's opinion – the Supreme Court washed its hands of the Cherokee nation. Powe (2009) (80) quotes Jackson as stating 'The authority of the Supreme Court must not, therefore, be permitted to control the Congress or the Executive when acting in their legislative capacities.' In fairness, Jackson was dealing with a nullification crisis under which states claimed that they could nullify federal laws they did not like.[12] In this context, the rights of the Cherokee, Choctaw, Seminole, Creek, and Chickasaw nations[13] to their traditional lands were not worth expending political capital for Jackson. Despite the alleged conflict between federal interests and state's rights, the way was cleared for Georgia to start their ethnic cleansing with the assistance of federal troops.

6.3.2   Regulating whites on Native American lands

Not all early Supreme Court decisions ruled against Native Americans. The earliest decision in Figure 6.5, *Johnson v. M'Intosh* (21US543), ruled that private (white) citizens could not purchase lands from Native Americans. Both the Oxford list and Wikipedia's list of landmark cases contain this case. Another early case in Figure 6.5 is *Worcester v. Georgia* (31US515), featured in Vile's (2010) discussion of essential Supreme Court decisions. Georgia enacted a criminal statute prohibiting non-Native Americans from being present on Native American lands without a state license. The Court ruled this law unconstitutional. In *The Kansas Indians* (72US737), the Court ruled Kansas could not tax lands of individual Native Americans of the Shawnee, Miami, and Wea nations.

6.3.3   Curtailing the authority of Native American courts

Ex parte Crow Dog (109US556) involved a conflict between two Sioux Nation members on reservation land. Crow Dog, a chief, killed another Sioux chief. Tried by a Native American council, he paid restitution: the case was settled under Native American law. However, a US court then tried Crow Dog for murder, sentencing him to death. A unanimous Supreme Court ruled federal courts had no jurisdiction after the Native American court had ruled on such cases. A writ of *habeas corpus* was issued. This decision, favorable to Native American sovereignty, led to white outrage. The Major Crimes Act of 1885 followed quickly. It placed murder, manslaughter, rape, assault with intent to commit murder, arson, burglary, and larceny under federal jurisdiction.[14] This greatly reduced Native American authority on their lands. Eroding this sovereignty has a very long history.

---

[11] According to Wikipedia, in 1800, the Cherokee owned about 140,000 km^2 of land in Tennessee, North Carolina, Georgia, and Alabama.

[12] This is still a live issue in the 21st century with Republican legislatures eager to nullify federal legislation enacted under President Obama.

[13] The areas from which Native American were removed covered a large area of the south-eastern USA even though the Supreme Court decision pertained to Georgia.

[14] This list was expanded to include 15 crimes. As a very modern postscript, Erdrich (2013) points out that rapes of Native American women by non-Native American men cannot be tried in Native American courts. She notes: 'More than 80% of sex crimes on reservations are committed by non-Native men who are immune from prosecution by tribal courts' and most of the offenders remain immune because US courts seldom prosecute these cases.

Reactions to *Crow Dog* are striking examples of the Court not always having the last word: any decision can prompt a law designed to undo it. Three years later, *United States v. Kagama* (187US294), also on the Oxford list, cited *Ex parte Crow Dog*. It concerned a Yurok Native American charged with murdering another Yurok. It was chosen to establish the constitutionality of the Major Crimes Act.[15] This decision cited *Cherokee Nation v. Georgia*, also in Figure 6.5.

A deeper intent lay behind the Major Crimes Act and *Kagama*: it was to affirm Congressional power to strip Native Americans of liberties and to reduce their sovereignty. They laid foundations for subsequent laws including the 1887 Dawes Act designed to eliminate Native American cultural identities and force their assimilation.

Figure 6.5 shows *Lone Wolf v. Hitchcock*, also on the Oxford list, citing *Cherokee Nation v. Georgia*. It affirmed Congress taking lands from Native Americans and its right to *abrogate all treaties* with them. From Hall (2005): 592: '*Lone Wolf* has permitted the United States to appropriate tribal lands and resources under the guise of fulfilling federal trust responsibilities' and noted that 'it is impossible for tribes to obtain judicial protection in disputes with the United States.'[16] This set of cases showed the Court and Congress moving jointly in subjugating Native American nations.

6.3.4 Taxing Native Americans and enforcing external laws

Vile (2010) combines Articles IV and V on federalism to help categorize Supreme Court decisions. He included decisions we have discussed (*Cherokee Nation v. Georgia* and *Worcester v. Georgia*) in this category. While focused on issues related to Native Americans, most of the decisions in this line island also deal with the conflict between federal and state authorities. *Carpenter v. Shaw* (280US363) dealt with an attempt by Oklahoma to tax Native Americans for gas and oil extracted from their land. The US Supreme Court overturned an Oklahoma Supreme Court decision affirming this tax in 1930. The 1959 Warren Court decision, *Williams v. Lee* (258US217), featured a dispute involving a Navajo man not paying at a trading post. An Arizona state court tried to compel payment. The Warren Court ruled Arizona had no jurisdiction. This case is in the Oxford list, and Wikipedia lists it as a landmark decision.

Fowler and Jeon (2008) note a sharp drop in outward citations to early precedents by the Warren Court. *Williams v. Lee* suggests that the Warren Court was more sympathetic to Native American populations and, as we show in Section 6.4, other outsider groups described by Irons (2006). It did not seek inconsistent decisions for precedent and reached back to some of the earliest decisions on this line island.

Kake v. Egan (369US60) concerned some incorporated Thlinget communities and the salmon traps they operated under permits issued by the US Army Corps of Engineers and the US Forest Service. Alaska passed a statute banning salmon traps. The Thlingets sued to prevent enforcement of this statute on them. The Alaska State Supreme Court's decision affirming this law was upheld by the US Supreme Court in 1962. About three years later, *Warren Trading Post v. Arizona Tax Commission* (380US685) overturned an Arizona State Supreme Court ruling allowing Arizona to tax income from a federally licensed trading post

---

[15] The local district attorney declined to prosecute citing *Crow Dog* but the District Attorney for Northern California decided to pursue the case.

[16] It quotes a federal judge in 1979 as describing *Lone Wolf* as the Native American's *Dred Scott* (see Section 6.6) and adds that it had 'not been repudiated by political events or judicial decisions.'

on Navajo land. The only decision in this line island it cited was *Cherokee Nation v. Georgia* in which the earlier Supreme Court decided that it did not have original jurisdiction.[17]

Two issues are prominent in the Burger Court (1969–1986) decisions on this island. One was state governments, some persistently, attempting to tax Native Americans and economic activities on their lands. The second was the sovereignty of Native American courts. On taxes, the Court ruled that New Mexico could not tax Apaches for land improvements on their reservation in *Mescalo Apache Tribe v. Jones* (411US145). But it did affirm a New Mexico lower court's ruling that taxes could be levied for an *off-reservation* ski resort. Arizona tried to tax a Navajo woman's income earned entirely on the reservation. A unanimous Court ruled in *McClanahan v. Arizona State Tax Commission* (411US164) that this was invalid.

6.3.5   The presence of non-Native Americans on Native American lands

Two cases were decided in 1976. Montana tried taxing cigarette sales at 'smoke shops' on the Flathead reservation. *Moe v. Confederated Salish and Kootenai Tribes on Flathead Reservation* (425US463) ruled these taxes could not be levied on Native Americans. But taxing for non-Native Americans was declared legal. Minnesota had sought to levy taxes on Native American property on their lands. *Bryan v. Itasca County* (426US373), another case in the Oxford listing, reversed a Minnesota State Supreme Court decision allowing these taxes. These rulings on 'smaller' issues favored Native Americans.

Clearly, states imposing taxes was complicated by the presence of non-Native Americans on reservations. Washington State tried to compel Indian retailers to collect its excise taxes on cigarette sales to non-Native Americans. *Washington v. Confederated Tribes of Colvill Indian Reservation* (447US134), in 1980, ruled that collecting this tax was valid.[18] Arizona attempted to tax a non-Native American working exclusively for Native Americans on their land. In the same year, the Court ruled in *White Mountain Apache Tribe v. Bracker* (448US136) that this state tax could not be collected. The underlying principle of these tax decisions was that Native Americans were protected against the impositions of taxes by states but non-members were not necessarily protected.

In 1975, the Court with *US v. Mazurie* reversed a 10th Circuit Court of Appeals decision by reinstating a non-Native American's conviction for introducing alcohol to Native Americans. The Court dealt with Montana's attempt to block a Cheyenne married couple from adopting a Native American child a year later. In *Fisher v. District Court of Montana* (424US382) the Court ruled that state courts had no standing: the Montana State Court decision interfered with 'tribal' self-government. In contrast, 1978's *Oliphant v. Suquamish Indian Tribe* (435US191), a 6-2 decision, ruled 'Indian tribal courts do not have inherent criminal jurisdiction to try

---

[17] The majority opinion quoted from Chief Justice Marshall's opinion in *Cherokee Nation v. Georgia*: 'From the commencement of our government, Congress has passed acts to regulate trade and intercourse with the Indians which treat them as nations, respect their rights, and manifest a firm purpose to afford that protection which treaties stipulate.' This statement seems decidedly at odds with allowing the Trail of Tears to occur. Because *Cherokee Nation v. Georgia* was decided in a context defined by conflicts between federal and state governments, as well as the potential problem of a civil war, Justice Marshall's opinion was complex, and different implications can be drawn from different parts of his opinion. This suggests that a citation to earlier Supreme Court decisions can be selective regarding its content as to what is cited.

[18] Washington State also wanted to apply its motor vehicle tax to mobile homes, campers, and trailers owned by Native Americans. The Supreme Court ruled that these taxes could not be imposed even if the vehicles also were used off the reservation.

and to punish non-Indians, and hence may not assume such jurisdiction unless specifically authorized to do so by Congress.'

After a Navajo man pled guilty to contributing to the delinquency of a minor, he was sentenced by a Native American court. A federal court then indicted him on the same charge. *US v. Wheeler* (435US313) reversed a 9th Circuit of Appeals decision by ruling double jeopardy did not apply in 1978. The Crow nation sought to limit fishing and hunting on their lands by non-Native Americans. In 1981, *Montana v. US* (450US544) ruled against them arguing 'exercise of tribal power beyond what is necessary to protect tribal self-government or to control internal relations is inconsistent with the dependent status of the tribes, and so cannot survive without express congressional delegation.' The Court reversing another 9th Circuit of Appeals decision to severely restrict Native American authority over non-Native Americans on their lands. These decisions are in stark contrast to more favorable decisions regarding taxes – and far more consequential.

6.3.6   Some later developments

While all the decisions on this line island dealt with issues featuring Native Americans, common themes kept emerging. Dominant among them was a concern to subjugate Native American populations in terms of forced relocations, eroding the authority of Native American courts and leaders, and imposing the will of state governments on these populations. The primary instigators of these cases were state and local governments, with the Supreme Court in the role of refereeing these conflicts. For the major issues with national implications, it tended to rule against Native Americans while on lesser, very local issues, it was more sympathetic to their concerns. This is the primary 'story' of the cases that we have considered. It was natural to ask if anything changed after the period covered by the decisions on this line island.

The decisions in Figure 6.5 were examined closely through *later decisions citing them*. Doing this expanded the number of decisions to 256. Half did *not* involve Native Americans. Some pertained to Peurto Rico and the Philippines ceded to the US by Spain in 1898 following its defeat in the Spanish–American War. These cases used the treatment of Native Americans in US courts as precedent for treating these Spanish-speaking populations similarly. Alcohol was a major national issue during the Prohibition Era.[19] Earlier cases about controlling alcohol on Native American reservations were cited despite not featuring Native Americans, again for precedent. Among the 256 decisions involving Native Americans, the issues of taxation on reservations, crimes by both Native Americans and non-Native Americans on reservations, extraction of mineral resources from reservations, and the sovereignty of Native American nations all appeared repeatedly. In that sense, little changed beyond adding additional groups to be treated in the same fashion.

6.3.7   A partial summary

An imbalance between Native Americans and white settlers from Europe existed from the start. In state and federal courts, this imbalance is reflected in so many decisions decided against Native Americans. It did not seem to matter if they were plaintiffs or defendants. Some

---

[19] Prohibition was a national ban on the manufacture, sale, and transportation of alcohol. It was initiated by the Eighteenth Amendment of 1920 and ended when the Twenty-First Amendment repealed it in 1933.

cases dealing with taxes and small-scale authority issues were settled in their favor. However, larger issues of taking land and curtailing Native American authority, including dealing with non-Native Americans, went against Native Americans. Looking closely at majority opinions of decisions in Figure 6.5 revealed selectivity in citations between Supreme Court decisions. Some decisions were cited while others, at face value equally relevant, were not. Also, some items *from* earlier decisions were selected while ignoring other parts. This suggests that *stare decisis* is not solely about citing previous cases but also what is *not* cited. This merits further attention for understanding the Court's decisions, despite doing so being a daunting task. Obtaining citations being present in majority opinions is an easy task compared to identifying missing citations. In Section 6.6 we do the latter with a focus on the *Dred Scott* case, a trivial task compared to doing this for over 30,000 decisions. Establishing and following precedent has been deemed important as a constraining force against arbitrary decisions by curtailing the authority of its Justices. Yet, if precedent has been established, even in part, by ignoring decisions that could have served as precedents because they do not support a decision being rendered, then precedent may be a mere chimera masking covert judicial and political intentions.

6.4   A 'Perceived Threats to Social Order' line island

The smaller line island considered in Section 6.2 illustrated what can be done by using this technique for understanding the Supreme Court. We now consider the largest co-cited island. Its span is 1919–1980, a much shorter period than for the Native American island. It has 95 decisions, 56 of which (59%) have been deemed important by legal scholars. The main constitutional items making this island coherent are the First Amendment (used in 75% cases) and the Fourteenth Amendment (used in 38% cases). Often, First Amendment protections were extended to the states under the Fourteenth Amendment.

Vile (2010): 207 observed 'The First Amendment is one of the most revered and important provisions of the US constitution.' The rights guaranteed under it concern political and religious rights. Political rights include freedom of speech, a free press, the right to assemble peacefully (with others), and the right to petition governments. Freedom of speech includes protections for subversive speech, symbolic speech, speech and conduct, and issues regarding the time, place, and manner of speech.

Religious rights guaranteed under the First Amendment come from the Establishment and Free Exercise Clauses. Under the former, the US government is prohibited from sponsoring or inhibiting specific religions. The idea's core is a 'wall of separation' between the state and the church. Under the second clause, citizens are free to practice their religion without governmental interference. However, this is not an absolute right guaranteeing *all* religiously motivated behavior.

The decisions considered here are far more diverse in their substantive issues than their underlying constitutional principles. This is not surprising: the Constituion is a *general* document designed to be applied across *many* substantive domains. We focus on the substantive issues appearing most often in the decisions in this island.

6.4.1   Perceived threats to social order

Our examination of the decisions on this island turned out to be fully consistent with the observations of Irons regarding the places of insiders and outsiders in the US legal system.

He noted how many cases were started by outsiders against some of the insiders. Based on considering the following decisions, we add that many cases are triggered *by* the insiders attempting to beat back the claims of outsiders or explicitly attacking selected outsiders. It seems that when outsiders mount challenges to a current social order they are seen as threats. This observation motivated our name choice for this island.

The legislative foundations for the early decisions in this island are three laws enacted by Congress before 1920. The Selective Service Act of 1917 came first and authorized the federal government's use of conscription to create a national army with the US entering WWI.[20] The 1917 Espionage Act was passed about a month later. Far more consequentially, it prohibited all attempts to: 1) interfere with military operations; 2) interfere with military recruitment, and 3) encourage insubordination within the military. It was designed to prevent any support being given to enemies of the US in wartime. *Schenck v. United States* (249US47), the earliest case in the island, was the first to challenge (unsuccessfully) this law.

The 1918 Sedition Act extended the Espionage Act by defining more 'offenses' meriting punishment. They included making negative statements about the US government or its war efforts. Also, statements encouraging others to view the US or its institutions negatively could be punished severely. These laws targeted behaviors threatening, or seen as threatening, the war time social order. While Germany was the war time enemy, 1917 also saw the Bolshevik overthrow of the Russian Tsars. The creation of a communist state, with the potential spread of communist ideas, was viewed as another, perhaps graver, threat to US interests. Activities of communists, socialists, and their sympathizers were viewed as threatening the social order. Many were *brought into* US courts by governmental agencies to face charges regarding their beliefs and actions.

Also among the earlier cases are Jehovah's Witnesses, whose practices were seen as potent threats to religious and social orders. Attempts by labor to organize collectively against capitalist interests were seen as threats to capitalism. Soon, unions were featured in cases coming before the Supreme Court. Attempts by labor to picket peacefully were targeted as threats to free commerce. This suppressive logic was extended to thwart attempts by African Americans organizing for civil rights. The National Association for the Advancement of Colored Peoples (NAACP) was targeted in multiple places. Often, the notion of 'communist infiltration' was used as a part of suppressing both types of collective organizing.

Cases involving freedom of speech and press freedom are in this island. This was not a huge leap because those challenging a social order speak out. But when the press reported events related to these threats or commented adversely on governmental responses they, too, were seen as threats. It was easy to see inflammatory speech, protests over race and residential exclusion, and protests against wars (especially the Vietnam War) as threats to the established social order. Frequently, they were prosecuted as 'breaches of the peace'.

Many of the decisions coming later in this island seem, at face value, to be far removed from the above issues. But the connection comes through freedom of speech. When obscenity, whether in the form of words, images, or movies, is presented publically (or even privately) it was taken as a threat (presumably, to controlling behaviors of people). Suits claiming defamation and libel became a way of protecting insiders. Additionally, and surprisingly, birth control and abortion were viewed as threats to social order. We consider decisions involving all these issues to support our contentions regarding threat to social orders. More

---

[20] The Supreme Court upheld this law in 1918. The law was cancelled at the end of WWI.

specifically, techniques learned for attacking certain outsiders were then applied to other outsiders. However, before doing so, we describe the network structures of this line island.

6.4.2   The structures of the threats to social order line island

Figure 6.6 has pairs of *heavily* co-cited decisions in the full citation network. Each vertex is labeled by its formal ID, and color coded by its primary substantive issue. The layout is designed, where possible, to group together decisions having the same issue content. (Placing decisions using substantive content, rather than locating them by time, leads to a much more readable image.) The color coding for the main issue of the decisions is at the foot of the figure.

Twelve decisions involving communists and socialists are located in the lower right of Figure 6.6 (with three more in the top right). The middle of this figure has 12 Jehovah's Witnesses (JW) decisions. Below them are five decisions involving labor organizing (with another one above the JW cases). Nine decisions involving African Americans organizing are in the top right of Figure 6.6. Slightly above and to the left of the the JW decisions are five breach of the peace cases. The levels of co-citation are highest among the decisions of these five issue areas.

The top left of Figure 6.6 has 13 obscenity cases (with one in the lower left). Immediately below them are six on defamation and libel. Below, slightly to the right of these cases, are four concerning press freedom (with two more below labor organizing cases). Below these four cases are five on birth control and abortion (plus another obscenity case). To the right of these are six miscellaneous cases (with three more in the upper right area) of Figure 6.6. To the right of the 13 obscenity cases are six decisions on commercial speech. We can account for discrepant placements of decisions away from others with the same substantive issue.

Figure 6.7 shows the *citation* network *within* the line island. The color coding of decisions is the same as for Figure 6.6. Decades are marked on the left – differences in the temporal 'scale' lead to a slightly more readable image, given the concentration of decisions in some periods. Decisions are located according to when they were made. Most substantive issues are distributed across this citation network. There are 1116 arcs from later decisions to earlier decisions. Additionally, nine edges in this citation subnetwork link decisions citing each other: they were cases decided jointly. This is a dense patch within the overall Court citation network. However, the co-citation of decisions implied by these links is a small part of the overall co-citation of decisions in this line island.

6.4.3   Decisions involving communists and socialists

The earliest decision is *Schenck v. United States* (249US47). It upheld the 1917 Espionage Act in 1919. Schenck, Secretary of the Socialist Party of America, distributed leaflets to potential draftees advocating resistance to the draft. He was indicted, convicted, and jailed for six months, a conviction upheld by a unanimous Supreme Court. As his actions were deemed 'a clear and present danger' to the USA and its wartime recruitment efforts, the Court ruled that they were not protected by the First Amendment. The 'clear and present danger test' of this decision helped lay foundations for determining conditions under which freedom of speech (as well as press freedom and peaceful assembly) was not protected and could be curtailed. This is reflected by the presence of such cases in the decisions occurring later in this line island.

Figure 6.6 Co-citation links among decisions in the threats to social order line island.

Colors: salmon (communist and socialist targets, e.g. 249US47); wild strawberry (restrictions on labor organizing, e.g. 307US496); dandelion (restrictions on black organizing, e.g. 357US449); light orange (Jehovah's Witness targets, e.g. 303US44); lavender (freedom of speech, e.g. 328US331); light cyan (defamation and libel, e.g. 376US254); light blue (press freedom, e.g. 328US331); light purple (obscenity, e.g. 333US507); green-yellow (commercial speech, e.g. 424US1); gray (breach of the peace, e.g. 336US77); light faded green (birth control and abortion, e.g. 410US113); white (seemingly idiosyncratic issues)

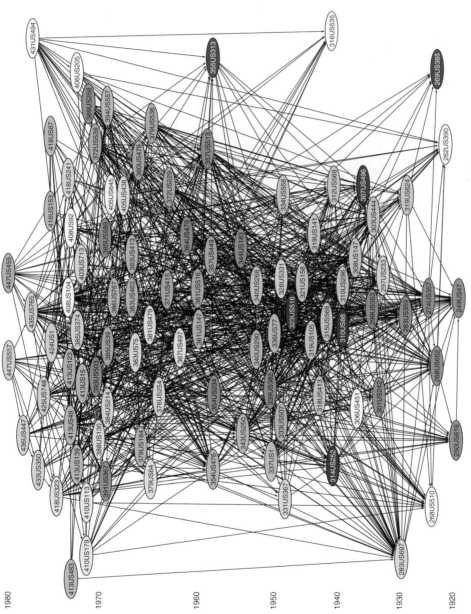

Figure 6.7 Citations among decisions in the threats to social order line island in time scale.

Colors: salmon (communist and socialist targets, e.g. 249US47); wild strawberry (restrictions on labor organizing, e.g. 307US496); dandelion (restrictions on black organizing, e.g. 357US449); light orange (Jehovah's Witness targets, e.g. 303US44); lavender (freedom of speech); light blue (press freedom, e.g. 328US331); light cyan (defamation and libel, e.g. 376US254); light purple (obscenity, e.g. 333US507); green-yellow (commercial speech, e.g. 424US1); gray (breach of the peace, e.g. 336US77); light faded green (birth control and abortion, e.g. 410US113); white (seemingly idiosyncratic issues)

The defendants in *Abrams v. United States* (250US616), another 1919 case in this island, were charged with inciting resistance to the war effort and urging labor strikes against plants manufacturing munitions. On conviction, they were sentenced to 20 years in jail. The 1918 Sedition Act, technically an amendment to the 1917 Espionage Act, made advocating restrictions on armaments production a criminal offense. The defendants, anarchists of Russian extraction, also protested the USA sending troops to Russia to aid the 'whites' against the 'reds' in opposing the Russian Revolution. The case against them was based on their actions being threats to the social order: indeed, they had advocated violent overthrow of the US government. A 7-2 decision upheld their convictions.[21]

Both of these cases were instances of outsiders being prosecuted, and they helped set the stage for the first 'Red Scare.'

6.4.3.1   The first Red Scare

The first Red Scare in the USA occurred in the early 1920s. It focused on denouncing worker revolutions and leftist radicals. *Gitlow v. New York* (286US652) was a part of this. Gitlow, a member of the Socialist Party of America, had distributed newspapers, starting in 1919, advocating the violent overthrow of the government. His conviction, under a New York criminal anarchy law, was upheld by the Court in 1925.

California had enacted its own criminal syndicalism act in 1919. Anita Whitney, a long-term women's rights activist, suffragist, and political activist who joined the Communist Party, was charged under this law in 1920 for helping create this political party. She was convicted and sentenced to a 1–14 year jail term. The Supreme Court upheld her conviction in 1927 with *Whitney v. California*.

Some states became very aggressive in their attacks on communists, so aggressive it seems that in the 1930s a change came in the nature of Supreme Court decisions regarding the 'threats' of communism and socialism. In 1919, California had passed a law banning the display of red flags, symbols of communism and socialism. This was struck down by the Court in 1931 in *Stromberg v. California*, a decision using the First and Fourteenth Amendments. The case was instigated by a group called the Better American Foundation wanting to purge California of threatening dissent.[22] They targeted a summer youth camp where Stromberg, a teenage member of the Young Communist League, was involved in a daily ceremony of raising a red flag.

Oregon went further by passing a law banning meetings of the Communist Party. This was struck down by a unanimous Supreme Court in *De Jonge v. Oregon* (299US353) in 1937. The basis was freedom of assembly guaranteed under the First Amendment. Also in 1937, *Herndon v. Lowry* overturned a conviction of an 18 month to 20 year jail sentence of Herndon for organizing public meetings of the Communist Party and the Young Communist League including both whites and blacks. It was a 5-4 decision with the minority composed of Justices Sutherland, McReynolds, Van Devanter, and Butler who became known as the 'Four Horsemen of Reaction.' According to Irons (2006: 264) they acquired this name by

---

[21] Justice Holmes, author of the unanimous *Schenck* opinion, dissented from *Abrams*. Irons (2006) argues that Holmes' views on the protections of the First Amendment shifted towards a broader view of freedom of speech. This is a reminder that both individual Justices and the Court can change over time regarding the nature of the constitutional principles underlying their decisions.

[22] There is a long history of groups using innocuous names to mask politically repressive goals.

'voting as a bloc against state and federal laws designed to rescue the American people from the ravages of the Great Depression.' The Supreme Court terms of these Justices were 1922–1938, 1914–1941, 1910–1937, and 1922–1939 respectively, with them consistently repelling perceived threats from outsiders.[23]

6.4.3.2 The second Red Scare

The second Red Scare following WWII lasted for at least a decade. It focused on preventing communists from infiltrating the federal government and influencing society as a result. Cases brought during this Scare are present in the next decisions in this island. Employers in government, education, and other sectors began using loyalty oaths. Potential employees were required to sign oaths declaring loyalty to the current social order as an employment condition. Some oaths required potential and actual employees to state they did not engage in political activities or belong to certain organizations, especially any associated with the Communist Party. Some even excluded educators from joining specified organizations. Not signing loyalty oaths could lead to dismissal or could deny applicants employment, especially in teaching positions. See Hyman (1959) for a fuller account of loyalty oaths.

Oklahoma enacted, in 1950, legislation requiring state officers and employees to take oaths pledging loyalty to the USA and to swear they did not support the overthrow of the US government. The primary target was support for alleged fronts of communist and other subversive organizations. A unanimous Court struck this law down in *Wiemann v. Updegraff* (344US183) in 1952. New Hampshire followed suit in 1951 by passing its subversive activities act defining any group advocating force to change the government as subversive. Such organizations could be dissolved under this law and their members barred from state employment. People employed in educational institutions were required to sign a statement declaring they were not subversives. *Sweezy v. New Hampshire* (354US234) struck down this law in 1957. It declared that Sweezy's academic freedom had been violated.[24] This decision invoked the protections of the First Amendment and the Bill of Rights.

Speiser v. Randall (375US513) added another wrinkle to loyalty oaths in 1958. California allowed some tax exemptions for war veterans who had served in WWII. Veterans claiming this benefit were required to complete a form. It was 'revised' to add a loyalty oath stating 'I do not advocate the overthrow of the Government of the United States or of the State of California by force or violence or other unlawful means, nor advocate the support of a foreign government against the United States in event of hostilities.' Those refusing to sign this statement, claiming that it was unconstitutional, were denied the tax exemptions and so sued the state. *Speiser* overturned this law in 1958 as violating First Amendment rights.

In 1960, with *Shelton v. Tucker* (364US479) the Court invalidated an Arkansas law requiring teachers at state-supported educational institutions to file an affidavit listing all of the organizations to which they had belonged in the previous five years. Further, they were

---

[23] They were staunchly laissez-faire and, on the Court, adamantly opposed programs enacted under the New Deal legislation of the 1930s.

[24] Sweezy, a Marxist and a member of the Progressive Party, had condemned the US use of force to support the capitalist social order. He had also given guest lectures at the University of New Hampshire. He was subpoenaed by the New Hampshire Attorney General to testify on multiple occasions. He answered most of the questions he was asked but refused to disclose the membership of the Progressive Party and the content of his lectures. Held in contempt, he was jailed. (For a fuller account, see http://lawhighereducation.com/124-sweezy-v-new-hampshire.html.)

required to list all of their contributions to organizations for the same period. When some teachers, including Shelton, refused to provide these lists, their contracts were not renewed.[25] The First Amendment (freedom to associate peacefully) and the Fourteenth Amendment (due protection) drove *Shelton* in 1960.

Keyishian v. Board of Regents, (385US589), another case involving education and communism, was decided in 1967. A New York law prohibited state employees from belonging to any organization advocating the overthrow of the US government.[26] The Regents of New York State University required employees to sign oaths stating they were not Communist Party members. Some faculty and staff had their contracts terminated after refusing to do so. Remnants of the second Red Scare were still active. In a narrow 5-4 decision, the Court overturned this law.

The final subversion case we consider had its origins in the 1950 Internal Security Act (also known as the Subversive Activities Control Act) enacted during the second Red Scare. Two features of this law surfaced in *United States v. Robel* (389US258), a case decided in 1967. One provision of this act was that the Communist Party was required to register with the US General Attorney. The second, more important, provision created the Subversive Activities Control Board. It had the power to investigate people suspected of conducting subversive activities. Robel was a communist and employed at a shipyard. He continued to work there even after it had been deemed a 'defense facility' by the Secretary of Defense. Under a section of the Subversive Activities Control Act, he was indicted for knowingly and unlawfully working at the shipyard while belonging the Communist Party. The Supreme Court ruled 7-2 in *Robel* that the US government cannot use 'national security' to deprive people of their constitutional rights.

6.4.3.3   The Warren Court on subversion

All of the *Sweezy* through *Robel* decisions discussed here were handed down by the Warren Court (1953–1969). Its stance firmly supported the freedoms granted under the First and Fourteenth Amendments. This court rebuffed concerted efforts designed to suppress, sometimes severely, activities deemed subversive. These repressive actions had been justified by the alleged need to face perceived threats to the nation articulated during the second Red Scare. The Warren Court did not have an accommodating view of these actions. Fowler and Jeon (2008) noted that the Warren Court's level of citations to prior decisions dropped compared to the historical record. Given its concerted effort to forge decisions more favorable to the outsiders described by Irons (2006), this drop is understandable.

These cases also point to an important feature of considering *co-cited* decisions: not all decisions in the same substantive domain will be included. For example, *Dennis v. United States* is absent. Dennis belonged to the Communist Party and was convicted for conspiring to overthrow the US government through force. The Vinson Court ruled, in 1951, that Dennis had no First Amendment rights of free speech, publication, and assembly if this involved plotting to overthrow the government. This decision was completely at odds with the subsequent Warren Court decisions. It is highly unlikely that it could be co-cited[27]

---

[25] He lost his case in the Arkansas courts even though the evidence established that he was not a communist. The case reached the Supreme Court on appeal.

[26] Nor were they allowed to belong to any organization deemed 'seditious' or 'treasonous'.

[27] *Dennis v. United States* was not co-cited enough to appear on any line island in Figure 6.3.

with diametrically opposed decisions made by the Warren Court. Co-citated decisions focus attention on consistent applications of constitutional principles and not only on substantive content.

This dramatic shift in handling the perceived threats of communists, socialists, and subversives, is a reminder of the potential and real impacts of unexpected events and changes in the composition of the Court. The sudden death of Chief Justice Vinson from a heart attack created a vacancy filled by the appointment of Earl Warren as Chief Justice. A very conservative Justice was replaced by a liberal Justice.

6.4.4   Restrictions of labor groups organizing

We next consider coordinated attempts to restrict organizing efforts of labor unions and African Americans. There are two connecting links between these decisions and those dealing with communists and socialists. One is that strategies and tactics used to deal with one set of perceived threats were refashioned to suppress other groups. The second connection was the use of communist threats in justifying the restriction of other organizing efforts seen as threats to social order.

The first decision on this island, *Hague v. Committee for Industrial Organization*[28] (307US496), came in 1939. A Jersey City ordinance forbade meetings of groups advocating 'unlawful' obstruction of government. Using it, the city disallowed a CIO meeting. The Jersey City mayor had called CIO members communists. The police also stopped the CIO from distributing its literature. The CIO filed suit under the First Amendment, prevailed in a District Court, and in the 3rd Circuit US Court of Appeals. *Hague* affirmed this decision striking down the city ordinance.

Thornhill v. Alabama (310US88) involved peaceful picketing of an employer during an authorized strike. However, an Alabama law prohibited all labor picketing. Thornhill, a local union president, was arrested, charged, and convicted. His conviction was reversed in 1940 using the First Amendment. Powe noted that this was labor's high point before the Court. 'The Court began to back away almost immediately, and a decade later a majority allowed a state to enjoin picketing for a lawful objective (2010: 229).' (See the *Douds* decision discussed below.)

Bridges v. California (314US252) was decided in 1941. Bridges, the leader of the long-shoreman's union, sent a telegram to the Secretary of Labor about a case pending at the Los Angeles County Superior Court. He implied that his union would strike if the Superior Court ruled unfavorably. Copies reached some West Coast papers, including the Los Angeles Times. The LA Times published an editorial on the court case. Both Bridges and the paper were held in contempt of court and fined. The Superior Court upheld the fines, a decision affirmed by Supreme Court of California. In a 5-4 decision, the US Supreme Court overturned the California contempt law and reversed both convictions using First Amendment free speech and free press protections.[29] Texas had a law requiring permits for pro-union speeches. Thomas, a labor organizer, was convicted after speaking at a union rally without a permit. In *Thomas v. Collins* (323US516) a unanimous Court reversed the conviction and overturned the Texas

---

[28] It was established as an umbrella organization for other unions in 1935. It soon became the Congress of Industrial Organization (CIO).

[29] See http://www.oyez.org/cases/1940-1949/1940/1940_1 for a fuller account.

law in 1944 ruling it inconsistent with First Amendment rights of freedom of speech and peaceful assembly.

The fate of labor unions plummeted with the Taft–Hartley Act (The Labor Managment Act) of 1947, an anti-union law designed to curb labor unions whose power was enhanced by the New Deal's 1935 National Labor Relations Act (NLRA). Business interests were bitterly opposed to the law creating the NLRA and sought to have it undone. Taft–Hartley was, in essence, designed for them. Its stated purpose was 'to promote the full flow of commerce, to prescribe the legitimate rights of both employees and employers in their relations affecting commerce, to provide orderly and peaceful procedures for preventing the interference by either with the legitimate rights of the other, to protect the rights of individual employees in their relations with labor organizations whose activities affect commerce, to define and proscribe practices on the part of labor and management which affect commerce and are inimical to the general welfare, and to protect the rights of the public in connection with labor disputes affecting commerce.' There were teeth under this seemingly vanilla legal language. The NLRA prohibited unfair labor practices by employers. Taft–Hartley added prohibitions of labor union actions included jurisdictional strikes (unions striking to get work assigned to their members), wildcat (unauthorized) strikes, solidarity strikes, picketing, and union contributions to political campaigns.[30] It laid foundations for efforts to break labor unions (that continue in the 21st Century).

This law also required union leaders to sign oaths and file them with the National Labor Relations Board (NLRB) stating that they were neither communists nor advocates of overthrowing governments. Unions whose leaders refused to sign this oath risked not being covered under the NLRA. Taft–Hartley was upheld by the Vinson Court in *American Communications Association (ACA) v. Douds* in 1950. The ACA represented telegraphists and radio operators. Many of its officers were communists who refused to sign the anti-communist oath. Douds, a New York regional director of the NLRA, blocked the ACA from taking part in an NLRB-supervised union organizing election, the first application of Taft–Hartley. The ACA sued, citing First Amendment rights. The Vinson Court ruled that the imposition of anti-communist oaths on union leaders did *not* violate the First Amendment!

Staub v. City of Baxley (355US313) was a 1958 Warren Court decision. Baxley enacted an ordinance targeting organizations, unions, or societies requiring member dues or fees. Organizations were required to get permits for approaching people within the city seeking members. The city mayor and council granted permits (or not) depending on assessing organizations seeking permits. Without one, Staub approached people in their homes requesting they join a union. She was convicted under the ordinance. The Court threw out the conviction ruling the city ordinance unconstitutional: First and Fourteenth Amendment constitutional freedoms could not depend on the arbitrary will of political authorities. It was a narrowly worded victory for labor compared to the far-reaching *Douds* decision.

6.4.5   Restrictions of African Americans organizing

The mid 1950s to the late 1960s saw civil rights for African Americans as a major issue: the Civil Rights Movement came of age on the national stage. Given the US history of

---

[30] It also allowed states to pass 'right-to-work' laws outlawing closed union shops (where employers were required to hire union members). The Taft–Hartley Act remains in force with states currently passing these laws.

slavery, strong and violent reactions followed. We focus on African Americans organizing and attempts to block them from doing so. Blacks organizing to confront the oppressive conditions of their lives was seen as a major threat to the established social order of the time, especially in the South. Not surprisingly, the five decisions we consider here all came from this region. The National Association for the Advancement of Colored People (NAACP) was prominent in this organizational effort. It was targeted by state and local authorities.

NAACP v. Alabama (357US449) started when the Alabama Attorney General filed suit against the NAACP. Alabama had a statute requiring 'foreign' (outside the state) corporations to qualify for doing business there. The NAACP was based in New York. It was involved in the 1955 Montgomery Bus Boycott, a historic event precipitated by Rosa Parks not yielding her bus seat to a white. It also supported black students seeking admission to the state university. Both efforts triggered widespread white outrage. Alabama's suit alleged that the NAACP was 'causing irreparable injury to the property and civil rights of the residents and citizens of the State of Alabama' and demanded its membership list. The NAACP refused to comply. It was held in contempt, fined $100,000, and banned from operating in Alabama. A unanimous Court ruled against Alabama: its courts' decisions were overturned and its law voided in 1958. The same tactic, demanding the NAACP membership list, was used in Arkansas. *Bates v. City of Little Rock* (316US516) voided an Arkansas law and a conviction based on it in 1960. First Amendment rights of free association implied that state authorities could not compel organizations to provide their membership lists.

The Virginia General Assembly enacted five statutes in 1956 aimed at curbing the NAACP. Segregationists saw it as 'stirring up' integrationist lawsuits against the state. The legislation's target was any group opposing or proposing legislation attempting to influence public opinion on behalf of any race. Virginia also required membership lists from these organizations. Their laws suffered the same fate as those from Alabama and Arkansas under *NAACP v. Button* (371US414).

Gibson v. Florida Legislative Investigation Committee (372US539) is listed as another important decision in a variety of sources and is present in this line island. Florida created an investigative committee to examine communist infiltration of organizations operating in the state. Gibson, president of the NAACP Miami branch, was subpoenaed to provide its membership list. He refused, was held in contempt, and found guilty. The Warren Court reversed this conviction on the same grounds as for the above decisions – but with a narrow 5-4 majority.

The Southern Conference Educational Fund (SCEF) was a group of Southern liberals fighting for African American civil rights. It was targeted and harassed as a result. Louisiana threatened also to prosecute the SCEF under anti-subversion laws. Dombrowski, who worked for the SCEF, was arrested. After his office was raided by state authorities, he wanted the seized materials returned and he sought damages. When his claim was dismissed in state courts, he appealed directly to the Supreme Court. In 1965, *Dombrowski v. Pfister* (380US479) ruled that the state conduct was unconstitutional. The wording of its opinion included 'The mere possibility of erroneous initial application of constitutional standards by a state court will not ordinarily constitute irreparable injury warranting federal interference with a good faith prosecution and the adjudication during its course of constitutional defenses. But equitable relief will be granted to prevent a substantial loss or impairment of freedoms of expression resulting from prosecution under an excessively broad statute regulating expression.' The Court noted the 'chilling effect' that the Louisiana court ruling could have on all First Amendment rights.

A consistency runs through suits against subversion (featuring communists and socialists), suits attempting to restrict labor organizing, and suits to obstruct organization by the Civil Rights Movement. All were seen as threats to social orders dominated by privileged elites, business interests, and whites: as a result, these were countered. The initial success in the courts against communists and socialists using 'clear and present danger' arguments appears to have motivated subsequent efforts blocking organization by other outsiders. Also, the subversive threat of communism could be used against unions and blacks. The Warren Court (1953–1969) stands out as a bulwark against some of these efforts, consistent with the decline of its citations to earlier decisions and the fewer citations to its decisions from the conservative Burger (1969–1986) and Rehnquist (1986–2005) Courts.

6.4.6   Jehovah's Witnesses as a perceived threat

All of the decisions considered thus far feature the First Amendment's protections of political rights. The decisions considered in this section are concerned with First Amendment protections of religious rights. Just as political actors can be seen as posing threats, so too can religious actors. In both groups of cases there has been selectivity as to which specific groups are seen as posing unacceptable threats.

Clashes between Christian groups – ranging from full-scale war through severe persecution to petty persecution – have persisted over the ages. The USA is no exception with regard to this, with repeated attacks on specific religions. One such group was the Jehovah's Witnesses who were targeted because of their religious doctrines, beliefs, and practices stemming from their beliefs.[31] Decisions on this island feature Jehovah's Witnesses in suits filed against them. Cox (1987): 189 argued that the Jehovah's Witnesses were 'the principal victims of religious persecution (after) they began to attract attention and provoke repression in the 1930s, when their proselytizing and numbers rapidly increased.' Clearly, they were seen as creating a threat that had to be neutralized and repressed. These cases can be group into 'nuisance' threats, direct threats to state interests, and 'petty' persecutions.

6.4.6.1   Nuisance threats

Lovell v. City of Griffin (303US444) was decided in 1938. Lovell, a Jehovah's Witness, distributed literature in the streets. She was charged under a city ordinance banning literature distribution deeming it a 'nuisance'. She was fined $50 and sentenced to jail for 50 days after refusing to pay the fine. Her appeals, including one to the Georgia Supreme Court, were denied. The US Supreme Court, using the First Amendment, ruled that the ordinance was too broad. The conviction was reversed. This issue appeared also in *Schneider v. New Jersey* (308US147), a decision made in 1939 under which a city ordinance was invalidated and a conviction under it reversed. This ordinance had gone further than Griffin's by banning people going door-to-door to distribute literature. Three more cases on this island – *Jamison v. Texas* (318US413), *Murdock v. Pennsylvania* (319US105), and *Martin v. City of Struthers, Ohio* (319US141) – all involving nuisance suits[32] had the same outcome in 1943. This geographic spread of using 'public nuisance' ordinances suggests a widespread tactic

---

[31] Mormons have been another targeted groups but their cases are not on this island.

[32] In all of these cases, it is not clear if the city ordinances were drafted with the Jehovah's Witnesses in mind or merely used selectively to target them.

used against Jehovah's Witnesses. All of the targeted individuals were acting peacefully in distributing information and seeking converts.

The 1940 case, *Cantwell v. Connecticut* (310US296), was complicated by a more provocative approach taken by Cantwell and his sons when proselytizing in a heavily Roman Catholic New Haven neighborhood. Going door-to-door, they explicitly attacked the Roman Catholic Church as an organized religion. Using the First Amendment's Free Exercise Clause and the Fourteenth Amendment, the Court ruled that this more aggressive, even offensive, behavior by some Jehovah's Witnesses was protected.

Cox v. New Hampshire (312US569) considered *collective* behavior of people proselytizing. Almost 70 Jehovah's Witnesses left their church to form smaller groups spreading their word. Groups with 15 to 20 people marched in single files impeding regular pedestrian flows. They did so without a parade permit claiming that their religious freedom for assembly and worship under the First Amendment meant that they did not have to get one. They were convicted for violating a state law and appealed. The Court ruled, in 1941, that while the government cannot regulate the content of speech, 'it can place reasonable time, place, and manner restrictions on speech for the public safety.' It stated also 'every parade or procession on public streets has to have a license and organizers have to pay a fee.' Even so, the Court cautioned that granting licenses must satisfy the First and Fourteenth Amendments.

Chaplinsky v. New Hampshire (315US568) also involved Jehovah's Witnesses using public spaces. Chaplinsky made a Rochester sidewalk his pulpit to call organized religion a racket. The town marshall warned him to 'keep it down' but a large crowd assembled and blocked roads. A police officer removed him from the sidewalk to a police station. There, Chaplinsky met the town marshall again and attacked him with 'fighting words'. He was convicted for breaching the peace. The Court upheld his conviction for attacking the town marshall. It seems his proselytizing was legal but for the large crowd assembling.[33] His removal from the street seems consistent with the principle of needing reasonable time, place, and manner restrictions on speech for public safety established in the *Cox* decision.

6.4.6.2   Direct threats to state interests

All of the nuisance cases (except *Schneider*) described above in various sources are listed as important Supreme Court decisions. While the next decision we discuss is listed also as important, it featured a completely different and more far-reaching issue. Pledges of allegiance (to the state) have been adopted often when the US was at war. Jehovah's Witnesses object to reciting these pledges on religious grounds. Also, they refuse to serve in the US military. Both stances helped increase antagonism towards them.

One *prior* Court decision, *Minersville School District v. Gobitis* is not in this island but it set the stage for the one we discuss next. *Gobitis* ruled, in 1940, that public schools could compel students, including Jehovah's Witnesses, to recite these pledges, despite their own or parental objections. Three years later, this issue was revisited in *West Virginia State Board of Education v. Barnette* (319US624) which overturned *Gobitis*.

The West Virginia Board of Education mandated saluting the flag in all public schools. Students not doing so were to be punished by expulsion and charged with delinquency by

---

[33] It is not clear if the crowd supported or opposed Chaplinsky. If it was the latter, a community response helped drive him off the street.

being 'voluntarily' absent. (Once expelled, students were readmitted only if they saluted the flag.) The Court ruled the West Virginia mandate unconstitutional on First Amendment grounds. However, this decision rested more on freedom of speech protections than on free exercise of religion. Justice Jackson wrote the majority opinion of this 6-3 decision. It included: 'If there is any fixed star in our constitutional constellation, it is that no official, high or petty, can prescribe what shall be orthodox in politics, nationalism, religion, or other matters of opinion or force citizens to confess by word or act their faith therein.' By reversing *Gobitis* this was a momentous decision.

Fowler and Jeon (2008) argue that overturned decisions are important. However, this does not imply that they will be co-cited frequently with the decisions overturning them. Indeed, this almost guarantees that they will not be co-cited. It is not surprising that *Gobitis* and *Barnette* were not co-cited enough to appear on this island because they rest on different constitutional principles despite being decided only three years apart. A case can be made for *not* treating the citation link between a decision and the one it reverses as just another citation tie.

Barnette was a major court victory for the Jehovah's Witnesses. Part of an explanation for why *Gobitis* was reversed lies in the changed composition of the Court. Justice Frankfurter wrote the majority opinion in *Gobitis*. He was joined by Justices Roberts, Black, Reed, Douglas, Murphy, and Chief Justice Hughes. Justice McReynolds, the last of the 'Four Horsemen of Reaction' still on the Court, wrote a concurring opinion. Only Justice Stone dissented. Three years later, Justice Jackson, a Court newcomer, wrote the majority opinion for *Barnette* joined by Justices Black, Douglas, Murphy,[34] and Rutledge (another new Court member) plus Chief Justice Stone (who had succeeded Hughes in this position). The dissenters were Justices Frankfurter, Roberts, and Reed – all from the *Gobitis* majority. Two other members of the *Gobitis* majority had been replaced by new Justices in the *Barnette* majority. Also, the lone dissenter to *Gobitis* had become Chief Justice. These compositional shifts seem decisive in this reversal, something not picked up in the receipt of citations to prior decisions.[35]

There is another profound contrast between these two decisions. Frankfurter's opinion and the *Gobitis* decision led directly to immense harm being done to the Jehovah's Witnesses. According to Irons (2006: 341), this decision 'unleashed a wave of attacks on the Witnesses across the country. Within two weeks of the Court's decision, two federal officials later wrote "hundreds of attacks upon the Witnesses were reported to the Department of Justice." ' These attacks, well outside the US court system, were seen as the terrorist actions they were. Irons quotes the same officials as claiming 'the files of the Department of Justice reflect an uninterrupted record of violence and persecution of the Witnesses. Almost, without exception, the flag and the flag salute can be found as the percussion cap that sets off these acts.' In order to constrain a peaceful perceived threat to state interests, the Court's decision helped trigger violence. Arguably, this was a greater threat to social order even though it does not appear that it was seen as such in the legal arena until the *Barnette* decision.

Irons, noting that Jackson was a close friend and judicial ally of Frankfurter, expressed surprise over Jackson writing the majority *Barnette* opinion. Irons (2006): 344 added: 'Most

---

[34] Justices Black, Douglas, and Murphy were also involved in writing concurring opinions.

[35] This was not picked up in co-cited counts either. However, examining co-cited decisions forces attention to their contexts by looking for the reasons behind Court decisions, especially reversals.

amazing of all, Jackson tore Frankfurter's *Gobitis* opinion to shreds: the Justice with one year of law school handed the Harvard professor a failing grade in front of the whole class.' Three Justices (Black, Douglas, and Murphy) had, in effect, switched sides on this issue. As an additional symbolic gesture, *Barnette* was handed down on Flag Day (June 14) as a potent reminder of its meaning. Jackson also cited the *Stromberg* decision discussed in Section 6.4.3 when a teenager had been convicted for raising a red flag.

6.4.6.3   Petty persecutions

The remaining decisions on this island feature petty persecutions of Jehovah's Witnesses. *Prince v. Massachusetts* (321US158), another listed important Court case, added a new twist. Prince, a Jehovah's Witness, was guardian for a nine-year old girl whom she brought to a downtown area while preaching. Literature was distributed in exchange for voluntary contributions. Prince was charged with violating Massachusetts child labor laws and convicted. The Court upheld this conviction in 1944 by viewing the case as a labor law issue rather than the free exercise of religion. In 1948, *Saia v. New York* struck down a New York law prohibiting using sound amplification equipment without the consent of a local police chief. Saia, a Jehovah's Witness minister, had obtained a permit from the Lockport Chief of Police to use sound amplification for his religious lectures at a fixed place on a Sunday. After the permit expired, he applied for a new one. He was denied one on the grounds that complaints had been received about noise.

The final Jehovah's Witnesses case is *Niemotko v. Maryland* (340US268), a decision handed down in 1951. Havre de Grace issued permits to civic groups to use a public park in the town. A Jehovah's Witnesses group sought a permit for one Sunday but was denied on the grounds of another group already having one. The City Council turned down all their permit applications for other Sundays. At a hearing, they were questioned about not saluting the flag and their opinions of Roman Catholics, topics irrelevant for permit applications. The Jehovah's Witnesses then held a meeting without a permit. Niemotko, their preacher, was arrested under a disorderly conduct statute despite there being no evidence of disorder at the meeting. A unanimous Court ruled that Havre de Grace had violated the First Amendment's free exercise of religion clause by denying permits for Jehovah's Witnesses when other groups had been granted permits.

6.4.6.4   A partial summary

Most, but not all, of the cases involving Jehovah's Witnesses resulted in decisions rejecting efforts by *local* authorities to restrict their activities. Two that did not, turned on other issues specific to those cases. In contrast, most of the decisions regarding communists (Section 6.4.3), labor organizations (Section 6.4.4), and black organizations (Section 6.4.5) involved state and federal authorities. When the Jehovah's Witnesses were seen as posing threats to state interests, higher levels of government became involved. Despite the involvement of very different levels of government in these cases, the same constitutional principles were involved in rendering decisions. The decisions regarding Jehovah's Witnesses[36] rejected

---

[36] According to Kosmin et al. (2001: Exhibit 13) and Mayer et al. (2001), in 1990, the Jehovah's Witnesses had the highest percentage (37%) of African Americans of all of the identified religious groups. This could well have been an additional motivation for seeing this religious group as a threat.

restrictions on targeted groups thought to threaten social order. Most of the rulings considered here affirmed First Amendment protections holding for all, including outsider groups.

6.4.7   Obscenity as a threat to social order

We consider another very different perceived threat in this section. Obscenity is very hard to define. According to the Merriam-Webster dictionary, components of this term include being abhorrent to morality or virtue, designed to incite lust or depravity or having language regarded as taboo in polite usage. Most cases in this island focus on the second item, especially in the form of pornography. Definitions of obscenity and pornography vary over time, between cultures and across places. If obscenity offends morality, virtue; or polite conversation, or incites lust and depravity, it can be seen as threatening social order. Most often, obscenity referred to explicitly sexual depictions in words and/or visual images. Reactions included attempts to remove the depictions from public view (and even private places). Laws outlawing obscenity, however defined, when enforced lead to censorship inhibiting free expression: civil liberties issues brought obscenity into the US court system where freedom of speech under the First Amendment was a primary consideration.

All definitions of legal standards for discerning obscenity were problematic due to its inherent ambiguity. Justice Stewart, writing an opinion for one of the cases that we consider here, noted that hard-core pornography was hard to define but added 'I know it when I see it.' Yet, different individuals will see different things as pornographic. This inevitable subjectivity complicates attempts to use the Constitution in a uniform fashion. Eight of the decisions about obscenity in this line island were made by the Warren Court (1953–1969) and seven by the Burger Court (1969–1986). Both Courts created 'tests' for defining obscenity.

6.4.7.1   The Warren Court on obscenity

Roth v. United States (354US476), decided in 1957, created a new legal definition of obscenity. Hitherto, the 'Hicklin test' (resulting from an 1868 case in England) held material that could 'deprave and corrupt those whose minds were open to such immoral impulses' was obscene.[37] Roth was convicted under a federal statute criminalizing using mail to send materials that were 'obscene, lewd, lascivious or filthy' and publishing documents with literary erotica and nude photographs. The Court upheld his conviction. Its new standard held obscenity was material whose 'dominant theme *taken as a whole* appeals to the prurient interest.' To make this more precise, the Court added a definition and a prescription: an 'average person, applying contemporary community standards' would assess the depiction of sexual matters. This new test included obscenity being 'utterly without redeeming value.' Obscenity, as defined by the Roth Test, was not protected speech under the First Amendment.[38]

A California law criminalized possession of obscene books. It was struck down by the Court in *Smith v. California* (361US147) in 1959. The Court objected to the law not requiring proof that anyone having such a book knew of its content. The First and Fourteenth

---

[37] Books including D.H. Lawrence's *Lady Chatterly's Lover* were banned because *isolated* passages in them could have an effect on, among others, children.

[38] Chief Justive Warren (in a concurring opinion) included a concern that the 'broad language used (in the majority opinion) may eventually be applied to the arts and sciences and freedom of communication generally.'

Amendments safeguarded freedom of expression and prevented such intrusions by the state. *Jacobellis v. Ohio* (378US184), another listed important decision, reversed a conviction of a theater manager for showing *Les Amants* (*The Lovers*) in Cleveland Heights. The Court ruled the film was not obscene. Therefore, showing it was Constitutionally protected. This 6-3 decision had five opinions supporting it with none having more than two signatures: the majority agreed on the outcome but without a fully shared rationale.

A Book Named 'John Cleland's Memoirs of a Woman of Pleasure' v. Attorney General of Massachusetts (383US413) was decided by the Warren Court in 1966. The book is better known as *Fanny Hill*. The Massachusetts Supreme Court agreed with its Attorney General: *Fanny Hill* was 'patently offensive' and obscene under the Roth test. Reversing this ruling, the US Supreme Court put stress on *utterly* in 'utterly without redeeming value' claiming its Roth test had been applied in error. The outcomes of *Smith* and *Fanny Hill* suggest that the Roth test was hard to apply in a consistent fashion across different cases.

Ginzburg v. United States (383US463), another case deemed to be important on this island, has additional interest value. Ginzberg was convicted for mailing circulars detailing how obscene publications could be obtained. The circulars were not obscene. The Court upheld his conviction because the sole purpose of the circulars was to exploit commercially erotic material with prurient intent. This seems at odds with prior Warren Court decisions. It confused matters further by claiming that erotic materials, if discussed within the context of art, literature, or even science, were protected by the First Amendment if human understanding was promoted. This suggests the presence of an implicit bias regarding what can be seen as not being obscene, one reinforced later by the Burger Court.

Another obscenity issue was raised in *Redrup v. New York* (386US767). Redrup, a Times Square news-stand clerk sold two paperback sex novels to a plainclothes police officer. He was arrested, charged, and convicted for selling obscene materials in 1965. The Court reversed his conviction by a 7-2 decision in 1967. The majority's rationale was twofold. The books: 1) were not sold to minors; and 2) were not foisted on unwilling audiences. As such, these sales were protected constitutionally. While this decision does not appear on the accepted lists of important decisions, an argument can be made for it effectively ending US censorship of written 'obscene' materials.

The final Warren Court decision considered here, *Stanley v. Georgia* (394US557), unanimously reversed the Georgia Supreme Court's upholding Stanley's conviction. Previously convicted for illegal bookmaking (betting), he was suspected of having continued his bookmaking activities. Police, with a federal warrant, searched his house for betting materials. They found none but they did find reels of pornographic film in a bedroom. He was convicted for possessing obscene materials. The Court ruled private possession of obscene materials was not a crime.

6.4.7.2   The Burger Court on obscenity

The Burger Court, dissatisfied with the Roth Test, created a new obscenity test in *Miller v. California* (413US15) in 1973. The Miller Test had three components: 1) the arbiter of obscenity was an 'average person applying community standards' when deciding if some work, as a whole, appealed to the prurient; 2) applicable state law was a foundation for deciding if some work is clearly offensive when describing sexual conduct; and 3) the work, as a whole, is devoid of 'serious literary, artistic, political, or scientific value.' To be considered obscene, all three components had to apply to a work.

Miller was the first of five 1973 Burger-authored 5-4 opinions on obscenity. Miller had a large mail-order business dealing with pornographic materials. He was convicted under a California law for mailing unsolicited materials with explicitly sexual content. The Burger Court upheld the conviction under its new test. The next opinion came in *Paris Adult Theatre I v. Slaton* (413US49) when the Court upheld a Georgia civil injunction preventing the showing of two films thought to be obscene. The theater operators had warned potential audiences of potentially offensive material and all required viewers to be at least 21. A lower trial judge ruled these provisos allowed showing the films. The Georgia Supreme Court overruled the judge. The US Supreme Court agreed. Both *Miller* and *Slaton* appear on lists of important decisions, presumably because they helped set a new course for dealing with the perceived threat of obscenity.

The Burger Court's more restrictive approach towards obscenity continued with *United States v. 12 200-Ft. Reels of Super 8mm Film* (413US123), *United States v. Orito* (413US139) and *Hamling v. United States* (418US87). In the first case, the defendant, Paladini, imported obscene materials from abroad for his private home use. Convicted under a California law, his materials were confiscated. The *Stanley* decision of the Warren Court had ruled having obscene material in the home was protected under the First Amendment. While the Burger Court accepted this argument, it ruled that the protection did not extend to importing obscene materials. *Orito* extended this logic to transporting obscene materials by interstate commerce. The Court upheld Orito's conviction and the national law underlying it: Congress had the power to prevent obscene materials not protected by the First Amendment from 'entering the stream of commerce.' *Hamling* upheld another conviction for mailing obscene materials. All of the decisions in this island made by the Burger Court upheld obscenity convictions.

The final obscenity decision we consider, *Jenkins v. Georgia* (418US153), also came from Georgia's Supreme Court, this time affirming a conviction for showing an 'obscene' film. However, a unanimous US Supreme Court reversed this decision arguing its Miller test had been applied incorrectly. The film, *Carnal Knowledge*, was shown in Albany, Georgia in 1972. Police seized the film, and the theater manager, Jenkins, was convicted for 'distributing obscene material.' Justice Rehnquist's majority opinion includes 'Our own viewing of the film satisfies us that *Carnal Knowledge* could not be found under the Miller standards to depict sexual conduct in a patently offensive way. Nothing in the movie falls within either of the two examples given in Miller of material which may constitutionally be found to meet the ''patently offensive'' element of those standards, nor is there anything sufficiently similar to such material to justify similar treatment. While the subject matter of the picture is, in a broader sense, sex, and there are scenes in which sexual conduct including ''ultimate sexual acts'' is to be understood to be taking place, the camera does not focus on the bodies of the actors at such times. There is no exhibition whatever of the actors' genitals, lewd or otherwise, during these scenes. There are occasional scenes of nudity, but nudity alone is not enough to make material legally obscene under the Miller standards.'

At face value, this decision is at odds with the Burger Court's stance on obscenity. Yet the Court used its Miller standard even though it reversed an obscenity conviction. The social background to this decision includes changes in moral values growing out of the 'swinging '60s' and the 1970s which led to a greater acceptance of public discussions of sexual conduct. *Jenkins* also revealed some differences existing between local community and national standards for obscenity. Of the three components of the Miller test, greater weight was placed on an implicit national standard. While the Burger Court took a more restrictive approach to obscenity than the Warren Court, both courts shared a bias in seeing potentially obscene depictions as not being obscene if they were made for some higher purpose, in this case, art. Other depictions were seen as a threat to community standards, and the implied social order, against which defenses were needed.

6.5   Other perceived threats

There are other substantive issues identified in this island. We deal with some briefly in this section because they show that perceived threats changed their form towards the end of the time period we cover. While these cases are held together by the First and Fourteenth Amendments, the perceived threats changed with the times even though some of the legal tactics for dealing with perceived threats did not.

6.5.1.1   Birth control and abortion

It is a short step from obscenity to issues of sexual freedom as perceived threats to social order. The first two cases considered here are listed as important Supreme Court decisions. The Warren Court decided both. *Poe v. Ullman* (367US497) considered an old Connecticut law banning contraceptive devices and providing advice regarding them. It applied even to married couples! A woman who had gone through a life-threatening pregnancy was advised that the next one could prove fatal. She challenged this law after Connecticut's Attorney General threatened to enforce it. The Court ruled, in 1961, that the time was not ripe to strike down this law: the plaintiffs had no standing. The Court did strike this law down in 1965 with *Griswold v. Connecticut* (381US479). Woven through the majority opinion of Justice Douglas was the idea of a 'right to privacy' with an emphasis on marital privacy. We note that the issue of privacy also was a part of some of the obscenity cases considered in Section 6.4.7.

The Burger Court went a step further in *Eisenstadt v. Baird* (405US438), another important case, by striking down a Massachusetts law – concerned with 'crimes against chastity' – criminalizing the use of contraceptives by unmarried couples in 1972. This 6-1 decision – two new Justices had not yet been sworn in – extended the *Griswold* right to privacy logic to unmarried couples. Chief Justice Burger was the lone dissenter. The majority's rationale lay in First Amendment protections. While *Griswold* and *Eisenstadt* were widely popular decisions, the next highly important decision, *Roe v. Wade* (410US113), on this island regarding a woman's right to an abortion set off a firestorm of criticism leading to the formation of the 'right to life' movement. The effects of this decision, decided simultaneously with *Doe v. Bolton* (410US179) in 1973, reverberate to this day.[39]

6.5.1.2   Press freedom and free speech as perceived threats

Freedom of the press and free speech are essential to a democratic society. Both are guaranteed under the Constitution. Yet they have been seen often as threats to institutionalized agencies. The first two cases we consider involve courts and the press. *Pennekamp v. Florida* (328US331) featured a contempt citation against a newspaper publisher, and an associate

---

[39] *Roe v. Wade* has been under attack since it was decided. Currently, US states under Republican control are passing ever more stringent laws with the intent of banning abortions altogether, seemingly without regard for rape, incest, or even threats to the life and health of the potential mother. The position of nominees to the Supreme Court regarding abortion became the most prominent 'litmus test' during their hearings before the Senate for confirmation (or not) since 1973.

editor, for printing two editorials and a cartoon critical of a Florida trial court. An 8-0 majority ruled in 1946 that this public press commentary was protected. Two decades later, in *Mills v. Alabama* (384US214), the Warren Court unanimously voided an Alabama law and a conviction of newspaper editor under it. One section of the law prohibited editorials advocating that people to vote in a particular way on an election day. The editor, Mills, published an editorial urging people vote for a mayor-council form of government. He was arrested, charged, and convicted. Justice Black's opinion included a stinging assessment: 'The Alabama Corrupt Practices Act, by providing criminal penalties for publishing editorials such as the one here, silences the press at a time when it can be most effective. It is difficult to conceive of a more obvious and flagrant abridgment of the constitutionally guaranteed freedom of the press.'

New York Times Co. v. United States (403US713) is a landmark case better known as The Pentagon Papers.[40] The First Amendment was the foundation for a 6-3 decision dismissing the Nixon White House claim – based on The Espionage Act discussed in Section 6.4.3 – that it had the executive authority to bar publication of the Pentagon Papers. This decision allowed the *New York Times* and *Washington Post* to publish the then-classified Pentagon Papers without fear of government censorship or punishment. While it was a ringing endorsement of press freedom, it was a *per curiam* decision with six concurring opinions. The three dissenters included the Chief Justice. Two years later, a unanimous Court in *Miami Herald Publishing Co. v. Tornillo* (418US241) struck down a Florida law requiring newspapers to grant political candidates equal access whenever a paper published a political editorial or an endorsement. Burger's opinion included 'the statute ... fails to clear the First Amendment's barriers because of its intrusion into the function of editors in choosing what material goes into a newspaper and in deciding on the size and content of the paper and the treatment of public issues and officials.'

These decisions were firmly for press freedom. However, when the issue shades into areas involving defamation and libel in publications, the outcomes are less clear. They depend on content and the accuracy of what is written. The first case we consider is *New York Times Co. v. Sullivan* (376US254). The context for this case is the Civil Rights movement touched upon in Section 6.4.5. At the time, newspapers reporting on events about civil rights in the South had to be very careful. There were many pending court cases in which newspapers had been sued over statements deemed libelous, especially as they applied to police actions. The New York Times published a one-page advertisement in 1960 soliciting funds to help defend Martin Luther King Jr. against a perjury indictment in Alabama. It described actions taken against civil rights protesters by the Montgomery police force. Some of the descriptions were inaccurate. The advertisement also stated that King had been arrested seven times by the Alabama State Police, another inaccuracy – he had been arrested four times when the advertisement appeared. The *New York Times* was sued by Sullivan, the Montgomery Public Safety commissioner, for libel. An Alabama Court awarded him $500,000 and the Alabama Supreme Court upheld the award. A unanimous US Supreme Court reversed this decision.

More importantly, this decision created a new criterion for assessing libel with its 'actual malice standard': plaintiffs have to prove that the publisher of a potentially libelous statement knew it was false or had acted in *reckless disregard* for the truth. False statements, alone, are not enough to warrant a guilty libel verdict. Motives for filing libel suits are relevant. The *New*

---

[40] This document resulted from the Secretary of Defense commissioning a large-scale top-secret history of the US role in Indochina. The context was rising public opposition to the Vietnam War.

York Times argued that the motivation for the suit against it was press intimidation, echoing the concerns of Justice Black in his *Mills* opinion. *Sullivan*'s affirmation of press freedom allowed other newspapers to report on civil rights matters including repressive actions by law enforcement and public officials. Force, both official and unofficial, was used across the south to thwart civil rights actions, and civil disorder resulted. Yet the perceived threat was the *reporting* of these events rather than the events themselves.

Garrison v. Louisiana (379US64) explicitly extended the logic of *Sullivan* to a case not involving the press. Quoting from the case, 'Appellant is the District Attorney of Orleans Parish, Louisiana. During a dispute with the eight judges of the Criminal District Court of the Parish, he held a press conference at which he issued a statement disparaging their judicial conduct. As a result, he was tried without a jury before a judge from another parish and convicted of criminal defamation under the Louisiana Criminal Defamation Statute.' Garrison was upset by a large backlog of pending criminal cases. He blamed this on the laziness and excessive vacations of the judges. He added, 'The judges have now made it eloquently clear where their sympathies lie in regard to aggressive vice investigations by refusing to authorize use of the DA's funds to pay for the cost of closing down the Canal Street clip joints. . . ' and '. . . This raises interesting questions about the racketeer influences on our eight vacation-minded judges.' The Court struck down Louisiana's criminal defamation statute and the conviction based on it: the Constitution limits state power to impose sanctions for criticism of the *public conduct of officials*.

6.5.1.3   Commercial speech and threats restricting it

In the free speech cases we have considered, the primary issue involved individuals charged for threatening the social order through their speech and the Court protecting their free speech rights under the First Amendment. *Buckley v. Valeo* (424US1), another listed important case in this line island, considered commercial speech. It centered on the Federal Election Campaign Act (FECA) of 1971, especially its amended form passed in 1974. This is another instance of federal laws and Supreme Court decisions affecting each other. The 'threats' concern the role of money in elections, an issue with a long legislative history.

President Theodore Roosevelt expressed concern over corporate spending as a corrupting force influencing elections. He sought campaign finance reform through legislation to ban corporate contributions designed to affect federal electoral outcomes. The Tilman Act of 1907 did this. The Federal Corrupt Practices Act followed three year later: it imposed campaign spending limits on political parties for House general elections. For the first time, public disclosure by political parties regarding their contributions to candidate campaigns was required. Amended in 1911 to include Senate races and primary elections, it established limits on campaign expenditures. These laws angered political parties, corporations, and rich individuals: they saw grave threats to their First Amendment free speech rights. The prime movers in the cases over commercial speech were insiders rather than outsiders. FECA was the latest round in this tussle between competing perceived threats. Legislation attempting to deal with this appeared often between 1911 and 1971.

The 1974 amended form of FECA, passed over the veto of President Ford, had many components. They included: 1) limits on campaign contributions to candidates for federal office and requirement for public disclosure of these contributions; 2) limited expenditures by candidates and their committees, limited independent expenditures, and limited candidate expenditures from personal funds; and 3) provided for the public financing of presidential elections. It also created the Federal Election Commission to oversee the process. Under the 1976 *per curiam* decision, *Buckley*, only parts of this law survived. The Burger Court upheld limits on individual contributions, public disclosure requirements, and public financing of presidential elections. However, restraints on candidates using their private funds, limitations of campaign expenditures, and limits on the 'independent' contributions of other groups were all struck down: the Court ruled that spending money to influence elections was constitutionally protected free speech. This decision was another major invalidation of a federal law, this time ruling in favor if insider (elite) interests.

The logic of *Buckley* was present in *First National Bank of Boston v. Bellotti* (435US765), another important decision striking down a Massachusetts law in 1978. This law had prohibited corporate expenditures to influence elections. In a 5-4 decision, the Court held that corporations had a First Amendment right to make contributions to influence political processes. The dissent by Justice Rehnquist, a very conservative jurist, contained: 'It cannot be so readily concluded that the right of political expression is equally necessary to carry out the functions of a corporation organized for commercial purposes. A State grants to a business corporation the blessings of potentially perpetual life and limited liability to enhance its efficiency as an economic entity. It might reasonably be concluded that those properties, so beneficial in the economic sphere, pose special dangers in the political sphere.' He was prescient regarding the flood of undocumented money flowing into the financing of political campaigns following *Citizens United* in 2010.[41]

The Burger Court applied its freedom of commercial speech approach in politics to commercial matters. Virginia had a statute banning pharmacists from advertising prescription drug prices. It further argued that pharmacists who did this acted unprofessionally. As a result, drug prices varied throughout the state. The Virginia Citizens Consumer Council challenged this law. In 1976, the Supreme Court struck it down with *Virginia State Board of Pharmacy v. Virginia Citizens Consumer Council* (425US748): commercial speech for profit was protected. It argued that the Board of Pharmacy could act to protect professional conduct among its members but not at the expense of the public. A year later, an Arizona law suffered a similar fate in *Bates v. State Bar of Arizona* (433US350) when the Court allowed lawyers to advertise their services. Finally, the Court invalidated New York Public Services Commission rules banning electrical utilities from advertising in *Central Hudson Gas and Electric Corp. v. Public Service Commission* (447US557) in 1980. All decisions rested on First Amendment rights and their application to states under the Fourteenth Amendment. As a result, commercial speech became explicitly protected, and attempts to restrict it were rendered null.

6.5.1.4   The coherence of the threats to social order line island

In terms of the substantive issues of the decisions in this line island, there is great diversity in the nature of the perceived threats. Some concerned 'national security' when communists and socialists were targeted. Others concerned threats to dominant economic interests posed by labor groups organizing. Whites saw threats in the organization of the black population in the pursuit of civil rights and racial equality. The Jehovah's Witnesses were targeted because of their beliefs. Obscenity was perceived as a threat to a dominant moral order as were birth

---

[41] It is not surprising that *Citizens United v. Federal Elections Commission*, a 2010 decision in which the Roberts Court held (5-4) that the First Amendment prohibited the government from restricting *any* independent political expenditures by corporations and unions cited *Buckley*.

control and abortion. Government officials saw threats when they were criticized and sought protections. Economic elites saw threats in attempts to rein in their use of economic resources to influence election outcomes. These diverse cases were held together by three features: 1) attempts to control the behaviors of outsiders seen to be posing threats; 2) privileged insiders seeking to extinguish these threats; and 3) the use by the Supreme Court of the First and Fourteenth Amemdments to resolve the rival claims.

One advantage of focusing on frequently co-cited decisions is the bringing together of diverse substantive issues that, otherwise, would be treated as separate domains. Identifying the commonalities across issues is an important part of seeing a bigger picture for the Supreme Court, and identifying line islands is an effective way of doing this. Unfortunately, space precludes discussion of decisions in other line islands. The Native-American and threats to social order islands serve as demonstrations of what can be done.

We consider next a particular decision known as the *Dred Scott* decision for two reasons: 1) it is invisible in the Fowler and Jeon (2008) analysis due to having few outgoing citations and receiving less incoming citations than many other decisions, and 2) it raises a serious data collection issue meriting further attention.

6.6   The Dred Scott decision

Scott v. Sandford[42] (60US393) was handed down in 1856. Dred Scott was an African-American slave, one of about four million in America at the time. The states were divided into slave states and free states (although, despite the name, slavery was present in them also). He was held by his 'master' in Missouri, a slave state. His master (owner), an officer in the US Army, took him to the free state of Illinois. Next, Dred Scott was taken to the free territory of Wisconsin. As a result, he lived in free areas for many years. When the Army ordered this slave owner back to Missouri, Dred Scott was returned to a slave state. After this army officer died, Dred Scott, helped by Abolitionist[43] lawyers, filed a suit claiming his freedom based on his stays in Illinois and Wisconsin. This claim was rejected by Missouri courts. On appeal, the case reached the Supreme Court.

The *Dred Scott* decision was a relatively early decision that came with multiple opinions, each signed by a Justice. Chief Justice Taney wrote the main opinion but it was *not* the majority opinion of the 7-2 decision against Scott. Taney decided three constitutional issues Powe (2009): 106. First, neither African slaves nor their descendants could become US citizens. Second, as the territories, not yet states nor included in states, were held in common for all Americans, Taney argued that Congress lacked the power to ban slavery in the territories. Third, if slaves were freed anywhere, this would be taking of property from slave owners without the due process of law.

This decision (especially Taney's opinion) has gone down as one of the worst decisions ever made by the Supreme Court. According to Irons (2006): 176, Taney 'misread history, twisted legal precedent, and bent the Constitution out of shape, all to achieve his predetermined goal of promoting the extension of slavery into the territories.' We look more closely at this twisting of legal precedent in Section 6.6.1.

---

[42] The actual name was Sanford but was misspelled in the court documents. We have stayed with the convention of using Sandford.

[43] The Abolitionist movement, dedicated to the ending of the slavery of all people of African descent, was active in the 18th and 19th centuries in the USA, Europe, and elsewhere.

While much has been written on this case, we do not pursue the substantive issues here despite their compelling importance. Many other sources can be consulted for discussions of them and the severe consequences of this decision for blacks. Methodologically, this case is instructive about constructing and analyzing Supreme Court citation data. One seemingly simple question is: What is a Supreme Court decision? Alas, the answer is ambiguous, which is illustrated by *Dred Scott*. While there have been changing conventions regarding the construction of decisions over the years, most often, when a decision is not unanimous, it can contain four items: 1) the majority opinion written by one Justice (which other Justices can join); 2) concurring opinions where one Justice (who can be joined by other Justices) writes a separate opinion providing a different rationale for the decision; 3) concurring and dissenting opinions where a Justice agrees, in part, with the majority but disagrees with another part of the majority opinion. (This could be with regard to the substance of a decision, with regard to the argument leading to the majority opinion, or both. Other Justices can join this opinion.); and iv) dissenting opinions where a Justice rejects the majority opinion completely stating why. Other Justices can join a dissent.

Dred Scott had multiple parts: the Taney opinion, six concurring opinions, and two dissenting opinions. This raises the question of what is included when constructing a citation network linking decisions. There are only two unambiguous decision rules for choosing from the above four items. One is to use only the majority opinion (as was done by Fowler and Jeon) and the second is to use all four items. Simplicity argues for the first option. However, some Supreme Court decisions have no majority opinion. Instead, a set of opinions, when combined, support the decision. Citations to and from such decisions were included, appropriately, in the data constructed by Fowler and Jeon (2008) for some decisions. For consistency, this suggests that using only the second decision rule is the better option. Yet, this was not done for *Dred Scott*, a stunning omission.

All Supreme Court decisions (available from any of the sources listed in Appendix A.3) provide all their opinions. It is the least ambiguous approach to obtaining citations between Supreme Court decisions. Doing this for *Dred Scott* exposes serious problems in the data provided by Fowler and Jeon. Dealing with these problems leads to potentially important methodological tasks when constructing the Supreme Court citation network.

6.6.1   Citations from *Dred Scott*

According Fowler and Jeon's data, Taney made four citations to prior Supreme Court decisions. In fact, he made six such citations pointing to data omissions.[44] Historically, the Taney opinion received most of the attention, especially immediately after *Dred Scott* was handed down. Powe (2009): 106–7 argued 'history is written by winners, and Taney's result and his reasoning have been condemned by history. But at the time and until the Civil War, Southerners had been winners, and Taney was busy writing Southern history.' Irons (2006): 170 added 'Taney decided to throw moderation to the winds and write a ''fire eating'' opinion that would serve as a pro-slavery manifesto.' Both supporters of slavery and their abolitionist opponents reacted strongly, perhaps as Taney intended for the former. 'Taney's opinion dominated public attention and shaped the debate over slavery as the nation plunged

---

[44] In fairness, when documenting their data, Fowler and Jeon concede mistakes in their data – as there are in most, if not all, large network datasets.

into *the most serious constitutional crisis in its history* (Irons (2006): 177, emphasis added).'
This being the case, it is hard to envision a more consequential decision – yet, as noted, it is
invisible in the analysis of Fowler and Jeon (2008).

The two dissents to *Dred Scott* came from Justices Curtis and McLean. Together, they
rebutted every claim made by Taney. We went through all of the *Dred Scott* opinions to
find citations to prior Supreme Court decisions. Curtis alone cited 33 prior Supreme Court
decisions as precedents.[45] McLean made 22 such citations (and also cited other lower court
decisions). Justice Daniel, in a concurring decision, cited more prior Supreme Court and
lower court decisions than Taney. All these relevant decisions were ignored by Taney. Irons
(2006) argued that Taney sought only precedents supporting his claims.[46]

To examine this network with the inserted ties and the ties between decisions, some steps
are required in Pajek. Preliminarily, two objects, a network and a cluster, are needed. The
`network` is created by inserting the additional ties into the entire Supreme Court network
and saving it. The `cluster` object has 0s everywhere and a single 1 for the Dred Scott
vertex. Having read this network and this cluster into Pajek, the steps are:

```
Operations/Network+Partition/Expand Partition/Greedy Partition/Input
Depth to go [1]
Network/Create Partition/Degree/Input
Operations/Network+Partition/Extract Subnetwork [Cluster 1]
```

The result of these commands is a network, shown in Figure 6.8, with $n = 53$ and $m = 102$,
one that is strikingly different from the network in the Fowler and Jeon dataset where $n = 5$
and $m = 4$.

Figure 6.8 shows citations to prior Supreme Court decisions in this island plus citations
between them (extracted using Pajek). Decades are marked on the left. Decisions are located
roughly according to when they were made. Arc colors represent citations from different
Justices. In Figure 6.8, other than citations by Curtis and Taney, only one line from 60US393
to cited earlier decisions is shown, even if Justices cited them in agreement. When Curtis
and Taney cited the same decision, red and blue lines link decisions: Taney cited them as
supporting his opinion while Curtis argued the reverse. Over time, *Dred Scott* lost its appeal
as a constitutional imperative. Curtis's dissent, ignored immediately after this decision, had
great relevance subsequently. Taney had been Chief Justice of the Supreme Court since
March 1836. It is remarkable how many decisions (22) of his own court he ignored when
other members found them relevant, especially in the dissenting opinions (see all non-red
arcs in Figure 6.8). Figure 6.8 differs greatly from a part of the source data.

The take-away message from this figure is how few citations Taney made to prior relevant
decisions that were identified by other members of his court, and provides support for the
notion of precedent existing not only in the form of cited earlier decisions but also by relevant
decisions, that could serve as precedent, being ignored.

---

[45] He cited also 36 decisions from state and local courts in rebutting Taney.

[46] Taney wrote in his opinion 'It is difficult at this day to realize the state of public opinion in relation to that
unfortunate race, which prevailed in the civilized and enlightened portions of the world at the time of the Declaration
of Independence, and when the Constitution of the United States was framed and adopted. But the public history
of every European nation displays it in a manner too plain to be mistaken.' In his dissent, Curtis cited *Munroe v.
Douglas*, a case from England, showing that this sweeping claim was unsupported empirically.

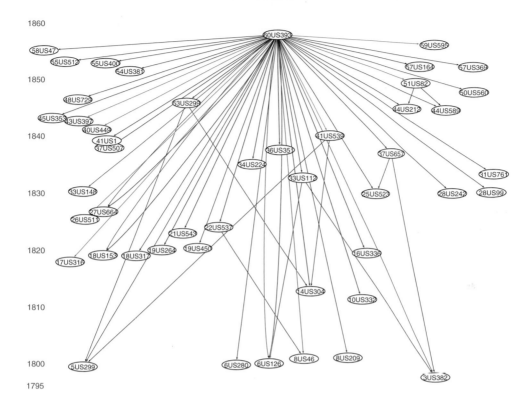

Figure 6.8   Citations from the *Dred Scott* decision.

Taney cites are in red; Curtis cites are in blue; McLean cites are in green; cites by both Curtis and McLean are in cyan; and cites from other Justices are in black. At the top, 60US393 is the *Dred Scott* decision.

6.6.2   Citations to *Dred Scott*

The change in outdegree from 4 to 46 for *Dred Scott*, documented in Section 6.6.1, suggested that checking Fowler and Jeon's (2008) reported indegree of 14 would be prudent. ARTL (2010) provided a list of Supreme Court cases citing *Dred Scott*.[47] Implicit in the construction of citations between Supreme Court decisions is the idea that 'a citation is a citation is a citation.' Considering the expanded list of citations to *Dred Scott* suggests that this assumption is far too restrictive: citations vary greatly with regard to establishing precedent (or not).

---

[47] The document can be found at http://www.issues4life.org/pdfs/artldredscottshepardized.pdf. It is an extremist screed in support of the 'right to life' movement and contains an outright attack on the Supreme Court concerning its composition (in 2010) and its historical refusal to reverse *Dred Scott* in subsequent decisions. This suggests solid support for *Dred Scott* through its subsequent decisions. Both this and the additional assertion that the Court is loathe to reverse itself or criticize earlier decisions is nonsense, as we argue below. We used this document *solely* for the list it provided – but only after checking the text of all its listed decisions to make sure they did 'cite' *Dred Scott* by reading them closely.

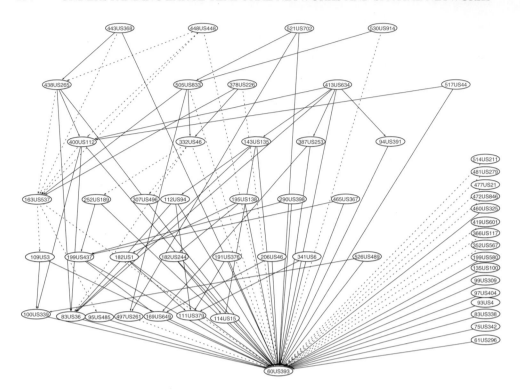

Figure 6.9   Citations to the *Dred Scott* decision.

Lines represent positive ties and dashed lines represent negative ties. 60US393 is the *Dred Scott* decision. Unlike Figure 6.8, this figure is drawn in layers only.

Citations to *Dred Scott* take a variety of forms including: 1) positive citations affirming and using it to inform a subsequent decision;[48] 2) outright critiques;[49] 3) rejections of *Dred Scott* as a precedent; and 4) utterly trivial and tangential citations.[50] These differences suggest that it would be useful to distinguish positive citations (item 1) and negative citations (items 2 and 3) to earlier decisions. Figure 6.9 shows citations to *Dred Scott* and citation links between decisions citing it. We read all these cases to assess whether the links were positive or negative. There are 52 cites to *Dred Scott* of which 38 are positive and 14 are negative. These 52 citations are far more than the 14 recoded by Fowler and Jeon and include negative citations. Overall, there are 76 positive ties and 25 negative ties in Figure 6.9. The decisions

---

[48] The Dred Scott decision is very long with 111,327 words. Even though many regard Taney's opinion as dreadful, it contained positive points. For example, Justice Souter in *Seminole Tribe of Florida v. Florida* (1996) noted, in a dissent, 'regardless of its other faults, Chief Justice Taney's opinion … recognized as a structural matter … ' before approving a *very specific* point in that opinion.

[49] For example, Justice Scalia in a dissent to *Planned Parenthood of Pennsylvania v. Casey* (1992), cited the Curtis dissent to *Dred Scott* and wrote: 'In my history book, the Court was covered with dishonor and deprived of legitimacy by *Dred Scott v. Sandford*, an erroneous (and wildly opposed) opinion that it did not abandon.'

[50] *Schiavone v. Fortune* (1986) referred to *Dred Scott* regarding the misspelling of Sanford as Sandford while discussing the incompetence of a court-appointed lawyer.

four balanced triples

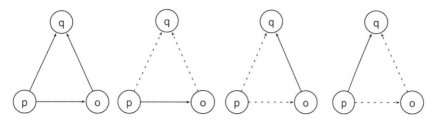

four imbalanced triples

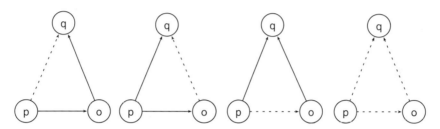

Figure 6.10 Balanced and imbalanced signed triple.

The top row of balanced triples are labeled B_1, B_2, B_3, and B_4. The leftmost imbalanced triple on the bottom row is labeled U_1.

on the right of Figure 6.9 cite *Dred Scott* but no other decisions in this subnetwork. Seven negative citations, drawn with dashed lines, are distributed among the decisions left and seven are from the decisions shown on the right. The presence of negative citations is concentrated in the more recent decisions, reflecting the changing evaluation of *Dred Scott* over time.

The network in Figure 6.9 is a signed network suggesting that structural balance (Cartwright and Harary 1956; Doreian and Mrvar 1996; Heider 1946) could play a role in understanding the structure of this subnetwork. A key feature of signed networks is the configuration of the signs in a signed triple. In Figure 6.10, vertices are denoted by p, q, and o. The sign of a triple is the product of the signs in the triple. If the product is positive, the triple is balanced[51] while a triple with a negative sign is deemed imbalanced. All eight possible complete directed signed triples are shown in Figure 6.10.

Rather than using the partitioning approach of Doreian and Mrvar (1996) to assess balance in Figure 6.9, it was more useful to identify network fragments as implemented in Pajek. The fragments we use are defined by the triples in Figure 6.10. These triples have to be identified and counted. The steps for doing this are: 1) create small network files containing the fragments; 2) extract the networks with only the fragments (separately) from the overall signed network; and 3) count them. To do the first step, it most useful to create, save, and

---

[51] Using the language of friends and enemies, the four triples in the top row of Figure 6.10 can be expressed, reading from the left, as: a friend of a friend is a friend; an enemy of a friend is an enemy; a friend of an enemy is an enemy and an enemy of an enemy is a friend.

read a Pajek `project file`. The project file we used contains each of the eight fragments as defined in Figure 6.10 is:

```
*Network OnlyPositive
*Vertices 3
*Arcs
1 2 1
1 3 1
2 3 1

*Network Two Negative /1
*Vertices 3
*Arcs
1 2 1
1 3 -1
2 3 -1

*Network Two Negative /2
*Vertices 3
*Arcs
1 2 -1
1 3 -1
2 3 1

*Network Two Negative /3
*Vertices 3
*Arcs
1 2 -1
1 3 1
2 3 -1

*Network One Negative/1
*Vertices 3
*Arcs
1 2 1
1 3 1
2 3 -1

*Network One Negative /2
*Vertices 3
*Arcs
1 2 1
1 3 -1
2 3 1

*Network One Negative/3
*Vertices 3
*Arcs
1 2 -1
1 3 1
2 3 1
```

```
*Network Only Negative
*Vertices 3
*Arcs
1 2 -1
1 3 -1
2 3 -1
```

The extraction of the fragments requires selecting a fragment as the first network and the full signed network as the second network. With these selections, the Pajek command is

```
Networks/Fragments (First in Second) [Find]
```

There are 35 B_1 triples, eight B_2 triples, eight B_3 triples, and three B_4 triples. In contrast, there are two U_1 imbalanced triples and no other imbalanced triples. The proportion of balanced triples in Figure 6.9 is 54/56 = 0.96 suggesting great consistency with classical balance theory in the signed subnetwork of citations to *Dred Scott*.

Figure 6.11 displays two subnetworks, one with B_1 triples (on the left) and the other with B_2 triples (on the right). The network with B_1 triples has 30 vertices and 59 positive ties. The network with B_2 triples has ten vertices with eight positive ties and nine negative ties. The large subnetwork on the right of Figure 6.11 supports the notion of the Supreme Court not reversing itself with regard to *Dred Scott*. Decisions positively citing *Dred Scott* positively cite other decisions that also cite it positively. However, the subnetwork on the right does not support this contention. In B_2 triples, decisions citing *Dred Scott* negatively also cite positively decisions citing *Dred Scott* negatively. There are three decisions linked by positive cites that all negatively cited 161US537 which cited *Dred Scott* negatively. All triples are consistent with classical structural balance.

Figure 6.12 contains networks composed of B_3 triples (on the left) and B_4 triples (on the right). The network with B_3 triples has 11 vertices with eight positive ties and 10 negative ties. The network with B_4 triples is tiny with five vertices, three positive ties, and four negative ties. In B_3 triples, all decisions citing *Dred Scott* negatively also negatively cite decisions that positively cited *Dred Scott*. In the B_4 triples, decisions negatively citing *Dred Scott* positively cited 163US537 which cited *Dred Scott* negatively. Both subnetworks have both positive and negative citations to *Dred Scott* in ways consistent with classical structural balance.

While it is clear that *Dred Scott* has not been completely repudiated by the Supreme Court in subsequent decisions by a reversal, the Court has been aware of its deficiencies. There are several reasons for *Dred Scott* not being reversed by the Court.

The most consequential stems from the Fourteenth Amendment adopted in 1869 which explicitly overturned *Dred Scott*. Taney had declared no slave or any descendent of a slave could become a US citizen. In contrast, Section 1 of the 14th Amendment states 'All persons born or naturalized in the United States, and subject to the jurisdiction thereof, are citizens of the United States and of the State wherein they reside. No State shall make or enforce any law which shall abridge the privileges or immunities of citizens of the United States; nor shall any State deprive any person of life, liberty, or property, without due process of law; nor deny to any person within its jurisdiction the equal protection of the laws.' The due process clause of the Fourteenth Amendment prohibited state and local governments from taking life, liberty, or property from people without ensuring fairness. The Court did not need to reverse *Dred Scott* because this had been done through a Constitutional amendment. Indeed, some of the subsequent decisions 'citing' *Dred Scott* explicitly stated its irrelevance with regard to

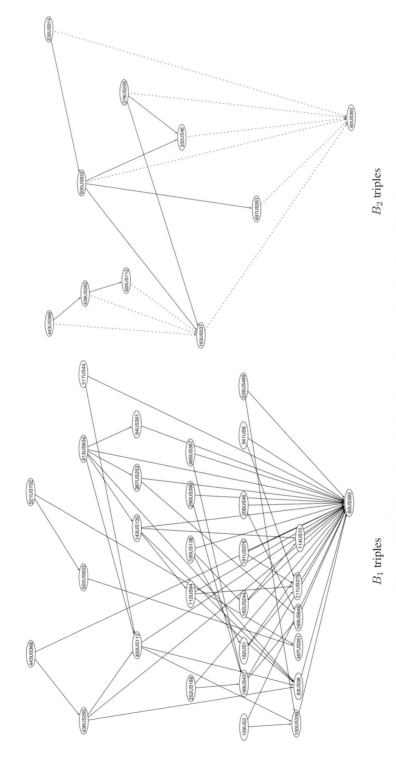

B_1 triples

B_2 triples

Figure 6.11 Two balanced signed networks organized by B_1 and B_2 triples.

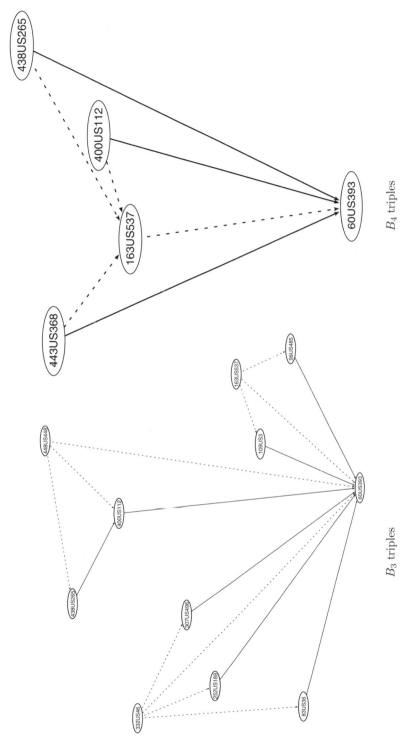

B_4 triples

B_3 triples

Figure 6.12 Two balanced signed networks organized by B_3 and B_4 triples.

precedent because of the Fourteenth Amendment. We took these as negative citations when constructing Figure 6.9.

Two issues cropped up frequently in decisions where *Dred Scott* was cited. One concerned citizenship for people who were neither slaves nor African Americans. *Dred Scott* was cited both positively and negatively regarding attempts to deprive people of citizenship and the rights of citizenship. This occurred when the Court considered, in 1970, amendments to the Voting Rights Acts of the 1960s. A second recurring issue is property. At face value, cases stemming from collisions of ships (e.g. *Jackson v. The Magnolia* (1857) and *Merrill v. Petty* (1872)) have nothing to do with slavery. Yet they cite *Dred Scott*. Taney declared slaves as property, and subsequent cases focusing on different types of property still cited *Dred Scott* regarding the disposition of property or providing remedies for lost property.

A constant tension exists over federal courts intervening in state level matters and states asserting their rights preemptively. Taney argued that the federal government had no Constitutional basis for imposing anything on Missouri in *Dred Scott*. As noted earlier, federalism has been a contentious issue, one inherent to the constitution's prose. It was never resolved. Attempts denying federal authority can appeal to *Dred Scott* to justify 'state's rights' as taking precedence over perceived federal intrusions.

6.6.3   Methodological implications of *Dred Scott*

Several important implications follow from the information and arguments in Sections 6.6.1 and 6.6.2 including:

1. Great care is needed when recording citations from one Supreme Court decision to prior Court decisions. A strong case can be advanced for considering the *whole decision*. If so, all opinions contained in a decision merit attention, especially dissents if they have been cited frequently in subsequent opinions. This has the potential for imposing severe data collection burdens on researchers.

2. The increase of citations by *Dred Scott* from 4 to 46 (a 1050% increase) and the increase of citations to *Dred Scott* from 14 to 52 (a 278% increase) imply a major problem in the Fowler and Jeon (2008) dataset constructed from only majority opinions. This problem hinges on what is a Supreme Court decision. At a minimum, clear rules are needed about this, and these rules have to be applied *consistently*. In the main, Fowler and Jeon were consistent but not completely so.

3. A citation is not just a citation. Citations differ greatly in their purpose and content. This must be addressed. We have shown that citations to *Dred Scott* can be coded as positive or negative, an important distinction because positive and negative citations have very different meanings. This implies that the *content* of a citation has significance. Simply counting citations to summarize the importance of decisions seems too limited: citations have characteristics meriting attention also. This also has the potential for imposing data collection costs.

We are not claiming that Fowler and Jeon's data are worthless for they can have great value. However, interpreting the results for a citation network if many citations are missing, and differences in the nature of citations have been ignored, must be done with caution. We simply do not know the extent of this problem. It could be that *Dred Scott* is an exceptional and egregious case with its problems not being reflective of the dataset as a whole. At a minimum, this needs to be checked.

6.7   Further reflections on the Supreme Court citation network

At face value, a set of Supreme Court decisions, together with citations between them, is just another citation network to be studied on its own. This is not the case if there have been major changes in the contexts within which decisions were made. *Dred Scott*, handed down in 1857, was informed by the Missouri Compromise of 1820 and the Great Compromise of 1850. At the time of the former, major conflicts existed between northern and southern states over regulating new territories in the west, especially over slavery. There was a concern to 'balance' the number of free and slave states. Rival bills introduced in the Senate and the House reflected the difficulties inherent in resolving these conflicts. Maine, formerly a part of (then) Massachusetts, was admitted as a free state while Missouri entered the Union as a slave state. Slavery was prohibited on the Great Plains but permitted in the Arkansas Territory.

The Great Compromise of 1850 tried to diffuse another major conflict between southern slave states and northern free states. It centered on the status of territories acquired after the 1846–48 Mexican–American War. Five mini-bills were passed and cobbled together because none of the big issues could be resolved in one piece of legislation. California was admitted as a free state (opposed in the south) while the New Mexico and Utah Territories could decide if they would be slave states (opposed in the north). Yet Utah and a northern fragment of New Mexico were in areas where the Missouri Compromise had prohibited slavery, suggesting that the 'compromise' greatly favored the interests of the South.

Before the Great Compromise, slaves were escaping to free states in the north, a phenomenon of concern to southern slave-owners. The 1850 Compromise contained The Fugitive Slave Act declaring all runaway slaves, once captured, were to be returned to their owners. This created a lucrative bounty-hunting industry where both runaway slaves and free blacks could be tracked down and returned to slavery regardless of where they resided or their legal status as free in the north. This compromise was thought to have prevented secession by the south. If so, it was only in the short term. Taney's *Dred Scott* opinion built on these legal 'compromises' with the intent of expanding slavery to all territories. These Acts, no matter how abhorrent they were seen in retrospect, cannot be ignored when considering Supreme Court decisions regarding slavery. Taney wanted to resolve these conflicts to favor the south. One direct response to *Dred Scott* was the passage of the Fourteenth Amendment – another example of laws and Supreme Court decisions impacting each other: one set cannot be understood without considering the other.

As noted earlier, the Constitution is an ambiguous document. Efforts to create the United States had to resolve deep conflicts to form a union. One central conflict concerned the institution of slavery, integral to the 'southern way of life', but present also in the north. Another featured resistance by individual states to federal control, especially if the federal republic had real nation state authority. These conflicts could not be resolved to the satisfaction of all the participants drafting the Constitution. It would not have been acceptable had it been written unambiguously. Ambiguity was there by design. Different meanings were attached to the same words permitting enough signatures to the document to provide authority for a federal government. It follows that the Constitution cannot be interpreted, once and for all, in a definitive way. Justices on the Supreme Court strive to base their decisions in terms of the Constitution. But they do so on the basis of *their own* understandings of this document. As Taney's opinion and Curtis's dissent in *Dred Scott* show, such understandings can be starkly different among members of the same court and across different courts with different compositions of the political interests and judicial philosophies of Justices.

Moreover, economic, political, and social differences, together with the conflicting interests following from them, have always been in play. The Supreme Court was a political institution when it was created. The rival interests that were revealed at the founding convention were present on this Court from its inception. The proportions in which they have been represented on the Court changed over the years after short-term battles over who 'should' or 'should not' be on the Court were settled. As a result, neither an objective understanding nor interpretation of the Constitution is possible. Indeed, understandings must change over time because the decisions of the Supreme Court reflect both the times in which they were made and its composition when cases before it were decided. Further, some state and federal laws prompt decisions that either affirm or overturn them. Finally, some Supreme Court decisions prompted Congressional lawmakers to enact new laws when they strongly opposed them. Supreme Court decisions and the citations between are best understood when their social, legislative, and legal contexts are considered. This includes considering the changing composition of this court. Examining co-citations of decisions provides a compelling way of establishing the foundations for doing this.

These arguments have implications for the collection and analysis of citation data for Supreme Court decisions because the Supreme Court citation network does not stand alone. It follows that it is more fruitful to consider how decisions are grounded in changing historical contexts. Examining line islands compels this. It implies that information about these contexts must be present in the information base to better understand this citation network. Additionally, as noted above, the notion of 'a citation is a citation is a citation' does not hold here. Citations to earlier decisions have contents that go beyond their mere existence. We distinguished positive and negative citations to prior decisions in Section 6.6.2 by looking closely at the texts of decisions.

This idea can be extended to consider explicit Supreme Court *reversals* of prior decisions because such reversals undo and repudiate the earlier decisions being reversed. There have been enough of them to merit their examination. Having both positive and negative citations of earlier decisions be parts of citation count as 'following precedent' is problematic. If anything, a reversal of the prior decision destroys its relevance as precedent – yet, paradoxically, they can appear to be 'important' decisions Fowler and Jeon (2008): 25. We think it would be useful to modify part of the underlying Supreme Court citation network to a signed form, use the kinds of fragments defined in Figure 6.10, and apply structural balance theory to create a more nuanced understanding of precedent.[52] Given this, we fully endorse one of Fowler and Jeon (2008): 29 concluding remarks: 'We believe that a contextual exploration of the positive and negative nature of each citation … may yield additional insights into the network of precedent and its effect on the relative importance of cited decisions.' The results in this chapter, while not focused explicitly or exclusively on precedent, are a step in this direction, especially with regard to signed citation ties.

---

[52] This will be the topic of future work because extant listings of reversals are incomplete and inconsistent. Indeed, the Fowler and Jeon (2008) data miss some reversals as citations. To treat a signed version of this network requires further checking for reversals and modifying the current database.

7

Football as the world's game

Football (soccer[1]) is now a global phenomenon as a sport played by over 250 million players in more than 200 countries with football leagues administered by national football associations (McGovern 2002). Countries have national teams playing international matches against each other. Football players (footballers[2]) have long moved over the globe to play 'the beautiful game'. In doing so, they move from club to club and between countries. Player movements generate the club-to-club and country-to-country networks we study to describe the organizational structures of football, the processes generating and changing these structures, plus some resulting outcomes. The organized system of world football, for all of the football clubs and footballers, is vast. There is no complete database with data on all players and all clubs through time. In response to this, we defined and assembled a more manageable database comprising *all* clubs that played in the English Premier League (EPL) together with *all* players on the first team squads of those clubs for the first 15 seasons since this league was formed for the 1992/3 season. This database and its construction are described in Appendix A.4. We characterize each player as having a career trajectory made up of an alternating sequence of stays and moves.[3] The stays are at clubs, with varying lengths of time, with the moves between clubs distributed over time. The rationale for using these data is provided in this chapter, together with the questions we consider and the results of some of our preliminary analyses.

A brief historical overview is provided in Section 7.1. Section 7.2 discusses some general features of clubs, while Section 7.3 discusses some general features of football players. We then describe some features of the organization of football in England (given our focus on the

---

[1] Following the codification of the game of football, the term 'association football' was introduced to distinguish it from other football games, especially rugby. The word 'soccer' was obtained from the word 'association' in the 1880s. We use football and soccer interchangeably.

[2] We use the terms 'football players', 'footballers' and 'soccer players' interchangeably.

[3] The number of stays in these players' careers in our data varies from 1 to 25. The number of player moves ranges from 0 to 24.

Understanding Large Temporal Networks and Spatial Networks: Exploration, Pattern Searching, Visualization and Network Evolution, First Edition. Vladimir Batagelj, Patrick Doreian, Anuška Ferligoj and Nataša Kejžar.
© 2014 John Wiley & Sons, Ltd. Published 2014 by John Wiley & Sons, Ltd.

EPL) in Section 7.4. This is followed by a characterization of player migrations in Section 7.5. We provide some details of the institutional arrangements together with the organization of football within which the player movements take place in Section 7.6. Both have clear – but general – impacts on how these player movements occur. Section 7.7 describes some court rulings that are thought to have had impacts on player migrations. Factors having more specific impacts on individual footballer migrations are presented in Section 7.8. Section 7.9 contains some hypotheses based on arguments drawn from the literature we survey, together with preliminary remarks about operationalizations we need to test these hypotheses. Some preliminary analyses regarding the flow of football players and the impacts of these flows are provided in Section 7.8. A partial summary of these results is in Section 7.12, and our main network results are reported in Chapter 8.

7.1   A brief historical overview

Initially, there were variants of (what was to become) football before the game was codified in the 1860s. The first version came in 1863 (Magee and Sugden 2002; Murray 1996). The men involved came from Britain's public schools who wanted to continue playing the game after leaving school. While the initial codification of football was not done by men in the working class, working-class men, especially in the industrialized areas of the Midlands, the north of England, and the industrialized part of Scotland, were quickly drawn to the game as spectators and participants. Further, colonial powers inadvertently took the game to their colonies. There are debates about which individuals were mainly responsible for spreading the game. Taylor (2006) argues that Britain's export of football involved the migration of many skilled and highly qualified workers. Magee and Sugden (2002) claim that the diffusion of football was driven by the recreation activities of English and Scottish manual workers, tradesmen, and railway engineers. They explicitly ruled out colonial administrators and those involved in administering the formal rule of the British Empire as being relevant. Alegi (2010) allows such administrators a role in spreading football as a way of 'civilizing' local populations, along with missionaries using football to instil a 'muscular Christianity.' Ben-Porat and Ben-Porat (2004) include merchants, bureaucrats, executives of private enterprises,[4] and engineers as carriers of football. Despite these differences about carriers, it is hard to disagree with Hobsbawn's (1995) claim that football became the world's sport as a result of the global economic presence of Britain.

7.2   Football clubs

Football teams are groups of players playing together against other teams. Football clubs are more formal arrangements as legal entities. At the time of their formation, they were located in, and identified with, specific *local* communities. Most often, they are formal organizations with officers and a variety of specialists supporting the club players. Clubs can range from village teams to giant clubs whose names are known the world over.[5] Clubs

---

[4] Some companies formed football teams for recreation and to instill discipline in their workers.

[5] Prime examples of such clubs are the members of the former G-14. It was formed by the executives of 14 big clubs in 2000 to lobby FIFA and UEFA. These clubs were: FC Barcelona and Real Madrid (Spain); Liverpool and Manchester United (England); AC Milan, Inter Milano, and Juventus (Italy); Olympique de Marseille and Paris Saint-Germain (France); Bayern München and Borussia Dortmund (Germany); Ajax and PSV Eindhoven (the

vary also from professional to semi-professional to amateur depending on their resources and ambitions.

Clubs are, most often, organized into a hierarchy of leagues coupled by promotion and relegation regimes under which, loosely, the bottom clubs of a higher league and the top clubs from the next lower level exchange places. The precise details of promotion and relegation have changed over time and can vary between different league systems at one point in time. However, the existence of promotion and relegation regimes is present in nearly all football leagues.[6] In this sense, success can be handsomely rewarded while failure is severely punished. Avoiding failure, including financial failure, becomes a major concern for all clubs. Some clubs are good enough, relative to their level, to seek success in the form of silverware (league titles and cup trophies). Clubs need to survive. This means that the financial costs for survival must be covered. Clubs have responded to this generic problem in multiple ways. Some formed football academies or other training arrangements as a way of recruiting and training young players. Another response was to put players on professional squads under contractual arrangements[7] with their club whereby the club owns their services for a specified length of time. Another response was to obtain players from other clubs via transfers of players (where, in essence, both the player and the contract move between clubs). Players can also be loaned from one club to another where the player moves temporarily to the receiving club but the ownership of the player is retained by the loaning club.[8]

Club success is measured by the outcomes of specific games, outcomes of games within a season, and the overall standing at the end of a season. This puts a premium on immediate success regardless of the level of a club. Arguably, the stakes get higher for clubs in the higher realms of league systems. This implies that clubs in higher leagues require higher-quality players relative to other clubs.

There is considerable instability in player composition of club squads. This results from multiple processes including: 1) clubs transferring players to other clubs; 2) players loaned between clubs; 3) clubs discarding players;[9] and 4) clubs introducing players into their squads from their football academies. In general, clubs seek players that are cheaper to buy and are self-disciplined (Maguire and Stead 1998; Poli 2010). Clubs face a potential dilemma in having to choose between recruiting 'foreign players' and players from the UK and Ireland.[10]

Netherlands), and FC Porto (Portugal). Four clubs were added in 2002: Arsenal (England); Bayer Leverkusen (Germany); Olympique Lyonnais (France); and Valencia (Spain). The G-14 was disbanded after reaching an agreement with UEFA. (Our source is http://en.wikipedia.org/wiki/G-14, accessed 9 May 2012.)

[6] Exceptions include the USA and Australia but they are few in number.

[7] There are instances of footballers at a club playing 'without a contract'. This applies mainly to players nearing the end of their careers.

[8] The periods of these loans can be as short as a single game (in emergency situations) to years. There are loans 'with the intention to buy' where the expected outcome is that, at the end of the loan, the player will be transferred to the receiving club.

[9] This includes players 'being released', players declared 'surplus to club requirements,' and players leaving their clubs 'by mutual consent.' Players can be put on a transfer list (and so be available to other clubs) unilaterally by a club. Players can request that they be placed on their club's transfer list. We do not distinguish these in our treatment of player transfers. We ignore trials that players can have with clubs because data on trials are seriously incomplete and inconsistent.

[10] We use the terms 'foreign' and 'non-English' interchangeably, even though there is a difference. Given the historical free movement of footballers between clubs in the UK, we also use the term 'non-British' on occasion. 'Foreign' has a pejorative sense that we not intend, even though it serves as a convenient shorthand. However, we

McGovern (2002) claims that clubs pursuing the former prefer to recruit players similar to indigenous players. If the price of local talent is thought to rise too high then there is a pressure to recruit foreign players. Clubs are partially changing their squads each year and this introduces uncertainty because they cannot predict accurately how the players they retain, or the players they recruit, will perform in the future. As a result, clubs when purchasing players are thought to prefer repeated transactions with other clubs (McGovern 2002). If correct, this practice helps introduce systematic patterns to player movements.

Managing a football club is fraught with hazard. If a club is performing badly, one frequent response is to fire the manager. Managers doing well with their clubs can become desirable commodities sought after by other clubs. As a result, uncertainty is introduced by the firing and recruitment of managers by clubs: managers have different preferences regarding playing systems and the kinds of players fitting those systems. Managerial movements induce player movements when managers recruit players with whom they have worked before.

7.3 Football players

Players have attributes directly related to football performance including their: 1) footballing skills; 2) ages; 3) career histories (including the reputations they have acquired); and 4) fitness levels (including injury status). Their services are contractually owned by clubs. As a result, throughout most of the history of the organized game, players have not been free to change clubs at will. The restrictions are less onerous now than in the past. (Some details regarding these changes are provided in Section 7.7.) Players bring intrinsic motivations to the game: 1) wanting to play top level football; 2) wanting regular first team playing opportunities; and 3) wanting to be challenged and to develop as players. They have extrinsic motivations: 1) to earn a living; 2) be on winning clubs; 3) to sample life in different places; and 4) to escape harsh restrictions, when they exist.

Magee and Sugden (2002) present a (not mutually exclusive) categorization of player types that has some value even though it cuts across intrinsic and extrinsic motivations. A *mercenary* is motivated primarily by maximizing his earnings and winning titles.[11] A *settler* is a player who has remained in a specific country for a substantial period (four or more seasons) and sees his economic reward as being coupled to lifestyle issues. An *ambitionist* is one who wants a professional football career, sees a specific country as a desirable place, and wants to improve as a player by being in a higher-quality league. An *exile* is one who for football-related, personal, or political reasons chooses to leave his country to play football.[12] A *nomadic cosmopolitan* is one whose playing career is motivated by wanting to experience different natures and cultures.[13] An *expelled* player is one who is forced to leave his country

---

note that in debates about the impact of the presence of non-English players on the game in England, the pejorative meaning of 'foreign' is clear.

[11] The Swedish international Zlatan Ibrahimović is often cited as such a player given his moves from Ajax to Juventus to Inter Milano to FC Barcelona to AC Milan to Paris Saint-Germain in less than a decade. All these teams won titles or were title contenders during his stays.

[12] The examples they provide to illustrate this category are: war (e.g. Slaven Bilić and Saša Ilić from the Balkans), political turmoil (e.g. George Weah from Liberia), or a complete lack of playing opportunity at home (e.g. Shaun Goater from Bermuda).

[13] Their examples include Jürgen Klinsman (Germany), Ruud Gullit (the Netherlands), and Andrei Kanchelskis (Latvia).

due to conflict with football authorities and/or faces implicit bans.[14] The *celebrity superstar* is one who gets 'front page' media coverage, for example, David Beckham. From Poli (2010) it is possible to add also *pioneers* who are the first players to move from a homeland to another country to play football[15] and *returnees* who go back to their homeland,[16] often when their playing careers are drawing to a close.

Given that clubs can succeed or fail, it becomes important for clubs to measure the contributions that players make. However, this can be done well only for some categories of players and some items.[17] There are intangibles such as organizing ability on the field, general support for other players on the field and 'locker room contributions'. This makes it difficult for clubs to assess player performances and the consequences of trading players and acquiring other players. Even so, players can be, and are, often traded between employers as if they were commodities like pieces of machinery or tracts of land (Ben-Porat 2004; McGovern 2002) or discarded as being obsolete or worthless.

7.4 Football in England

Every country has its own football history. However, we focus here on some historical details for England because of our study of the EPL. The league system was created in 1888. From the outset, England dominated the British scene by recruiting players heavily from Scotland, Wales, Ireland, and – after Ireland was partitioned in 1916 – from Northern Ireland. The within-Britain migratory patterns of players moving between countries were present at the beginning, especially with Scottish players moving south of the border to play in England. There is nothing new about between-country movements of footballers except, subsequently, such volumes became much larger and covered greater distances to include most of the world.

One consequence of England being seen as the birthplace of football is that officials of both the Football League (FL) and Football Association (FA), together with most club owners, officers, managers, and players, believed they had nothing to learn about football from others. This explains why, for many decades, new ideas about strategies and tactics adopted elsewhere were ignored.[18] As Murray (1996) (p. 41) puts it: 'When the nations of Europe and South America grafted their own skills on to the British plant and fed the result with their own flair, they created a game that the British ignored at their own peril.'

---

[14] Examples include Eric Cantona (France), Duncan Ferguson (Scotland), and Paul Gascoigne (England).

[15] Scattered throughout the narrative of Harris (2006) are instances of the first player of a given non-English nationality playing in England.

[16] Nearly all databases kept for particular countries ignore most returnees once they move from the country where one (restricted) version of their playing record is kept.

[17] For all players, but especially strikers, counts are kept of the goals they score. More recently, for all positions, assists are counted where at most one assist is possible for every goal scored. Records can be kept of the number of passes attempted (and the percentage that were completed). Clean sheets are counted for goalkeepers (even if they make no saves in a game). A more sophisticated measure is the winning percentages of a team with and without a player in the line-up.

[18] When the USA defeated England 1-0 in the 1950 World Cup, it was the shock of the tournament. According to Murray (1996) those running the game in England learned nothing from this. Only when the 'Golden Team' of Hungary thrashed England 6-3 at home in 1953 and 7-1 in Hungary the next year was their complacency jolted.

Coupled to this condescending attitude to football played in the rest of the world, a systematic effort to keep foreign players out of English football was employed. 'Foreign' meant not being born in the Home Countries and, implicitly, not coming from the (white) English-speaking part of the British Commonwealth. Imports of foreign footballers were restricted to refugees and amateurs (Harris 2006). The FA drafted and adopted an amendment in 1931 stating 'A professional player who is not a British-born subject is not eligible to take part in any competition under the jurisdiction of this Association unless he possesses a two years' residential qualification within the jurisdiction of Association' (quoted in Harris (2006): 36). This was enforced stringently. The wording was quite subtle by not stating an outright ban. However, requiring a young footballer, or one in his prime, to forgo earning a wage from playing football for two years deterred in-migration of non-British players as effectively as a ban. It was reinforced in 1952 when even temporary transfers of well known non-English players were banned under League rules (Harris 2006). As with most rules, there were loopholes because players with certain colonial ties were allowed into England to play football. Prior to the 1960s, the majority of these players came from South Africa[19] (Maguire and Stead 1998; Murray 1996). This de facto ban remained in place until 1978.

This insularity until 1978 meant that England was unlike other major leagues in Europe that were more open to recruiting 'foreign' players. France included players from their colonies (regardless of race) from the 1930s, as did Portugal in the 1950s. Italy and Spain drew more on players from South America (Harris 2006; Lanfranchi 1994; Maguire and Stead 1998; Taylor 2006). England remained solidly 'English' and isolated from the new football currents circulating elsewhere.

The men responsible for codifying football, as well as the FL and FA administrators who followed them, believed that football should be played for its own sake. While men of means could play within this rubric, the influx of working-class players triggered the need to pay them. These payments frequently were made under the table. Players playing for money were professionals while those playing without payments were amateurs. There has been a long history of conflict in many countries regarding professionalism, but in England this was resolved early when professionalism was accepted in 1885, albeit with heavy restrictions that became unworkable and were quickly abandoned (Murray 1996). However, the increased player salaries came with contracts whose terms were onerous. 'For most of the past century, a combination of club and community loyalty and draconian contractual arrangements meant that the relation between club owners and players could be characterized as master and serf' (Magee and Sugden (2002): 434).

7.5 Player migrations

Footballer movements are fundamental by forming the basic network structure of the organized game (McGovern 2002; Poli 2010). Present since the beginnings of organized football, it has a long and complicated history (Taylor 2006). Flows of players are parts of a much broader flow of people, goods, services, ideas, and information (Magee and Sugden 2002). These recruitment networks are built upon personal connections crossing national boundaries

---

[19] Murray (1996) states that the number of South African footballers playing in England through this era was more than the combined total for players from Australia, Canada, and New Zealand.

(Taylor 2006) and are shaped by these social ties (McGovern 2002). They involve multiple stakeholders (Poli 2010) including clubs, scouts, leagues, associations, and player agents. Such varied stakeholders also creates uncertainty, especially when there are shady deals with illicit payments and murky ownership of players.

Player migrations are based on coupled decisions of players and clubs creating recruitment patterns (Taylor 2006). These decisions and migration ties are formed within extant football networks and, at the same time, when cumulated across players, clubs, and nations can change the form of these networks over time. Expressed differently, player migration is network generated and is a network-dependent process (McGovern 2002), one that is both dynamic and cumulative. We study the structure of these networks as they operate and change over time with a view to studying some of their dynamics.

7.6 Institutional arrangements and the organization of football

As noted earlier, every country has a body that organizes the game within its borders. The names of these organizational bodies feature the descriptors 'football association' or 'league' (or translations of them). In 1904, representatives of just seven national organizations formed the Fédération Internationale de Football Association (FIFA). By 1914 there were 24 members. The British FAs remained aloof (Taylor 2006), consistent with their disdainful view of football elsewhere. By 2012, FIFA had 208 member associations including all nations in the UK and Ireland.[20] The confederations making up FIFA are: the Asian Football Confederation (AFC); the Confédération Africaine de Football (CAF); the Confederation of North, Central American, and Caribbean Association Football (CONCACAF); the Confederación Sudamericana de Fútbol (CSA); the Oceania Football Confederation (OFC); and the Union des Associations Européenes de Football (UEFA). This organizational structure is hierarchical: FIFA dominates its confederations, which dominate their national FAs which, in turn, dominate their clubs. All this is done by enforcing rules. The interests of these entities are not always consistent and can diverge starkly.[21]

The global system of football is Eurocentric with the members of UEFA long being dominant. Countries belonging to UEFA have always been over-represented in World Cup Finals, the preeminent tournament of the game organized by FIFA. Europe is dominant also by having the top five leagues in the world. Many players from around the globe reach clubs in these leagues: the EPL (England), Ligue 1 (France), the Bundesliga (Germany), Serie A (Italy) and La Liga (Spain). In short, Western Europe forms the core economy of football (Maguire and Stead 1998).

Media companies have had an interest in sport in general and, more specifically, football. Newspapers encouraged football from the earliest days. They benefitted also from the growth of this game (Murray 1996). In the 1930s, radio became a new media form by providing live accounts of games. It became the dominant source for news about clubs, players, football

---

[20] The source for this is: http://www.fifa.com/aboutfifa/index.html (accessed 4 May 2012)

[21] This was acute when CAF wanted more World Cup Final slots and seats on FIFA's board once they were decolonized. UEFA saw this as a serious threat. FIFA compensating clubs whose players were injured while playing for their national teams remains a contentious issue.

matches, and match outcomes.[22] Television provided a new market for use in commercial activities. It became an integral part of modern life (Murray 1996). As such, its impact on the organization of football was dramatic even though, as noted by Murray, it has not affected directly – yet – the way the game is played on the field in terms of team selection and the tactics employed on the field.[23]

Creating private TV companies created competition for the rights to show live matches. This resulted in bidding wars for these rights. One outcome was a flood of money into football with FIFA, its confederations, football associations, and some football clubs all acquiring huge new revenue sources. Murray (1996) (Chapter 8) provides an extended examination of some consequences of these media revenue sources. For our purposes, only some impacts are relevant. The foremost was the formation of the EPL when the richest clubs broke away from the rest of the English League to tap disproportionately into the new TV money source, while avoiding sharing their riches with smaller clubs. Another was creating cup competitions involving the top clubs of different countries as another lucrative revenue source.

The European Cup, the most prestigious of the UEFA competitions, started in 1955 as a knockout competition for which only the top national league champions qualified. It was expanded to include additional high-ranking top-league clubs. UEFA regulates its competitions, holds their media rights, and controls the prize monies. In 1992, the (reorganized) European Cup became the Champions League. It is very lucrative for successful clubs.[24] A second UEFA competition, the Europa League, was formed for the 2009/10 season with lower prize monies. It also has forerunners. It was known as the UEFA Cup after UEFA incorporated it in 1971. Before that, from 1955 to 1971, it was known as the Inter-Cities Fairs Cup. These competitions meant greater potential earnings and created new opportunities for some clubs along with increased risks. Footballers want to play in these competitions. This desire motivates player migration to European leagues.

It is well known that English clubs take home more than clubs elsewhere due to the £400m that Sky and ITV paid for Champions League rights. As a result, in England, UEFA competitions magnified differences between the 'big' clubs and other clubs: the largest clubs garner more resources and become even more dominant. They can pay higher salaries to their players, play before larger crowds, and have the opportunity of playing in UEFA competitions (McGovern 2002). At this top level, there was a rapid rise in the price of local players (reflected by increased transfer fees) in the early 1990s. In response, many EPL clubs

---

[22] Murray (1996) notes that cinema became the most popular indoor leisure activity in the 1930s and their newsreels nearly always had extracts from major football matches.

[23] There is some very recent evidence of very wealthy owners of Premiership clubs trying to exert influence in this direction.

[24] For example, in 2010/11, clubs reaching the playoff round received €2.1 million. The payment for reaching the round-robin group stage was €3.9 million. In addition, the clubs received €550,000 for each game played in the group stage. Further, a win nets a club €800,000 and, for a draw, the two clubs involved receive €400,000. Clubs reaching the first knockout round received €3 million, reaching the quarter finals netted €3.3 million, and reaching the semi finals added another €4.2 million for clubs. The runners-up received another €5.6 million and the Champions League winners walked away with another €9 million.These rewards continue to rise. As evidence of this the BBC (http://www.bbc.co.uk/sport/0/football/24958189, accessed 17 November 2013) reported Manchester United earned £30m for getting to the Champions League's last 16 in 2013. Arsenal earned £26m, and Manchester City £24m. Chelsea earned £53m for winning the Champions League in 2012.

sought cheaper foreign players instead of investing in local talent (Taylor 2006). Some of the bigger clubs were able to do both.

Some key features of football distinguish it from many other commercial ventures. Clubs are fixed in their locations (McGovern 2002). Even when they build new stadiums, these stadiums are very close to the locations of their old stadiums.[25] This means *capital is fixed in geographic space* for football. It is labor that moves geographically to generate the fundamental networks linking football clubs. Arrangements between labor and capital have been, and continue to be, inherently conflictual in capitalist societies. Within football, the tight control of players worked to the advantage of club owners. When not settled by force, most often, disputes are resolved in courts. Football was not immune, with some major labor issues settled there when players were able to file suits. Four cases are thought noteworthy for studying football networks.

7.7   Court rulings

From a player's standpoint, England's 'retain and transfer system' was a major problem: club owners valued highly this form of coercive control of players. It was abolished as a result of the George Eastman case (against Newcastle United in 1964), the first time a player had been able to get a hearing in the courts.[26] A judge ruled that Eastman's contract was an illegal restraint of trade (Murray 1996).

In February 1978, the EU handed down a decision that, in essence, ended England's ban of non-British players playing football in England. All Football Associations within the EU had to change their labor rules to make them consistent with those of the Treaty of Rome. We do not explore football fan behavior, but we note the UEFA decision to ban indefinitely English teams in 1985 from participating in UEFA sponsored club competitions due to off-the-field behavior of English fans.[27] This was triggered primarily by the disaster at the 1985 European Cup Final when rioting by Liverpool fans led to the deaths of 39 Juventus supporters. This ban was lifted for the 1990/91 season, providing major incentives for non-British players to play in England (Hirst, 2004).

In December, 1995, the EU Court of Justice handed down another decision,[28] for a consolation of three cases involving one player, which came to be known as the Bosman Ruling. The decision concerned freedom of movement for workers and so affected all clubs in national leagues within the EU. By allowing players of these clubs to move, without restraint, to another club, once their contract had expired, it introduced freedom of contracts for players and, equally important, it also abolished player quotas by nationality. Another decision by the European Court of Justice, in May 2003, in effect extended the Bosman

---

[25] Examples from the EPL include Arsenal's move from its Highbury stadium to the Emirates Stadium, Southampton's move from The Dell to St Mary's, and Manchester City's move from Maine Road to the City of Manchester (Etihad) Stadium. AFC Wimbledon moving 90 km to Milton Keynes and becoming the MK Dons was a rare (and conflictual) exception. Moves of (American) football teams in the USA where the Los Angeles Rams became the St Louis Rams and the Baltimore Colts became the Indianapolis Colts are unthinkable for most soccer leagues. Even more unthinkable are movements like the Kansas City Scouts (an ice hockey team) to Denver to become the Colorado Rockies before moving to Newark as the New Jersey Devils.

[26] Other footballers before him had tried to break this control by owners but both the FA and the FL had managed to keep players and their potential suits out of the courts.

[27] http://news.bbc.co.uk/onthisday/hi/dates/stories/june/2/newsid_2494000/2494963.stm (accessed 6 May 2012).

[28] Union Royale Belge des Sociétés de Football Association v. Jean-Marc Bosman (1995).

Ruling to apply to citizens of countries having Association Agreements with the EU and having valid work permits. They had to be treated, as far as employment was concerned, as if they were citizens of an EU country. These decisions are seen as greatly facilitating freedom of movement for footballers to pursue their football careers.

7.8   Specific factors impacting football migration

We consider some general factors affecting football player migrations. Colonial histories plus their created interdependencies had great impact (Maguire and Stead 1998). Geography, language, religion, colonial history, and post-colonial ties all play a role in player movements (Taylor 2006). More sharply, migration reflects historical contexts including exploitation of weaker nations by stronger nations (McGovern 2002). While migrations between all FIFA confederation regions occur, the key migrations at the top level involve movements to and from Europe and within Europe. These movements prompt Poli (2010) to view player migrations as creating the 'Europeanization of sport' including soccer.

Specific concrete mechanisms directly affect player movements including labor rights, salary caps, recruitment and retention strategies, player nationalities, residing in other countries, and cultural dislocation before and after moves (Maguire and Stead 1998). The EU rulings stripped most of them of their power to create general systematic patterns, even though they have impacts on individual players.

The struggle between professionalism and amateurism within organized football, while having a long conflictual history in many places, ended with football being professional at the top levels. From the beginnings of professionalism in Britain, the professional clubs were supported by a vast network of amateur teams (Murray 1996). This has not changed. Immediately after WWII, the Scandinavian countries and the Netherlands were fiercely amateur. They lost many top players to professional clubs elsewhere in Europe (Lanfranchi 1994). For between-nation movements, these dynamics finished well before our study's time frame.

Turmoil in specific regions can occur at any time, and such turmoil triggers migrations of many people including footballers (Duke 1994; Maguire and Stead 1998). Of these, the dissolution of the Soviet Union with the fall of the Berlin Wall, and the Balkan wars leading to the dissolution of Yugoslavia, have had a dramatic impact on the movement of footballers from these regions.

7.9   Some arguments and propositions

We outline some general issues that influenced our selection of problems, and state specific hypotheses for testing, the results of which are reported at the end of this chapter and in Chapter 8. Organized football has existed since 1888. We have divided the period between this starting date and the 2010/11 season into three distinct eras: The *Old Era* is from 1888 to the beginning of WWII (but we spend very little time considering this period). In essence, and especially for Europe, organized football ceased during WWII. We define the *Pre-Modern Era* as starting with the 1946/47 season and finishing with the 1989/90 season. The, so called,

Modern Era starts with the 1990/91 season and ends, for us, with the 2010/11 season.[29] Of course, this era has continued after our study period.

Our hypotheses vary from being very simple to being rather complex either in their statement or in the data and methods required to assess them. Given the legal forces that have acted upon English football authorities, and the changes in the organization of football around the globe, together with Europe being a magnet for non-European footballers, the first (obvious descriptive) hypothesis is easy to state.

Proposition 7.9.1. *The regional composition of the EPL has changed throughout its short history.*

We use the FIFA confederations to operationalize regions. However, we split the countries of UEFA into Western Europe (WEU) and Eastern Europe (EEU) because there are systematic differences in the flows for these two broad areas (Duke 1994; Maguire and Stead 1998). Given Australia's recent move to AFC, we have classified it there throughout, even though the move was recent (in 2006). New Zealand is the only other source of players in the data we have from the Oceania Football Confederation and we have grouped it with the AFC (for simpler visual summaries only).

If EPL clubs were faced with demands for short-term success and clubs responded by recruiting cheaper foreign players, this makes Proposition 7.9.1 trivially true. It can be made less obvious by arguing that this applies to all clubs and not just to the EPL as a whole.

Proposition 7.9.2. *All clubs in the EPL have squads with increased diversity of the regional and national origins of players.*

Underlying the hypotheses stated thus far is the presumption that clubs have been changing their recruitment strategies and preferences for players. If so, this has an effect on which types of players are recruited and the routes over which they reach the EPL. Proposition 7.9.3 comes from McGovern (2002) and is rather subtle.

Proposition 7.9.3. *Early in the history of the EPL, clubs recruited players who most resembled indigenous players.*

Prior to 1978 when the EU required club compliance with EU labor laws, Proposition 7.9.3 held trivially: recruitment was within the UK and Ireland, allowing for some player recruitment from the British Commonwealth. Following 1978, testing this hypothesis becomes more difficult due to the inherent ambiguity of 'resemble' and 'indigenous players'. The latter becomes more ambiguous when its composition changes over time. To address Proposition 7.9.3 more fully, we consider some more specific hypotheses. If non-English players are not the same as indigenous (English) players then Proposition 7.9.4 follows.

Proposition 7.9.4. *The proportion of English players in the squads of the EPL has diminished through time.*

---

[29] There are three conventions for reporting specific seasons. Given that, in Europe, most football seasons start in one calendar year and end in the following year, one convention uses both years for labeling a season. An alternative is to use the starting year and another is to use the ending year. When we do not use both years in our narratives, we employ the convention of using the starting year of a season as the label for that season.

One vigorous debate in England concerns the impact of having non-English players in EPL squads. Clearly, this diminishes playing opportunities for English players[30] (Magee and Sugden 2002). Since the start of professional football in England, English clubs recruited heavily from the Celtic Fringe (a term coined by McGovern (2002)). It is reasonable to think that the shrinking presence of English players has been matched by a reduced presence of Celtic Fringe players. Further, given a 'loophole' in the effective ban on foreign players in the English League, some non-English players from the Commonwealth were admitted. McGovern (2002) notes that the presence of players from the British Commonwealth declined and this may have continued into the EPL.

Proposition 7.9.5. *The proportion of players from the Celtic Fringe in the squads of the EPL has diminished through time.*

Proposition 7.9.6. *The proportion of players from the British Commonwealth in the squads of the EPL has diminished through time.*

Following WWII, labor shortages existed in England. In response, people were admitted from Commonwealth countries, including African colonies (before and after they became independent), the West Indies, India, and Pakistan. In addition to a strong bias against foreign players, English football had one against recruiting black players[31] (Harris 2006). Yet many black players are indigenous by the second generation. Increasing numbers of black players, both English and non-English, in British football changed the composition of the indigenous player pool. We have not yet obtained complete race data for all players who were in EPL squads for the first 15 years of the EPL. So we use a crude surrogate. If Proposition 7.9.3 is correct, then the expansion of non-English players would have been through players recruited first from Northern European nations at the beginning of the EPL. Further, this recruitment would expand and be followed by increased recruitment from Eastern Europe given that both areas are largely 'white'.

Proposition 7.9.7. *The largest part of the recruitment of non-English players was from Western Europe and this has expanded over the course of the EPL.*

Proposition 7.9.8. *The recruitment of players from countries of Eastern Europe has expanded over time.*

Comparisons between English and non-English players playing in the EPL can be made. The football style in England has been characterized as being a 'long game' which meant, despite some exceptions, dribbling and short passing moves were less valued. There has been discussion also that managers and footballers were not dedicated to training and fitness regimes. Non-English players such as Jürgen Klinsman and Eric Cantona are thought to

---

[30] Authors such as Exall (2011) argue that this lowered the quality of England's national team. However, Kuper and Szymanski (2009) offer an alternative view, that it has had little impact on England's performance because playing with and against good non-English players raises the quality of English players. However, we do not explore this issue here.

[31] Viv Anderson was the first black player to play for England, and his first call-up did not come until 1978. A 'let's kick racism out of football' campaign started in 1993. Recent (2012) bans and fines of two players in the EPL for 'racial abuse' have become focal points in this campaign.

have inspired their team-mates to take fitness and training seriously. This raises an obvious question: do non-English and English players differ in fitness levels? If so, has this changed over time with English and non-English players converging in fitness levels? In assessing this issue, playing positions must be taken into consideration.

Proposition 7.9.9. *At the beginning of the EPL, non-English players were fitter than English players and over time, the two groups are converging with regard to physical characteristics.*

We argued that there are two mechanisms for players entering the EPL. One is recruiting very young players and training them. Another is obtaining players through transfers or loans. For clubs, both strategies carry risks. Kuper and Szymanski (2009) discuss the recruitment strategy of Olympique Lyonnais, a club that won seven consecutive Ligue 1 titles in France starting with the 2001/2 season. It avoided recruiting very young players (their potential might not be realized), former star players thought to be beyond their prime, and star players. This recruiting approach seems consistent with another hypothesis of McGovern (2002).

Proposition 7.9.10. *EPL clubs avoided top-end international players in their early to mid-twenties especially from Argentina, Brazil, Germany, Italy, and Spain.*

Assessing this hypothesis requires the operationalization of 'top-end' and 'country'. The latter is easy to resolve. We use the nationality of the players. Another option is to use the countries of the clubs from which players are recruited. We use both meanings in separate analyses. The notion of 'top-end' is operationalized as having played for their national teams.

All clubs are compelled to 'strengthen their squads' through recruitment. Does the recruitment of non-English players have any bearing of league success? If all clubs are improved then the improvements across squads would cancel each other out. Also, clubs may be imitating each other when seeking players from a worldwide pool. The distribution of nationalities across clubs need not have any relationship to club performance. One variable known to be an effective predictor of league success in England is level of player salaries (Kuper and Szymanski 2009).

Proposition 7.9.11. *Controlling for salary levels, the proportion of non-English players in EPL squads is unrelated to success within the EPL.*

If the number of non-English players in the EPL has increased then it is reasonable to expect that the actual routes to the EPL have multiplied. Maguire and Stead (1998) provide detailed information on general player routes: there were 422 migration routes for footballers with 296 routes within Western Europe, 147 routes from Eastern Europe, and another 126 routes into Europe during the late 1980s and the 1990s. This detailed information is hard to digest broadly. Each player can generate a new (idiosyncratic) route and this alone can increase the reported number of routes. Nor is there any information regarding the composition of these routes.[32]

---

[32] For example, if one player has the migration route Australia to Croatia to Germany to England and another travels from Australia to Serbia to Germany to England, they do count as different routes. However, they could be viewed as the same generic (or fundamental) route from Australia to England in the country-to-country network.

Question 1 Is it possible to establish a set of basic types of generic routes of footballers to the EPL?

In the network of player moves between countries (and also between clubs), we define a *trail* of a player as an alternating sequence of country (or club) stays and moves between countries (or clubs) from the start of a career (age 15 for us) and ending when the player hangs up his boots. In the country-to-country network, this will entail a consideration of the number of steps in the player trails for specific countries through which players pass. The same holds for the club-to-club network. In essence, answering Question 1 requires a consideration of all trails in a graph made up of individual player moves. We have one simple hypothesis, one that is more methodological.

Proposition 7.9.12. *There is a set of fundamental player routes to the EPL.*

At face value, this seems trivial. However, if these fundamental routes, representing large sets of players, exist they will have great value. Once this has been established, it may be possible to consider if the distribution across them has changed. We think of these fundamental routes less in terms of geographical features, in contrast to Maguire and Stead, but more in terms of the overall ranking of clubs. Given the EPL's presence as a top football league, we think of player migrations to the EPL as movements between clubs within *a stratified system of clubs*.

(McGovern 2002): 24 argues 'employers prefer to engage in repeated transactions with reliable and known sources as a means of reducing the uncertainties that characterize labour as a commodity' and applies this notion to the recruitment of footballers. McGovern argues that EPL clubs, and English clubs in general, prefer repeated transactions with other clubs when acquiring footballers.

Proposition 7.9.13. *EPL clubs, and English clubs in general, prefer repeated transactions with other clubs for transfers and loans.*

To the extent that transfers and loans are risky transactions for both clubs and players, it makes sense that the 'provenance' of a club source for players has value. Yet, given the high turnover of players and managers of football clubs, this hypothesis, if confirmed, is likely to be a short-term phenomenon. To the extent that player movements are driven by immediate-term contingencies, the hypothesis will not hold.

In principle, any club could appear in the top level of a nation's league system. However, in practice, this is a fiction. This leads to another descriptive hypothesis:

Proposition 7.9.14. *The top level of the league systems of the top leagues in Europe is inaccessible for a huge majority of clubs in their league systems.*

We have argued, in neutral terms, that player moves between clubs define the fundamental network structure of football (as does Poli (2010)). However, there may be more to this network in terms of its dynamics, and it raises the following question.

Question 2 Is this fundamental network also an integral part of the processes leading to the almost closed nature of the top league in national league systems?

This question implies looking at the network of player moves as more than a description of a worldwide football structure. Many of the foregoing hypotheses and the arguments for them point to the motivations of clubs coupled to those of players. They require a closer look at the nature of relations between clubs.

Proposition 7.9.15. *The fundamental relation between clubs is predation where stronger clubs prey on weaker clubs when acquiring players.*

Supporting arguments for this hypothesis include: 1) leading clubs buy the best players;[33] 2) lower-level clubs cannot retain their best players when the players want to move, especially if their club has just been relegated from the EPL;[34] and 3) lower-level clubs must often sell their best players in order to survive. Put differently, stronger clubs dominate the transfer market (McGovern 2002). We noted that young players can enter the world of organized football when they are recruited to a football academy operated by a club. McGovern argues that the top clubs also dominate the training system.

Proposition 7.9.16. *The majority of English players entering into the EPL do so through football academies controlled by EPL clubs.*

Players move, or are forced to move, when clubs upgrade their squads. From the vantage point of players, the downside of clubs making these upgrades is that players who were performing adequately no longer have value for the club. They are sold or are released. Such players usually move to another club if they are not approaching the age when their careers would end. An obvious question is: Where do they go to? In response, we propose:

Proposition 7.9.17. *English players who have started with EPL squads have longer and more successful careers when they start in the EPL compared to English players who start their careers outside the EPL.*

The Bosman Decision is controversial regarding its impact on player flows in general and the flows of players into and out of the top level of English football. One view is represented by Poli (2010) who argues that the Bosman Decision did not increase the spatial diversification of players and merely reinforced older channels. In contrast, Maguire and Stead (1998) argue that the Bosman Decision was powerful in increasing the flow of non-English players into British football. The wordings of these two claims differ as do the terms of reference as to what the authors were focusing upon. We confine attention to the EPL and suggest the following hypothesis.

Proposition 7.9.18. *The Bosman Decision had no effect on the flow of non-English footballers into the EPL beyond reinforcing older patterns.*

---

[33] Much of the evidence supporting this takes an anecdotal form. For example, Chelsea bought Fernando Torres from Liverpool for £50 million – then an English record transfer fee – and Liverpool bought Andy Carroll from Newcastle United for £35 million (both trades were in January 2011). Neither of the selling clubs could resist these sums of money. More recently (May 2012), BBC Newcastle had an interview with the Newcastle United manager, Alan Pardew, who claimed that it is impossible to prevent Real Madrid, FC Barcelona, Chelsea, and Manchester City from coming for his club's players. Newcastle had players on lower salaries than those of the super-clubs. Pardew insisted that Newcastle's predicament was the same as that faced by Arsenal: it was unable to keep Samir Nasri and Cesc Fàbregas from signing for Manchester City and FC Barcelona respectively in 2011. (Source: http://www.bbc.co.uk/sport/0/football/18042729). Two moves from EPL clubs, both to Real Madrid, are consistent with this escalating trend, consistent with the observation of Newcastle United's manager. Gareth Bale became the world's most expensive football player when he moved from Tottenham Hotspur to Real Madrid for £88m in September 2013. In 2009, the Spanish giants had paid £80m to Manchester United for Cristiano Ronaldo.

[34] For example, when Blackpool was relegated from the Premier League at the end of the 2010/11 season, Liverpool was able to get their captain and star player, Charlie Adam.

We have focused, thus far, on movements to the EPL. Footballers reach this league through flows within the UK or flows into the UK from elsewhere. They can be known players with a good reputation or unknown players yet to make their mark. Regardless of how they arrive, once they leave the EPL, in the main, they never return. Of course, some can return with clubs promoted to the EPL. Others leave relegated EPL clubs to be recruited to other (struggling) EPL clubs or clubs promoted to the league. The main factor driving this proposition is age: professional football in England is a young man's profession. As discussed earlier, clubs discard players once they have no further use for them.

Proposition 7.9.19. *Once footballers leave the EPL, they are highly unlikely to return.*

An additional conjecture, Proposition 7.10.1 is added in the next section.

7.10 Some preliminary results

Players from 148 countries were in the EPL between 1992 and 2006. Europe had 55 countries with players in the EPL – 30 in Western Europe and 25 in Eastern Europe. Africa was represented by 43 countries. All 10 countries from South America had players in the EPL. A total of 23 countries in North America – including Central America and the Caribbean – (using NAM as a label instead of CONCACAF) were represented along with 18 countries from Asia.

Figure 7.1 shows percentages of players from the Home Countries (England (ENG), Ireland (IRL), Scotland (SCO), Wales (WAL), and Northern Ireland (NIR)) and six regional

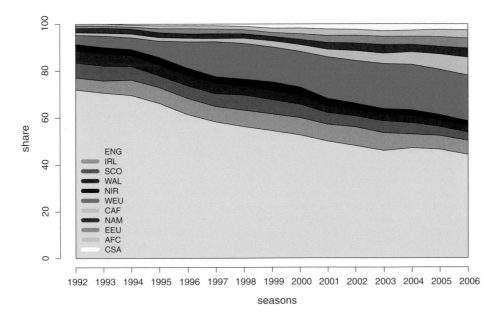

Figure 7.1 Player presence in the EPL by UK countries, Ireland, and FIFA confederations: 1992–2006.

groupings: WEU (actually, the rest of Western Europe excluding the Home Countries), CAF, NAM, EEU, AFC, and CSA over time. The colors depict Home Countries and world regions. A legend coupling places to colors is on the lower left of Figure 7.1. Shades of gray represent the Home Countries with the lightest shade used for England. The horizontal axis is marked by the seasons and percentages are on the vertical axis.

7.10.1   The non-English presence in the EPL

The shapes of the overall colored plots support Proposition 7.9.1: the regional composition of the EPL changed over its first 15 years. The major detailed features are well known also: 1) a steady decline in the percentage of English players in the EPL (consistent with Proposition 7.9.4); 2) a smaller but clear decline in the presence of players from Northern Ireland, Scotland, and Wales (all in the Celtic Fringe) supporting Proposition 7.9.5; 3) players from Ireland were present in slightly greater numbers in the middle of this period but their numbers then dropped to a level slightly above the 1992 level, largely consistent with Proposition 7.9.5; 4) a dramatic increase in players from the rest of Western Europe (WEU) supporting Proposition 7.9.7; and 5) a steady but smaller increase in the percentage of players from EEU supporting Proposition 7.9.8. The percentage of players from Africa was small in the first year of the EPL. After that, there was a modest increase in the number of CAF players, which accelerated at the end of our study period. This also provides support for Proposition 7.9.3. The presence of players from North America has fluctuated slightly, finishing at a higher level. Finally there is a lower but steady increase in the percentage of players from South America and Asia.

7.10.1.1   Club level variations

While Figure 7.1 confirms Proposition 7.9.7 for the EPL as a whole, it masks variations at the club level. To examine these variations, we focus attention on the seven clubs that were present throughout our study period. Indeed, they have never been out of the EPL. They are Arsenal, Aston Villa, Chelsea, Everton, Liverpool, Manchester United, and Tottenham Hotspur. We label these as the 'top clubs' for this discussion. We note that Arsenal, Chelsea, Liverpool, and Manchester United were routinely described as the 'Big Four' until Manchester City won the EPL title in the 2011/2 season. By far, the two most successful clubs in this period were Manchester United and Arsenal.[35] Figure 7.2 shows the regional variation of player presence for these two clubs. The decline in the proportion of English players was far more dramatic for Arsenal than for Manchester United: in 1992 Arsenal had a higher proportion but in 2006 it had a lower proportion. Manchester United both started and finished with a higher proportion of players from the Celtic Fringe than did Arsenal. Apart from one Irish player, Arsenal eliminated Celtic Fringe players from its squad. Players from the Home Countries fared better, in terms of squad presence, under Manchester United's Scottish manager than under Arsenal's French manager.[36] Arsenal expanded the number of Western

---

[35] Manchester United won nine EPL titles and Arsenal won three. In addition, Arsenal won five FA Cups with Manchester United winning four. In this period, Chelsea won two EPL titles and two FA Cups. Other top club winners of the FA Cup were Liverpool (three), Chelsea (two), and Everton (one).

[36] Sir Alex Ferguson was present throughout the study period at Manchester United while Arsene Wenger was present after 1996 at Arsenal. Making comparisons of the trajectories of regional presence at these clubs for 1996-2006 is simpler, in contrast to the other top clubs where managerial changes occurred.

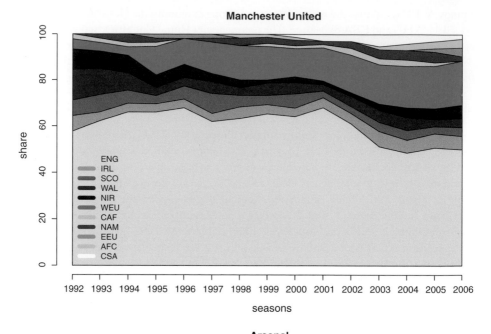

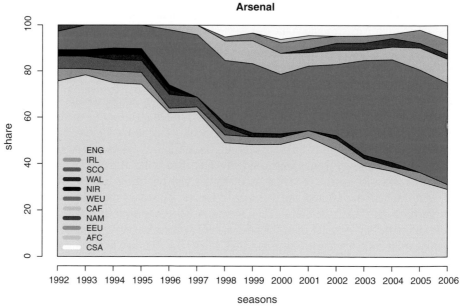

Figure 7.2   Regional presence at Manchester United and Arsenal: 1992–2006.

European (WEU) players earlier than Manchester United and it did so to a greater extent. Indeed, it had a much greater proportion of players from outside the Home Countries than local players in 2006.

The corresponding trajectories for Chelsea and Liverpool are shown in Figure 7.3. The proportion of Chelsea's English players declined through time until 2003 but then increased slightly. Overall, the proportion of English players at Liverpool dropped to a greater extent. Chelsea started with a large proportion of Celtic Fringe players in 1992. None remained in 2006. Liverpool also shed players from the Celtic Fringe although three were present in 2006. Both of these clubs had an expanding proportion of WEU players. They expanded also the number of players from other regions. These data also support Proposition 7.9.1. Propositions 7.9.4 (decline in the proportion of English players) and 7.9.5 (decline in Celtic Fringe player presence) are supported strongly by the club-specific data.[37]

Figure 7.4 provides further minor variation on the same overall theme. The trajectory for Everton has a pattern similar to Manchester United with a peak in 1999. Apart from two upward jumps between seasons, the trajectory for Tottenham Hotspur is downwards throughout. Even so, there were more English players at Manchester United and Tottenham Hotspur in 2006. A case can be made for the declines in numbers predicted in Proposition 7.9.4 coming only *after* 1999/2000.

Figure 7.5 shows the numbers of English (on the left) and Celtic Fringe (on the right) players. Overall, they show support for Propositions 7.9.1, 7.9.2, 7.9.4, and 7.9.5 using absolute numbers rather than proportions. They reveal also club-level differences. The trajectory for Manchester United's English players increases through 2000, contrary to Proposition 7.9.3. However, the drop in their numbers after 2000 is precipitous.

The differences between clubs are more dramatic for the numbers of Celtic Fringe players. Everton had managers from the Home Countries (Scotland 4, England 4 (with two of them being short term), and Wales 1) throughout this period. The trajectory for Everton through 1999 flatly contradicts Proposition 7.9.5. This club plays in the city of Liverpool, a short boat ride from Ireland and a prime destination for Irish immigrants over the years. The football club, Liverpool, had Home Country managers (with two from Scotland) through 1998. Its trajectory for Celtic Fringe player presence through 1997 also contradicts Proposition 7.9.5, but supports it after 1997. Except for 1997–2000, the trajectory for Aston Villa also supports this proposition. Throughout, the Arsenal and Chelsea trajectories support Proposition 7.9.5, as is the case for Tottenham Hotspur after 1999. The numbers for Manchester United show no obvious pattern. Overall, the safest conclusion is that the proposition regarding the declining presence of Celtic Fringe players holds only after 1999/2000. Also evident is that Manchester United and Everton are different from the other top clubs in retaining more players from the Celtic Fringe – from where two long-term managers hailed.

There is some modest evidence for the nationality of a club's manager having an impact on the regional composition of their squads. The number of WEU players jumped after Arsene Wenger, a French manager, arrived at Arsenal. Among them, the highest proportion came from France. The same happened at Liverpool following the arrival of Gerard Houllier, another French manager. He was succeeded by the Spanish manager, Rafael Benitez, in 2004. Thereafter, the number of French players declined while the number of Spanish players jumped. The arrival of Gianluca Vialli and Claudio Ranieri, both from Italy, at Chelsea led to an increase in the number of Italian players in its squads. All of these managers brought

---

[37] We suspect the same results hold for other clubs that played in the EPL. However, comparisons with these clubs are complicated by relegation of clubs from the EPL and the promotion of clubs to this league. Both types of events trigger additional discarding and acquiring of players.

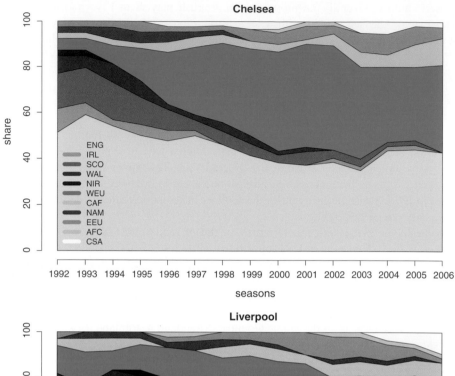

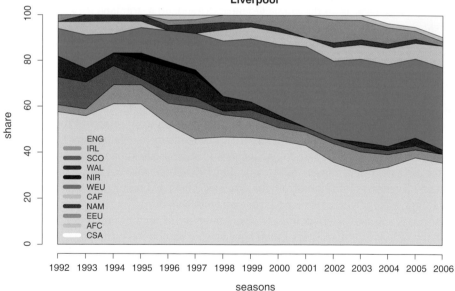

Figure 7.3   Regional Presence at Chelsea and Liverpool: 1992–2006.

not only players of their own nationalities to their new clubs but also a wide knowledge of the WEU football scene regarding the distribution of talent and the availability of players.

However, caution is merited when linking manager nationality to squad composition. None of the changes at Manchester United can be attributed to the nationality of its manager because he was present throughout this period (and beyond it). The decline in the number of Celtic Fringe players at Aston Villa continued during the tenure of the Irish manager, David O'Leary, at the club. While it is tempting to see the arrival of the Portuguese manager, Jose Mourinho, as triggering an increase in the presence of Portuguese players, this increase started just prior to his arrival. Also, when the managerial presences at clubs are short, it is

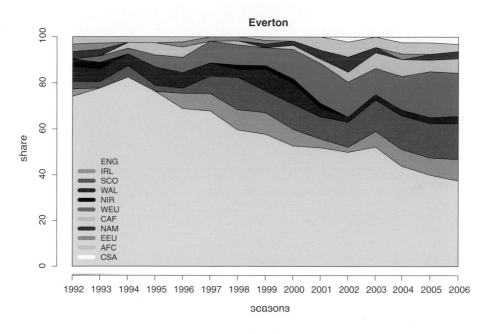

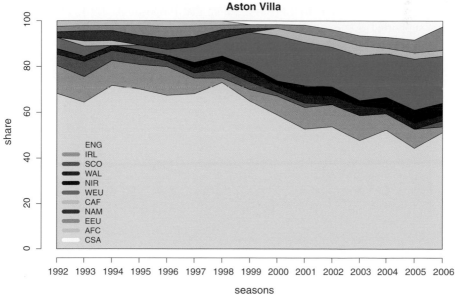

Figure 7.4   Regional presence at Everton, Aston Villa, and Tottenham Hotspur: 1992–2006.

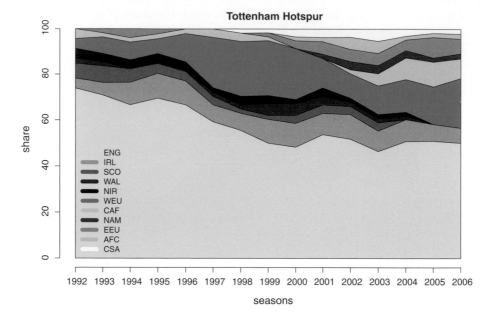

Figure 7.4   (*Continued*).

more difficult to discern the impacts of their tenures on squad composition. Finally, at a club level, the numbers are too small for a general definitive claim about nationality and regional composition of squads.

In general, it is clear for the EPL as a whole that Propositions 7.9.1, 7.9.2, 7.9.4, and 7.9.5 are supported. Some caveats regarding this conclusion for Propositions 7.9.4 and 7.9.5 come from several top clubs. Even so, the evidence supports them after 1999/2000. There is no denying that the EPL was far more diverse in 2006 than in 1992 regarding the regions from which football players came. According the Harris (2006) only ten 'foreign' players

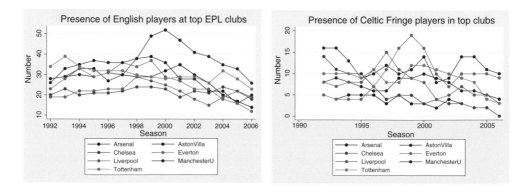

Figure 7.5   English and Celtic Fringe players at the top EPL clubs: 1992–2006.

were among the 242 players in the starting teams on the opening day of the EPL. He reports a steady decline of English players appearing in the league. Their numbers were: 391 in 1992/3; 262 in 2001/2, and close to 200 in 2005/6.

The results have varied at the club level also. Aston Villa fielded an all-English squad in February 1999 for an EPL game. This was the last time this happened in the league. On Boxing Day 1999, Chelsea had an all-foreign starting line-up, although some Home Country players were on the bench. Arsenal was the first club to name a complete squad with no Home Country players, in February 2005.

7.10.1.2   Effects of the Bosman decision

The results shown in Figure 7.1 for the EPL as a whole leave no doubt about the increasing presence of footballers in the league coming from elsewhere, a phenomenon present at the club level as shown in Figures 7.2, 7.3, and 7.5. However, beyond being potentially suggestive, the trajectories do not show whether the Bosman Decision or the later EU decision extending it to footballers from Eastern Europe had a specific impact on these flows.

There are two broad arguments regarding the impact of this decision on the flow of non-Home Country players to the EPL: 1) it had a dramatic impact on the flow of these players into the EPL (e.g. Maguire and Stead (1998)), and 2) it merely continued previous flows of players (e.g. Poli (2010)). We counted the transfers of players to the EPL from outside the Home Countries for three periods: 1992–1995; 1996–2001, and 2001–2006. This categorization was constructed because the Bosman Decision came in 1995 and the second EU decision in 2001. More importantly for our purposes, we counted the number of players involved in these transfers.

As background, we report the composition of the EPL by the nationality of players in the first season of the league. Table 7.1 shows the composition for the Home Countries and Table 7.2 does the same for the FIFA confederations. Clearly, the EPL at first was very much a Home Countries league, with 92% of the players coming from these places. And, from Table 7.1, of these footballers, 78% were English. Of the non-Home Country regions, players from WEU, EEU, and NAM dominate with 76% of the foreign players. The distribution in Table 7.2 provides a baseline for assessing the impact of the Bosman Decision. To the extent that the WEU players come from Northern Europe, EEU players are white and the NAM players come from North America and Canada, the numbers in Table 7.2 provide modest support for Proposition 7.9.3 concerning a preference for recruiting footballers who most resemble indigenous players.

Table 7.3 shows the counts for the presence of non-Home-Country players by regions defined by selected FIFA confederations. All transfers between clubs in the Home Countries were excluded, even those involving foreign players already present.

Table 7.1   The initial distribution of Home Country players.

Season	ENG	SCO	WAL	IRL	NIR	Total
1992/3	525	49	42	40	19	672

Labels denote: ENG (England), SCO (Scotland), WAL (Wales), IRL (Ireland), and NIR (Northern Ireland)

Table 7.2 The initial distribution of non-Home Country players.

Season	WEU	EEU	NAM	CAF	CSA	AFC	Total
1992/3	26	6	15	9	0	6	62

Labels denote: WEU (Western Europe), EEU (Eastern Europe), NAM (North America), CAF (Africa), CSA (South America), and AFC (Asia)

Summarizing the figures in Table 7.3:

1. The jump from 88 to 466 between the first and second periods shown on the left for all non-Home-Country players is fully consistent with the first argument regarding the impact of the Bosman Decision. The drop from 466 to 426 between the second and third periods does not seem consistent with this argument, suggesting that there was only a temporary, but still significant, jump in the second period.

2. All of the jumps in the numbers of players from all of the regions shown are consistent also with the argument of the Bosman Decision having a dramatic impact, especially for players coming from elsewhere in Western Europe.

3. There was a shift in the composition of the number of players coming from the FIFA confederations between the second and third periods compared to the changes between the first two periods. Compared to the earlier changes, the numbers coming from Western Europe *dropped* in the third period while the numbers coming from the other areas continued to rise, albeit to a much lesser extent. The second EU court decision expanding Bosman to Eastern Europe appears to have had a minimal impact. While the number of players coming from that area rose, so too did the numbers from Africa, South America, and North America by about the same amounts.

Some of the changes between the second two periods can be partially accounted for after we consider the changes in EPL squad sizes.

Another way of assessing the impact of the Bosman Decision is to consider the primary mechanism for recruitment of players to the EPL from countries other than the Home Countries. This mechanism is recruitment from clubs located in other countries in Western Europe. Of the 88 players listed in Table 7.3 as moving into the EPL during the 1992–1995 period (before Bosman) from elsewhere, 74 (84%) came from WEU clubs. The corresponding figures for 1996–2001 are 466 and 454: 87% of these players were transferred from WEU clubs. For 2002–2006, the figures are 426 and 359, with 84% for the percentage coming

Table 7.3 Flows of non-Home-Country players to the EPL from non-Home-Countries clubs.

Period	WEU	EEU	CONCACAF	CAF	CSA	Total
1992/1995	55	11	9	6	7	88
1996/2001	306	38	51	34	37	466
2002/2006	208	46	68	59	45	426

Labels denote: WEU (Western Europe), EEU (Eastern Europe), CONCACAF (North America, Caribbean and Central America), CAF (Africa), and CSA (South America)

Table 7.4 External flows of traditional source players to the EPL.

Period	HC	AUS, NZE, RSA	NAM	Total
1992/1995	9	12	2	23
1996/2001	45	16	14	75
2002/2006	32	11	22	65

Labels denote: HC (Home Countries), AUS, NZE, RSA (Australia, New Zealand South Africa), and NAM (North America)

from WEU clubs. Just considering these percentages, a (very weak) case can be made for these movements being no more than a continuation of extant patterns prior to the Bosman Decision. While this is consistent with Proposition 7.9.18, the jump of foreign players moving to the EPL from 88 (during 1992–95) to 466 (during 1996–2001), almost a 430% increase, is a far more potent signal of the dramatic impact of the Bosman Decision on recruitment to the EPL following it being handed down by the EU Court. Overall, Proposition 7.9.18 is not supported.

Table 7.4 restricts attention to moves from more 'traditional' places for players recruited to the top tier of English football from clubs outside the Home Countries. In summary:

1. The change in the number of players from the Home Countries across the three periods is similar to the pattern for non-Home-Country players coming to the EPL from clubs outside the Home Countries. If so, these returnees were caught up in the general flows of players from WEU clubs.
2. The increase in numbers from British Commonwealth countries (Australia and New Zealand) and South Africa was too small to claim that the Bosman Decision had any impact on them. Those movements seem a continuation of older patterns of recruitment, especially given the numbers entering the EPL from these three countries via WEU clubs being 3, 6, and 7 across the three periods.
3. It is unlikely that much of the movement from NAM, especially from the USA, can be seen as a secondary impact stemming from the Bosman Decision.

We note that the increase in the number of players from CONCACAF countries included players from former British colonies in the Caribbean, such as Jamaica, Trinidad and Tobago, and Bermuda whose movements would have been affected little by the Bosman Decision. Movement from these areas had been freer given their Commonwealth status, and some of these players were born in, or were already in, England due to migrations by their parents.

7.10.1.3 Squad sizes

One club characteristic thought to be important for success is the size of its playing squad. Clubs must allow for injuries to, suspensions of, and the loss of form of, their players. Squads have to be large enough to meet these contingencies should they arise. As shown on the left in Figure 7.6, containing box plots of squad size for each season during the first 15 seasons

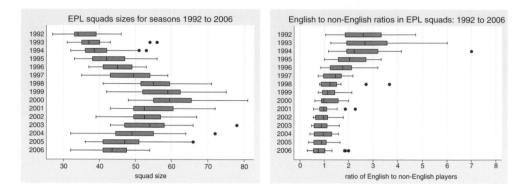

Figure 7.6   EPL squad sizes and proportions of English players: 1992–2006.

of the EPL, squad sizes varied greatly across teams.[38] There was systematic variation with increasing squad sizes from the 1992 season through to the 2000 season and, thereafter, declining squad sizes through to the 2006 season. Most, but not all, box plots show some slight skews on the high side in the form of some high outliers, medians closer to the lower quartiles than to the upper quartiles, and some longer whiskers at the upper end of the box plots. The median squad size for the 1992 season was 34 (with 32 and 39 as the lower and upper quartiles). For the 2002 season the median was much higher at 59.5 (with 55 and 62.5 as the lower and upper quartiles). Finally, the median for the 2006 season was down to 43.5 (with 41 and 54 as the lower and upper quartiles). We note that in these 15 seasons, the clubs with smallest squad were relegated in seven of those seasons. Manchester United had the largest squad for 13 of these 15 seasons and was the EPL champion for six of these occurrences. However, Sheffield United had the largest squad in the 2006 season and was relegated. While suggestive, these instances of success and failure, in relation to squad size, are only partially conclusive.

Figure 7.6 (right) shows box plots for the ratio of English to non-English players for the EPL over the first 15 seasons. A median ratio of 1 holds if the number of English and non-English are equal in the EPL. Median ratios less than 1 show more non-English players than English players in EPL squads and overall. Every EPL club had a ratio above 1 for the 1992 season. Indeed, the median EPL ratio was above 2 during the first four EPL seasons and dropped gradually to 1 for the 2000 and 2001 seasons. Thereafter it dropped below 1 and remained there. Indeed, the upper quartile is close to 1 for the 2006 season: three-quarters of EPL clubs had more non-English players in their squads than English players. This clear pattern is fully consistent with the overall pattern shown in Figure 7.1. It also provides support for Proposition 7.9.4 at the level of clubs.

Across these 15 seasons, for the proportion of English players, there are eight high outlier teams and six of them were relegated: Leicester City (1994); Charlton Athletic (1998); Ipswich Town (2001); Watford (2006); Middlesbrough (2006); and Sheffield United (2006).

---

[38] A small part of the variation may be accounted for by the ways in which clubs list their players on their websites but much of the variation is real. Players who are seriously injured will have played for the club but, in effect, are replaced by other players who are also recorded as belonging to the squads. The numbers reflect 'total' squad size over a season.

Table 7.5 Number of footballers in the EPL.

Region	1992–1995	1996–2001	2002–2007
England	854	1197	832
Celtic Fringe	244	430	294
Rest of the world	173	655	745

Being extremely high in the proportion of English players appears to pose some risk for clubs. However, the other two high outlier clubs were not relegated: Aston Villa (1998) and Manchester United (2001).

We noted in Table 7.3 a sharp drop in players coming from Western Europe during the 2002–2006 period compared to those entering during 1996–2001. The panel on the left of Figure 7.6 shows a general shrinkage in squad sizes, suggesting that EPL clubs cut back in the number of players following the sharp increase in non-Home Country presence during 1996–2001. Indirectly, this reinforces the result of the Bosman Decision having a great impact on the recruitment of players because it triggered a correction by the clubs in the EPL. This affected English and non-English players differently. There were drops in the total number of players from England and the Celtic Fringe from 1996–2001 to 2002–2007 while the number increased for players from elsewhere. Table 7.5 shows the number of players by regions, consistent with the expansion of squads during 1996–2001 and a contraction thereafter.[39]

7.10.2   Player fitness

Proposition 7.9.9 focused on player fitness over time, and possible differences between English players and those from elsewhere. The obvious variable to consider is the body mass index (BMI), defined as (weight/height^2), as a simple measure of fitness. However, BMI is a ratio of two random variables and using such ratios as dependent variables in regression models is problematic (Kronmal 1993). Also, treating height and weight separately does not get at fitness directly. Instead, we estimated a multiple linear regression model:

$$\text{weight} = \beta_0 + \beta_1 \text{position} + \beta_2 \text{start year} + \beta_3 \text{height}^2 + \beta_4 \text{position} \cdot \text{height}^2 + \epsilon$$

In essence, this relates weight to squared height while taking into account both time and position played. Time is operationalized as 0 for the initial EPL season (1992/3) and 1 for the remaining seasons. An explicit comparison is made between fitness at the beginning of the EPL and subsequently. It is clear that controls have to be imposed for position played,

---

[39] A more compelling measure of the English presence in the Premiership is the amount of time such players are on the field rather than the number of them in squads in the league. According to a report commissioned by the BBC (http://www.bbc.co.uk/sport/0/football/24467371), the percentage of time on the field for them was about 32.3% for 2012/3, down from 35.5% for 2007/8. In order for a striker to score goals he has to be on the field, on the whole, for long periods. The last time an English player was the top scorer was in 1999/2000. In the 14 seasons since then the top goal scorer has not been English. The distribution by country is France 5 (two players), the Netherlands 4 (three players), Ivory Coast (one player) and Portugal, Argentina, and Bulgaria once. In the 11 EPL seasons before 2001, an English player was the top scorer nine times.

Table 7.6 Estimated parameters for the implicit BMI model.

Coefficient	Variable	Estimate	Std Error	t-value	p-value
(b_0)	(Intercept)	7.850	2.160	3.64	0.0003
(b_{1m})	Midfielder	5.189	3.201	1.62	0.1051
(b_{1f})	Forward	4.110	3.196	1.29	0.1986
(b_{1g})	Goalkeeper	15.875	5.552	2.86	0.0043
(b_3)	Start year	−2.069	0.191	−10.84	< 0.0001
(b_2)	$Height^2$	21.558	0.650	33.19	< 0.0001
(b_{4m})	$Midfielder.Height^2$	−1.993	0.989	−2.02	0.0439
(b_{4f})	$Forward.Height^2$	−2.457	0.973	−1.50	0.1346
(b_{4g})	$Goalkeeper.Height^2$	−3.843	1.592	−2.41	0.0159

$R^2 = 0.547$; Adjusted $R^2 = 0.546$; $F(8,3612) = 544.8$, $p < 0.0001$

given the systematic differences in height and weight by position. Position is a categorical variable taking the form of four dummy variables – one each for football playing positions defenders, midfielders, forwards, and goalkeepers – of which only three can be used in the regression. Defender is used as the omitted (reference) category. The weight and height of players is recorded in most databases when they enter a league system. There does not appear to be a database with temporal data on weight and height for players. However, our specific comparison is not affected seriously by this even if there are temporal shifts in these physical features, because our attention is on the arrival of players to the EPL. We term this regression an implicit BMI model. Its estimated parameters are shown in Table 7.6.

The plot of the residuals against fitted values formed a dense ellipse showing no skewness and is devoid of pattern: there are no departures from a linear model. The regression accounts for about 55% of the variance and shows: a highly significant but unsurprising link between weight and height; a highly significant effect for start year in that, controlling for height, weights are lower by about 2 kg after the first EPL season; a significant difference for goalkeepers with them being about 16 kg heavier than defenders controlling for the other variables; and the interactions of both goalkeeper and midfield positions with squared height being significant with slightly lower increases in weight given height. Figure 7.7 shows the actual changes in average weight (on the left) and height (on the right) in the top two panels. The lower left panel shows the plot of weight against height (derived from the regression model) and, on the right, are the plots BMI over time. In each plot, the four basic playing positions are distinguished.

In general, there was a general decline in the weight of goalkeepers, forwards, and midfielders over the first 15 years of the EPL. For defenders there was only a slight decline but with a sharp upwards spike in the last season. The height of goalkeepers was steady, with fluctuations, and a spike upwards in the last season shown. Players in all other positions were getting taller across the 15 seasons. The trend lines for weight against height on the left of Figure 7.7 show the increase of weight given height, an obvious (trivial) result. The trend lines, drawn separately by position played are ordered clearly (from top to bottom) by goalkeepers, defenders, forwards, and midfielders. The rate of increase of weight for increased height is highest for defenders while being lowest for midfielders and goalkeepers, consistent with the parameter estimates in Table 7.6. These changes were driven by the recruitment of new players over these seasons. Both weights and heights are ordered in the

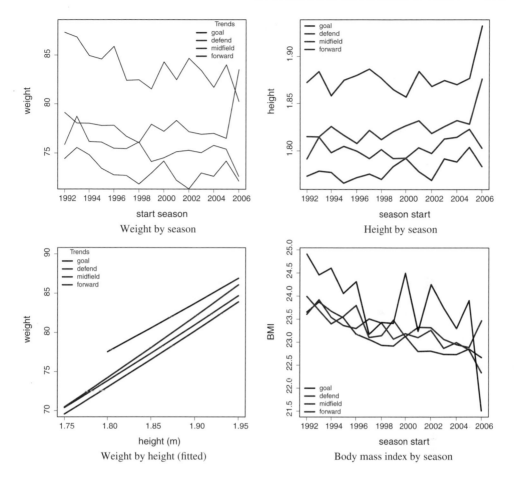

Figure 7.7   Plots of weight, height, and BMI over time: EPL.

same way by position (except for goalkeepers and defenders at the last season shown). At the extremes, having midfielders as the lightest and shortest with goalkeepers the tallest and heaviest makes considerable sense. Midfielders cover the greatest distances when playing in a game while goalkeepers cover the least.

The plot of the average BMI index over time in the lower right panel of Figure 7.7 shows dramatic drops (and corresponding increases in fitness) for forwards, midfielders, and defenders. Indeed, their BMI measures are close for all seasons. The BMI trajectory for goalkeepers had a different and more variable pattern. There is also a dramatic drop in the BMI of goalkeepers in the last season shown. Even so, as a group, goalkeepers were also fitter at the start of the 15th season than in the first. In order to assess Proposition 7.9.9, it is necessary to distinguish foreign-born players from local-born players. Including a dummy variable capturing this difference between English (coded 1) and players from elsewhere (coded 0) together with its interaction with start year the likelihood ratio test of the two nested models (with and without the new dummy variable) shows that the new more complicated model does *not* improve the fit to the data significantly. Further, the coefficient for foreign vs.

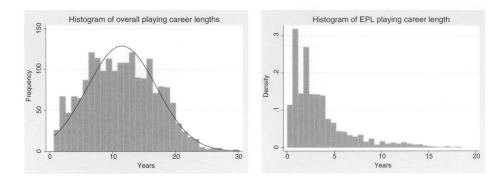

Figure 7.8   Length of overall and EPL careers for EPL English players: 1992–2006.

indigenous variable, was not significant. In short, while players became fitter over the first 15 seasons of the EPL, there were is no statistical evidence of the difference between native-born and foreign-born players regarding fitness. Proposition 7.9.9 is not supported. The anecdotal evidence concerning Jürgen Klinsman (from Germany) and Eric Cantona (from France) is just that - anecdotal. While they may have provided inspiration to their team mates initially, there are no *systematic* differences between native-born and foreign-born players: they were equally fit over all of the 15 EPL seasons considered. Instead of converging trajectories, the trajectories are the same, albeit with real increases in fitness. The latter is implicitly consistent with part of Proposition 7.9.9 but the former contradicts the main part of this proposition: recruiting foreign-born players does not, by itself, create a fitter player pool.

7.10.3   Starting clubs for English players

We consider Propositions 7.9.16 (the majority of English players entering into the EPL do so through football academies controlled by EPL clubs) and 7.9.17 (English players who have started with EPL clubs have longer and more successful careers when they start in the EPL compared to English players who start their careers outside the EPL) in this section. Figure 7.8 shows the distribution of the lengths of overall careers (left) and the lengths of EPL playing careers. The shapes of these distributions are dramatically different. One is reasonably close to a normal distribution while the distribution of EPL careers is highly skewed.

To examine these distributions in relation to the starting clubs of players, their starting clubs were coded into four categories: 1) starting at a top-four club (coded 3); 2) starting at another top-league club (coded 2); 3) starting at a club in the rest of the Football League (coded 1); and 4) starting at a non-league club (coded 0). The 'Big Four' clubs, traditionally, are Arsenal, Chelsea, Liverpool, and Manchester United. However, given the importance of a top-four finish in the top league, we added to these clubs, other clubs finishing in the top four places in each season. With players starting their careers before the formation of the EPL, we included seasons prior to its formation. As a result, 75 starts were added to the top four clubs. They are, with the number of times they were included depending on seasonal finishes: Aston Villa (7); Birmingham City (1); Blackburn Rovers (8), Crystal Palace (1); Derby County (1); Everton (6); Ipswich Town (2); Leeds United (15); Newcastle United (13); Norwich City (6); Nottingham Forest (9); Sheffield Wednesday (1); Tottenham Hotspur (4);

Table 7.7 Career starts and median lengths of playing careers.

Start level	Frequency	Percent	Median overall career	Median EPL career
Non-league	172	8.92	14.25	2.10
Rest of league	699	36.26	12.50	2.00
Top league	703	36.46	9.95	2.50
Top four club	354	18.36	9.50	3.06
Total	1928	100.00	11.41	2.40

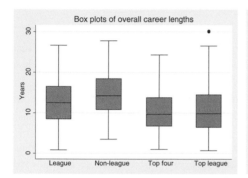 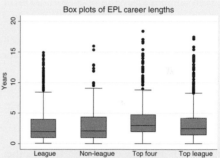

Figure 7.9 Box plots of career lengths by club level starts: 1992–2006.

and West Ham United (1). The overall rationale for this change is that being a 'top' club is historically defined in time.

With 1057 of the 1928 English players (54.82%) starting on EPL squads, Proposition 7.9.16 has support. Regarding the median length of overall playing careers, Proposition 7.9.17 is contradicted for overall playing career lengths. Indeed, the order of these medians drops the higher the starting club level. However, Proposition 7.9.17 seems supported to some extent for lengths of EPL playing careers. The next step was to see if these differences are significant. For Proposition 7.9.16, the 95% confidence interval around 0.548 (the proportion starting at a top league club) is (0.505, 0.591) so the proposition is supported. Dealing with Proposition 7.9.17 is more complex, especially for EPL career lengths. See Figure 7.9.

7.10.3.1 EPL career lengths

For the overall comparisons of median EPL career lengths across starting levels, the Pearson $\chi^2(3)$ is 38.929, $p < 0.001$. There are some differences. To see which are real, comparisons were made across the levels of career starts. The results are shown in Table 7.8. The results of these tests for median EPL playing career lengths are:

1. starting a playing career at one the top four clubs confers an advantage over all other starting club levels;

Table 7.8 Pearson $\chi^2(1)$ and (p-values) for start club levels for the EPL.

Start level	Rest of league	Top league	Top-four clubs
Non-league	0.2506	2.3961	8.8461
	($p = 0.617$)	($p = 0.122$)	($p = 0.003$)
Rest of league	—	9.9316	37.3329
	—	($p = 0.002$)	($p < 0.001$)
Top league	—	—	13.4794
	—	—	($p < 0.001$)

In terms of inference, the same results are obtained using the continuity corrected Pearson $\chi^2(1)$, the Fisher exact test, and the rank sum test. Given that six tests were made, with $\alpha = 0.001$, the Bonferroni inequality implies an overall p-value of 0.006 for the set of tests.

2. starting at another club in the top league confers an advantage of starting in the rest of the Football League *only*. There is no such advantage compared to starting at non-league clubs; and

3. starting at a club in any of the second, third, and fourth levels of the Football league confers no advantages at all relative to starting at a non-league club.

Without much surprise, starting their careers at a top-four club confers significant advantages for players compared to all other starting levels. The advantage is less evident for starting at other top league clubs but still present. Proposition 7.9.17 is confirmed. It is surprising that starting elsewhere in the Football League does not confer advantages over starting at a non-league club.

7.10.3.2 Overall playing careers

The basic conclusion of this section is that Proposition 7.9.17 is rejected decisively. Still focusing on medians for overall playing careers, the Pearson $\chi^2(3)$ is 80.942, $p < 0.001$. There are differences in the values reported in Table 7.7. The results of the pairwise comparisons across starting club levels are shown in Table 7.9.

Table 7.9 Pearson $\chi^2(1)$ and (p-values) for start club levels overall.

Start level	Rest of league	Top league	Top-four clubs
Non-league	7.3272	48.4248	44.7836
	($p = 0.007$)	($p < 0.001$)	($p < 0.001$)
Rest of league	—	37.7321	28.8489
	—	($p < 0.001$)	($p < 0.001$)
Top league	—	—	0.1705
	—	—	($p = 0.680$)

In terms of inference, the same results are obtained using the continuity corrected Pearson $\chi^2(1)$, the Fisher exact test, and the rank sum test. Given that six tests were made, with $\alpha = 0.001$, the Bonferroni inequality implies an overall p-value of 0.006 for the set of tests.

In summary, the results in Table 7.9 are:

1. the length of overall playing careers at both the top-four clubs and the rest of the top league do not differ;
2. overall playing careers for those starting in the top league – both for the top four and other top level clubs – are shorter than for players starting in the rest of the football league or with non-league clubs; and
3. players starting in the rest of the football leagues have shorter overall careers than those starting in non-league clubs.

The results are surprising (at least to us) and devastating for Proposition 7.9.17: it fails completely. In an admittedly ad hoc fashion, some plausible arguments for this result include the following. First, more games are played at the top level, especially for the top clubs, even before the formation of the EPL, and this carries greater risk of injury to players. So one reason for shorter overall playing careers could be having these careers curtailed by injury. Second, there is greater competition for places in top-level squads. This may discourage some players from staying at top-level squads in order to get more first-team playing experience at lower levels. Also, some players could quit the game as a result of not being successful at the top level. Third, players get cut from top teams more often and may seek more security at lower levels. It seems that getting into the academies of top football clubs is no guarantee of later success in terms of longer playing careers. Fourth, players can recognize the match between their talents and the levels of football teams for which they could play and so make adjustments. As players age, many become less valuable to their clubs, especially at the top level. The very few players who retire at the top level are truly exceptional. Finally, the non-league world includes professional clubs, semi-professional clubs, and amateur clubs. As a result, the motivations for playing there differ from playing in the Football League as do the pressures. This may help account for the longest playing careers there for players who once played in the EPL.

7.10.4   General features of the top five European leagues

Officials in charge of English football acted to keep as many foreign players out of England as possible. This made England different from the other top European leagues until 1978. There have also been extensive discussions of the corrupting influence of money in the modern game[40] as we focus on the EPL. The purchase of English clubs by non-English owners is part of this debate along with the claim that, with enough money, clubs can buy titles by spending lavishly on transfer fees and player salaries. To gain some perspective on this, we focus on the inequality of participation and winning titles in these top five leagues.

In principle, clubs can rise from a being a tiny local club and reach the top league. It would take a very long time in large league systems such as in England. As clubs rise in the levels of a such a system, the geographic range of where they play games broadens. Club stadiums have to satisfy increasingly stringent quality standards. Making the necessary upgrades is

---

[40] This is not a new concern. An official history of the FA, published in 1953, claimed that the organization had long stood against the evils of violence, women's football(!), and the corrupting power of money (Murray 1996). As no women have ever player in the EPL (or any of the top European leagues), our use of the masculine gender is reasonable despite the welcome increasing presence of women in football around the globe.

very expensive. As the costs of participation rise, chances are slim to nil for tiny local teams rising far. Much larger clubs are always better placed to be at, or near, the top of national league systems. Clubs can get into financial difficulties and descend though the levels of the league hierarchy. Even so, from a long-term perspective, overall membership in the top level of the English league system is *remarkably stable*. For the 111 seasons from 1888/9 to 2010/11, only 64 clubs[41] have played of the top tier (in its various forms) in England. Of the 12 founding clubs for the top tier of English football, seven played in the EPL during the 2010/11 season.[42]

The second tier of each league system acts as a filtering system: clubs enter the top tier only from the second tier. Together, the top two tiers define the elite clubs of all European league systems. The membership in the broader elite is even more concentrated. We split the post-WWII period into the pre-modern era and the modern era with a view to asking if anything has changed since the flood of money that came into English football at the top level in the late 1980s and early 1990s. More specifically, we ask if the inequality of participation has changed between the two eras. Of the 64 clubs that have played in the top tier since 1888, 59 have been present in the top tier at least once since 1946, another indicator of membership stability. Arsenal is the only team never to have been out of the top flight since WWII.[43] The clubs of the top tier can be split into three categories: 1) clubs consistently in the top tier; 2) clubs bouncing between the top two tiers; and 3) clubs whose appearance in the top tier was relatively brief. Given these features we add:

Proposition 7.10.1. *The level of inequality of participation in the top European Leagues is the same in the pre-modern and modern eras.*

We turn to assessing this proposition next.

7.10.4.1   Pre-modern era

Figure 7.10 shows the memberships for the pre-modern era in a formatted two-mode network array for clubs appearing in the top tier. Clubs are ordered by the number of times they have appeared in the top level. In addition to the black squares showing top-tier membership, the gray squares show second-tier membership. Note that in each row, a change in color of squares indicates either promotion or relegation, depending on the color sequence of the squares. The white squares show when clubs slipped into even lower tiers. By design, the density for the presence of clubs in the top tier is much higher at the top of the array than at the bottom where it is decidedly sparse. Arsenal, never relegated, is in the top row while, at the bottom, there are four clubs with one-year stays. There was great inequality in England in the pre-modern era.

---

[41] There are 67 clubs listed as having played in the top tier but three of these clubs are forerunners to other clubs in this list. Newton Heath was founded in 1878 and was renamed Manchester United in 1902. Similarly, Leicester Fosse, founded in 1884, became Leicester City in 1919. Small Heath, founded in 1873, went through a series of name changes before becoming Birmingham City in 1943.

[42] None of these seven clubs have been continuously in the top tier since its founding in 1888.

[43] Manchester United was relegated once but spent only one season in the second tier. Everton was relegated once and spent three seasons at the second level while their intra-city rivals Liverpool, was relegated once and spent eight seasons at the second level. Rounding out the top five clubs, in terms of participation at the top level, is Tottenham Hotspur. Relegated twice, it spent a total of five seasons in the second tier.

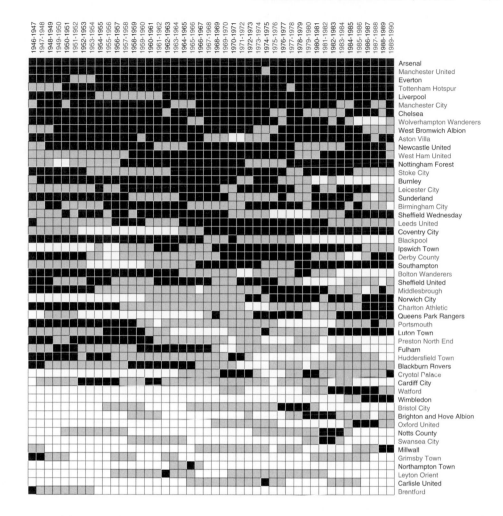

Figure 7.10   Top tier memberships of English football: pre-modern era.

The shading is: black squares, in the top tier; gray squares, in the second tier; and white squares, in a lower tier. Clubs are ordered from the top by the number of times they were in the top tier.

The English system was not unique in its concentrated membership. The formatted arrays for the other four top European leagues are very similar. In Germany, only Hamburger SV has a continuous presence in the Bundesliga. Inter Milano, Juventus, and Fiorentina were the only clubs that stayed in Italy's Serie A throughout the pre-modern era. In Spain, FC Barcelona, Real Madrid, Athletic Bilbao, and Atlético Madrid were never relegated from La Liga in this era. France shows a slight difference by having no team with a continuous presence in Ligue 1. The top five French teams, in terms of long-term presence, were all relegated on multiple occasions in the pre-modern era.[44]

---

[44] Bordeaux, St Etienne, and Olympique de Marseille were all relegated four times; RC Lens dropped out of the top flight five times (and even spent a season in the third tier), and FC Sochaux went down six times.

Italy and Spain show clearly the extremes where some clubs had long stays while others had short stays. Italy has the highest incidence of short stays. The medians for stay durations are: England, 19; France, 14; Spain 13; and Italy 9. They (or some of them) are significantly different (Pearsonian $\chi^2(3) = 9.21$, $p = 0.027$.) The visual and statistical results make it clear that some teams fared much better than others in the pre-modern period.

7.10.4.2   The modern era

Figure 7.11 shows the formatted array for participation in the top English league in the modern era. Visually, the concentrated (differentiated) pattern looks very similar to that of the pre-modern era shown in Figure 7.10. In terms of the overall pattern of participation, not a great deal changed in the modern era. Proposition 7.10.1 is supported for England. Figure 7.12 provides the pre-modern participation arrays for the other four top European leagues where, again, there is the same concentrated pattern of membership.

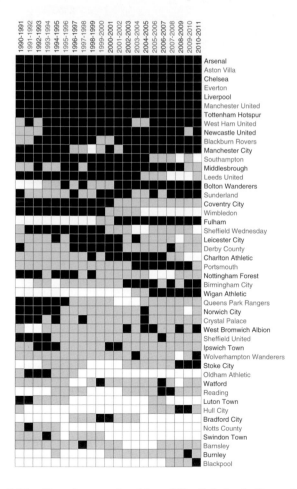

Figure 7.11   Top-tier memberships of English football: modern era.

Black squares: in the top tier; gray squares: in the second tier; white squares: in a lower tier. Clubs are ordered from the top by the number of times they were in the top tier.

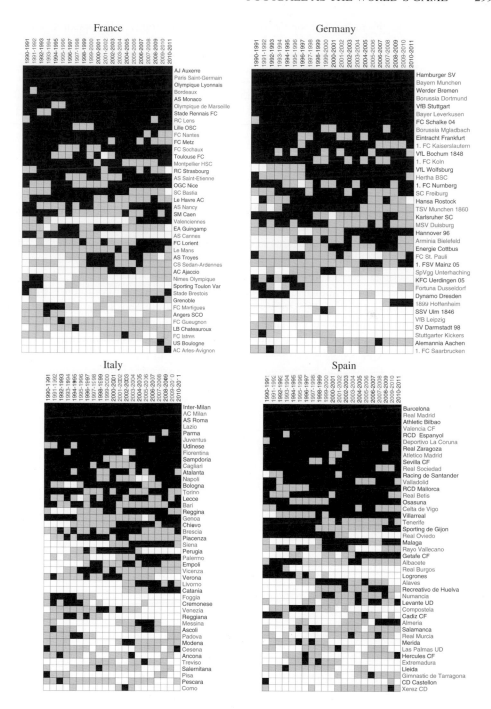

Figure 7.12   Top-tier memberships in the top continental European leagues: modern era.

Black squares: in the top tier; gray squares: in the second tier; white squares: in a lower tier. Clubs are ordered from the top by the number of times they were in the top tier.

Table 7.10 Inequality indexes of top tier membership for the top five European leagues in two eras.

Country of league:	England	France	Germany	Italy	Spain
Pre-modern era	0.390	0.425	—	0.490	0.492
Modern era	0.408	0.411	0.398	0.409	0.381

For Germany, our data go back only to 1964.

An obvious question, given the debate about the influence of money, is: did anything change in the modern era when the economic stakes for top-tier membership became much higher. To examine this more closely, we look at Table 7.10 and Figure 7.13. The bootstrapped 95% confidence intervals for the modern era Gini indexes suggest that the five leagues differ little with regard to inequality of club membership. Put differently, they are equally unequal. This provides considerable support for Proposition 7.9.14. This is confirmed by the Lorenz curves for the five leagues shown in Figure 7.13. The bootstrapped confidence intervals suggest also that drops in the values of the Gini index for Italy and Spain are significant: modern era membership inequality is smaller than the pre-modern era inequality for these two national leagues. See Table 7.10. Proposition 7.10.1 has provisional support.

Counting titles won provides further evidence of concentration. In England, they are: Manchester United 12; Arsenal 4; Chelsea 3; Blackburn Rovers 1; and Leeds United 1. The count of Bundesliga titles is: Bayern München 10; Borussia Dortmund 4; Kaiserslautern 2; VfB Stuttgart 2; Werder Bremen 2; and Wolfsburg 1. In Italy, the Serie A title count is: AC

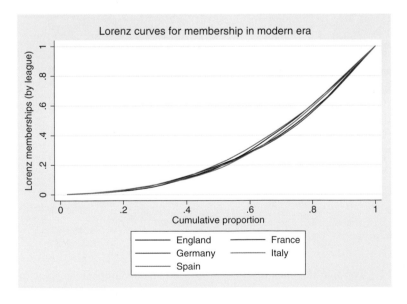

Figure 7.13 Lorenz curves for the top five European leagues: modern era.

Table 7.11   Inequality for titles won in the top five European leagues in the modern era.

Country of league:	England	France	Germany	Italy	Spain
Modern era	0.943	0.663	0.898	0.951	0.936
Maximum equality	0.544	0.500	0.400	0.522	0.537

Milan 7; Inter Milano 5; Juventus 5;[45] Lazio 1; AS Roma 1; and Sampdoria 1. The title counts in Spain are: FC Barcelona 11; Real Madrid 6; Valencia 2; Atlético Madrid 1; and Deportivo de la Coruña 1. For France it is: Olympique Lyonnais 7; Olympique de Marseille 3;[46] Nantes 2; Bordeaux 2; AS Monaco 2; AJ Auxerre 1; RC Lens 1; and OCG Lille 1.

The high values of the Gini indices reported in the first row of Table 7.11 cannot be taken at face value. They are severely inflated by having only 21 seasons. The 'most equal' distribution of champions would hold if a new champion won the title each season. So, for England, Germany, and Spain, an artificial distribution was constructed with 21 clubs each winning one title. The artificial distribution for France and Italy has 20 clubs winning one title (each had a season when no title was awarded due to corruption in the form of match fixing). Even these 'maximally equal' distributions are necessarily unequal. Given that the lowest possible values of the Gini indices are about the same, the computed indices are roughly comparable (without taking the actual values as the real measures of inequality). This suggests that France and Germany were less unequal in the distribution of titles won in this era. Even so, the top five leagues in Europe are all very unequal systems and have long had this form. *That this inequality has persisted suggests that this system-level impact of the inflow of huge sums of money has been minimal* regarding inequality. Of course, some clubs have benefitted far more than others at particular times but the underlying inequality has been affected very little by these flows. However, the primary mechanisms are different. Throughout most of the pre-modern era, clubs were able to keep their best players because of the onerous terms of the retain-and-transfer system. In the modern era, clubs can buy success through the purchase of (what they think are) the best players. Proposition 7.10.1 is fully supported for the EPL as well as in two of the other top European leagues in terms of the inequality in winning titles.

7.10.5   Flows of footballers into the top European leagues

Footballers migrate over routes in the club-to-club and country-to-country networks. For the other top four European leagues after WWII, foreign player flows were much larger than for England due to its restrictive labor practices until 1978. Players reaching England often do so through leagues in Europe. We are particularly interested in the roles of France, Germany, Italy, and Spain in these migrations.

---

[45] Juventus won a sixth title but was stripped of it for its involvement in match fixing. No title was awarded because the runner-up was also involved to a lesser extent in the scandal.

[46] Marseille initially won the 1992/3 title but were stripped of it after being found guilty of bribery. No title was awarded that year. Had it gone to the runner-up then Paris Saint-Germain (henceforth, PSG) would have been the champion.

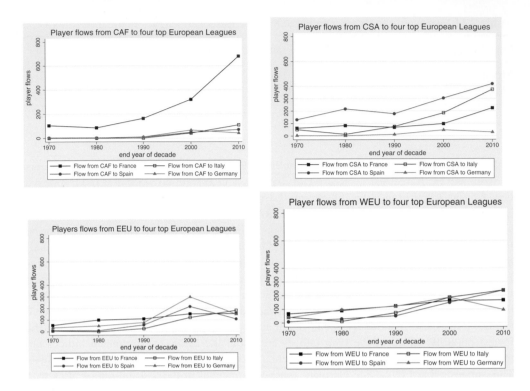

Figure 7.14   Four regional flows to Europe.

Figure 7.14 shows the flows of players from the four regions with the largest player presence in the EPL (other than the Home Countries) as shown in Figure 7.1. They are Africa (CAF), South America (CSA), Eastern Europe (EEU), and (the rest of) Western Europe (WEU). By far, France's Ligue 1 had the most players from Africa in every decade. Even for the early 1960–1970 decade, there were 105 African players in Ligue 1 compared to just four African players in Spain and none in the other two leagues. The number in France dipped in the 1970–1980 decade with the other counts remaining very low. For the 1980–1990 decade, there were eight times as many African players (163) in the Ligue 1 as in the other three top leagues combined. This number rose to 321 for 1990–2000, twice the sum for the other top leagues even though the African presence increased in all these leagues. For 2000–2010, the 680 African players in Ligue 1 was more than three times the sum for the Bundesliga, Serie A, and La Liga combined. France has long made greater use of its former colonies as sources for elite players. We anticipate that France will be a major conduit for African players to the EPL.

The presence of South American players was the highest in Spain for every decade (see the top right panel of Figure 7.14). The presence of CSA players in La Liga in the 2000–2010 decade (at 419) is the second highest across all panels of Figure 7.14. Players from CSA in La Liga, other than Brazilian players, all come from former Spanish colonies. Over the last three decades, the number of South American players in Serie A increased steadily to approach the level in La Liga for 2000–2010 (with 373 players). The presence of South

American players has always been lower in the Bundesliga. South American players have had a greater presence in France than in Germany, especially for 2000–2010. We expect La Liga, Serie A, and Ligue 1 will be important conduits to the EPL from CSA countries.

The number of players from Eastern Europe was the highest in France for the 1960–1970 through 1980–1990 decades. It continued to grow. There was an EEU jump for the other three leagues in the 1990–2000 decade, undoubtedly due to the wars in the Balkans that tore Yugoslavia apart and the break-up of the Soviet Union. The largest number (297) of Eastern European players were in the Bundesliga, followed by La Liga with 215 players and then Italy with 152 players from EEU. The number of these players dropped sharply during the 2000–2010 decade for the Bundesliga and La Liga. Some of them may have moved to the EPL in that decade while the level of their presence remained steady in Ligue 1.

The trajectories in the lower right panel of Figure 7.14 show little dramatic change over the decades shown. In general, there is a rise of (other) Western European players in these four top leagues except for a drop between 1960–1970 and 1970–1980 in Serie A and between 1990–2000 and 2000–2010 in the Bundesliga. There appear to be no obvious implications regarding movement to the EPL.

Our data source for the presence of non-national players in the top leagues (see Appendix A.4) had data for England only for the EPL for 1992–2000 and 2000–2010. The corresponding plots for the EPL cannot be drawn usefully for either period because the presence of players from WEU dwarfs those for all other regions. Even so, we can make some comparisons with the top leagues regarding the presence of footballers from other FIFA regions. Although the AFC had a minor presence in the top leagues for all periods, the largest presence of players from the AFC, for both 1992–2000 and 2000–2010, was in England by a wide margin. For both of these periods, the second highest presence was for footballers from CAF in England. The presence of players from CSA to England ranked last compared to the other top leagues in 1990–2000. CSA presence in England exceeded only its presence in Germany for 2000–2010. Among the flows from EEU, the flow to England ranked last in 1992–2000 but ranked second in 2000–2010. There was also a clear presence of footballers from NAM in the EPL.

7.11   Player ages when recruited to the EPL

Proposition 7.9.10 restated McGovern's (2002) claim about the EPL clubs having avoided recruiting top-end foreign players in their early years, especially from Argentina, Brazil, Germany, Italy, and Spain. This can be tested with the data we collected. However, one difficulty in testing this proposition is operationalizing 'top-end'. This could be done by using players appearing in their respective national teams as an indicator. This requires both locating the date of the first appearance of each player for his country and using the number of subsequent appearances for his country as a quantitative measure. While the former is fixed in time, the latter varies over time for many international players. Tracking these streams of information would be complex. Instead, we took a seemingly indirect but simpler approach by looking at the arrivals of players in the EPL by nationality. The players from outside the Home Countries were divided into two subsets. One was made up of all of the players from the above five nations (the top football countries) with all of the other non-English players (from all of the other countries) placed in the second subset.

A second difficulty with this proposition is that 'in their early years' is ambiguous. Instead of defining a threshold separating early years from later years, we worked with the

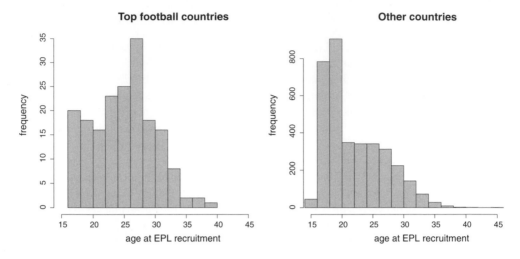

Figure 7.15   Distribution of ages for foreign players when recruited to the EPL: 1992–2006.

full age distribution to construct each player's age when he first appeared in the EPL.[47] In terms of testing the proposition, the distributions for the top football countries and other countries were compared. These histograms are shown in Figure 7.15 with the top countries on the left and the other countries on the right. Visually, these distributions are clearly different.[48]

For the other countries, the modal ages of recruitment to the EPL are between 16 and 20 while the modal ages for recruitment to the EPL are between 24 and 27 for the top football countries. As was the case for similar tests in the Chapter 5 on patents, we used empirical cumulative distribution functions for a two-sample Kolmogorov–Smirnov test. These cumulative distributions, one for the top countries and one for the other countries, are shown in Figure 7.16. The cumulative distribution for the top football countries is marked in red while the corresponding distribution for the other countries is marked in black. The p-value for the test is off the charts at $1.731 \cdot 10^{-14}$. Clearly, the two distributions are very different: players arriving from the top football countries do so at a much later age. Overall, EPL clubs avoid recruiting young players from Argentina, Brazil, Germany, Italy, and Spain. Proposition 7.9.10 is supported strongly. One caveat to this result is that we have shown only a difference on the age of arrival in the EPL for all players from the two subsets of countries. However, by definition, all of the top-end players from the top football countries are included in the set of players from these countries. If there is a tendency to recruit top-end players later in their careers, the test we have performed is a rather conservative test of the proposition. A more general claim would be that top-end players from outside the Home Countries are recruited later in their careers regardless of their nationalities. This would

---

[47] Players for whom there are no available birth dates were excluded from this analysis.

[48] In the data, we have the dates of birth and the dates of arrival in the EPL. To compute the corresponding ages in years we divided age in days by 365.25. As a result, there was only player whose age of arrival was 14.9, slightly below our cut-off of 15 for the start of professional careers. Hence, the leftmost box in the histogram for the other countries extends below 15.

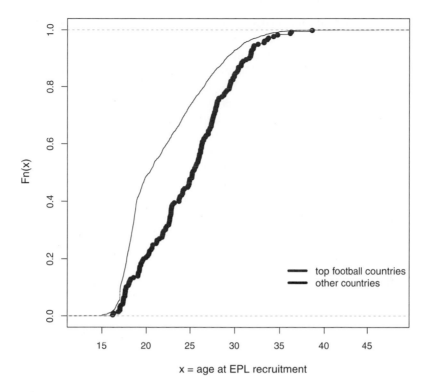

Figure 7.16   Cumulative distributions for the ages of arrival of foreign players in the EPL: 1992–2006.

require operationalizing the notion of top-end and constructing data from a widely scattered set of resources.

7.12   A partial summary of results

We have provided a general outline of the history of football in England, given our focus on the EPL, together with some partial descriptions of football clubs and football players. The idea of club-to-club and country-to-country networks was provided to motivate our study of these networks, their operation, and some consequences of these footballer migrations. We provided some propositions in Section 7.9 and presented results relating to some of them in Section 7.8.

The following propositions all received strong support: Proposition 7.9.1 (the regional composition of the EPL has changed throughout its short history); Proposition 7.9.4 (the proportion of English players in the squads of the EPL has diminished through time); Proposition 7.9.5 (the proportion of players from the Celtic Fringe in the squads of the EPL diminished through time); Proposition 7.9.7 (the largest part of the recruitment of non-English players was from Western Europe and this has expanded over the course of the EPL); Proposition 7.9.8 (the recruitment of players from countries of Eastern European has expanded over time); and Proposition 7.9.10 (the EPL clubs avoid recruiting young players from the top football countries). While many of these ideas are not new, testing them using data on *all* players appearing in the EPL for the first 15 years of the leagues existence is.

Proposition 7.9.14 (the top level of the league systems of the top leagues in Europe is inaccessible for a huge majority of clubs in their league systems) was supported. Proposition 7.9.3 (early in the history of the EPL, clubs recruited players who most resembled indigenous players) received indirect support. Additionally, Proposition 7.10.1 (the level of inequality of participation in the top European Leagues is the same in the pre-modern and modern eras) received strong support regarding participation in the top leagues. Proposition 7.9.9 (at the beginning of the EPL, non-English players were fitter than English players) was not supported. This rendered the second part of this proposition (regarding the convergence of the two groups of players regarding fitness) moot. However, players became fitter overall, regardless of position played, in the EPL. Undoubtably, this is true for all of the other top leagues.

Proposition 7.9.18 (about the Bosman Decision having no impact on the flow of players to the EPL) was rejected. There was a sharp jump in the number of players not originating in the Home Countries coming into the EPL in the period 1996–2001. The only way of establishing whether there was only a continuation of extant flows was to compare proportions from regions for the flows in 1992–1995 and 1996–2001 by regional sources. This alleged support pales in the face of the sharp jumps following Bosman. Of some interest is the contraction in squad sizes during 2002–2007 suggesting that there had been an unsustainable expansion following this decision. The lowering of squad sizes affected both native-born and foreign-born players with a more consequential drop for players from the Home Countries.

Propositions 7.9.16 and 7.9.17 concerned the career starts of English players appearing in the EPL. Proposition 7.9.16 (claiming that most English players in the EPL started with EPL clubs) was supported. However, while significant, having a proportion of 0.548 starting there was a mild surprise. While, at the club level, the top league is inaccessible for most lower level clubs, this is not the case for players. There is considerable opportunity for movement upwards, even from starts with non-league clubs. Given the primary relation between clubs is predation, these two results are coupled. Lower-level clubs frequently lose their best players to top-level clubs in order to survive and assuming a desire of these players to play at a league higher level.

Proposition 7.9.17 (concerning English players who started with EPL clubs having longer and more successful careers when they start in the EPL compared to English players who start their careers outside the EPL) was formulated with the length of overall careers in mind. This proposition was decisively rejected: the median lengths of the playing careers were shorter for players starting with EPL clubs than for players starting in the rest of the Football League. In turn, the median career lengths for players starting in the rest of the Football League were lower than for players having non-league starts. We caution that the data pertain only to footballers who appeared in the first 15 seasons of the EPL. There is support for Proposition 7.9.17 if attention is restricted to the length of playing careers in the EPL.

The results concerning the persistent inequality of the top league systems are important for several reasons. First, the playing field, as it were, has never been level: some clubs have been massively advantaged[49] while others were seriously disadvantaged, especially

---

[49] Reading through publications such as Dorling and Kindersly's *Football Yearbook, 2004/5* (edited by David Goldblatt) reavels that in all leagues a small handful of clubs have dominated the collection of trophies.

after WWII. All that changed are the mechanisms generating these inequalities. Second, arguments concerning the increasing flood of money flowing into English football and how this corrupts the game are beside the point if cumulative advantage (and disadvantage) is the central concern. While the current flow of money into the game can corrupt, especially with the administration of FIFA (Jennings 2006; Yallop 1999), the underlying inequality of football league systems has changed little. Third, the movements of players take place within this overall structure of the game.

The propositions examined in this chapter used data focused on attributes of players, clubs, and league systems. The remaining propositions deal more with the network structure of player movements as they link clubs and countries. These are examined in Chapter 8.

8

Networks of player movements to the EPL

Our primary focus in this chapter is on the *flows* of players from across the globe to European leagues, especially the EPL. We examine in Section 8.1 why the EPL is such a lure for playing talent as well as the primary determinant of success at the club level.

8.1 Success in the EPL

Here, we consider the predictors of success in the EPL. As noted in Chapter 7, the EPL, by intent, reaped most of the influx of funds in England during the modern era. Evidence supporting this claim is shown in Figure 8.1. Four plots of aggregate salaries are shown, one for each league in League Football.[1] These data cover the middle of the period for our core data (the first 15 seasons of the EPL). It is clear that the aggregate salaries paid in the EPL soared during this period while the other leagues lagged well behind, especially in League One and League Two. To the extent that earning higher salaries is an incentive, playing in the EPL has immense appeal for many footballers. Further, playing in the EPL is on a larger stage with the chance of playing in UEFA competitions if their clubs have a sufficiently high finish.

With clubs trying to improve their playing squads, there is considerable turnover in EPL squads: new players arrive and others depart. It is reasonable to think that acquiring and discarding of players could led to success. However, (Kuper and Szymanski (2009): Chapter 3) present compelling evidence for club salary levels being the primary determinant of success. Their Table 3 presents data for 58 clubs at the elite levels of English football for the period 1998–2007. The variables they constructed are the average league positions for

---

[1] http://www.footballeconomy.com/content/estimated-total-players-earnings-english-football-leagues-1996-2004 is our source of these data, accessed 14 May 2012.

Understanding Large Temporal Networks and Spatial Networks: Exploration, Pattern Searching, Visualization and Network Evolution, First Edition. Vladimir Batagelj, Patrick Doreian, Anuška Ferligoj and Nataša Kejžar.
© 2014 John Wiley & Sons, Ltd. Published 2014 by John Wiley & Sons, Ltd.

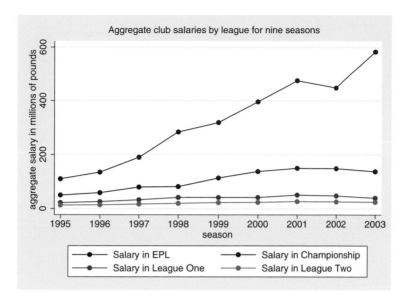

Figure 8.1 Salary levels in different English leagues.

Table 8.1 Estimated parameters: predicting league success from aggregate salary.

Coefficient	Variable	Estimate	Std error	t-value	p-value
(β_0)	(Intercept)	18.341	0.548	33.44	<0.0001
(β_1)	Aggregate salary	−14.914	0.923	−16.15	<0.0001

Adjusted $R^2 = 0.872$; $F(1,37) = 260.89$, $p < 0.0001$.

these clubs[2] and each club's wage spending relative to the average spending of all clubs. Aggregate salary accounts for about 89% of the variance in the average league position for these 10 seasons. Aggregate salary is a powerful predictor of success. As some of the clubs in their listing have never appeared in the EPL, we eliminated them[3] to focus attention on the clubs appearing in the EPL between 1998 and 2007. We used the (logged) club wage bill as the predictor but used seasonal position as the predicted variable: figures plotting position against salary are easier to digest visually. The resulting plot is shown in Figure 8.2. The fitted regression is reported in Table 8.1. The results shown in Table 8.1 result from estimating:

$$\text{(league) position} = \beta_0 + \beta_1 \text{aggregate salary} + \epsilon$$

---

[2] Their operationalization of average league position is $-log(p/45 - p)$ where p is the actual league position. In their data, the average league position varies from 2 to 42. The time that some clubs played in the second tier and a lower tier contributes to lower average league positions.

[3] Including the very low ranking clubs amplifies slightly the explained variance.

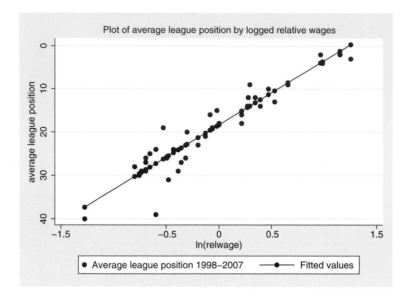

Figure 8.2   Average aggregate wages and average league position.

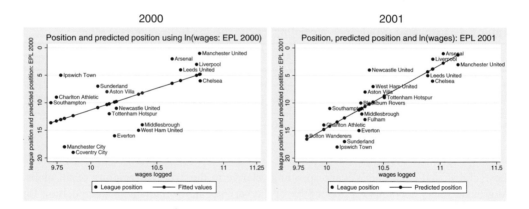

Figure 8.3   Club salary levels and league position.

We obtained seasonal total club wages for EPL clubs for only the 1993, 2000, and 2001 seasons. The fitted values from the regressions suggested by the Kuper and Szymanski analyses for the 2000 and 2001 seasons[4] are shown in Figure 8.3.

The regression results reported in Table 8.1 and Figure 8.3 leave some room for thinking that the arrivals and departures of players, together with squads sizes, could have some additional predictive value: clubs act on this presumption. However, the results reported in Table 8.2 dash these hopes. Using the three seasons for which we obtained complete data,

---

[4] For the 2000 regression, $R^2 = 0.34$. One season later, $R^2 = 0.71$ for 2001, consistent with the greater spread around the fitted line on the left in Figure 8.3.

Table 8.2 League success, wages of, and squad features.

Coefficient	Variable	Estimate	Std error	t-value	p-value
(β_0)	(Intercept)	−3895.380	842.896	−4.62	<0.001
(β_1)	Wages (logged)	−10.860	1.775	−6.12	<0.001
(β_2)	Season	2.004	0.428	4.68	<0.001
(β_3)	Arrivals	0.040	0.157	0.25	0.800
(β_4)	Departures	0.058	0.147	0.39	0.696
(β_5)	Squad size	−0.013	0.114	−0.12	0.908
(β_6)	Proportion foreign	9.315	7.108	1.31	0.196

$R^2 = 0.532$; Adjusted $R^2 = 0.476$; $F(6,50) = 9.47$, $p < 0.0001$.

the estimated equation is:

$$position = \beta_0 + \beta_1 \log(wages) + \beta_2 season + \beta_3 arrivals + \beta_4 transfersout + \beta_5 squad\ size + \beta_6 pforeign + \epsilon$$

The results are straightforward. First, higher club wages are associated with better league performance (1 is the top position and 20 the lowest position). Second, there is variation across seasons. In part, this is due to a large gap between the first season and the two later seasons. However, the turnover of clubs in the EPL due to promotion and relegation may also be a contributing factor. Neither the number of arrivals nor the number of departures have predictive value on league performance. Squad size has no relevance for success. Nor does the proportion of foreign-born players. In short, the sole substantive predictor for the EPL success of clubs is player wages.[5] Proposition 7.9.11 is supported strongly.

This result suggests that the trading and exchanging of players has no bearing on league success by clubs despite the pursuit of league success driving all of the trading. An economic argument is that the more talented players will get the higher wages, especially at the higher levels. However, buying and selling players amounts to gambling: player performances need not match the wages paid. The specters of injury and suspensions plus the intangibles of team chemistry all affect club success. Players do move from club to club and this movement is the focus of the rest of this chapter, despite its systemic irrelevance for club success. Indeed, this movement is *more* interesting as a general phenomenon.

8.2 The overall presence of other countries in the EPL

For this analysis, the 15-year period we study is divided into three 5-year slices: 1) 1992/3 through 1996/7 (time slice 1); 2) 1997/8 through 2001/2 (time slice 2); and 2002/3 through 2006/7 (time slice 3). Figure 8.4 shows world maps with country outlines. Various shades are used to represent the extent of foreign-born player presence in the EPL for three 5-year time periods. The measured level is the square root of the number of foreign players. Countries in white have no such presence in the EPL for each period.[6] Black

---

[5] Fitting a random effects model leads to the same result.

[6] We left England as white because our interest is to show the global reach of the EPL for attracting players.

countries have the highest such presence with varying shades of gray representing a lesser presence.[7]

We know from Figure 7.1 that the percentage of players from England and the Celtic Fringe declined[8] over the period we study while the presence from elsewhere increased. Figure 8.4 provides another visual image of this. The presence of players from the rest of Western Europe was high during 1992–1996 and became much higher in both 1997–2001 and 2002–2007. The increase for Spain on the Iberian Peninsula is noticeable: it was highest in the third period. Scandinavian players have long been present in English football, consistent with Proposition 7.9.3. A noticeable Finnish presence was added in the second period. There has always been a presence of footballers from Australia. It was highest in the middle period. The presence of players from the USA in all periods is clear. It was highest in the third period.[9] This is due to the founding of the Major League Soccer (MLS) in 1993, with the presence of players from the USA getting stronger through the three periods. There was a clear presence of Brazilian players in 1997–2001 which increased for 2002–2007. The Argentinian presence in football was strongest in 1997–2001. There was a notable Turkish presence for 1997–2001. For 2002–2007, the increased presence of players from Eastern Europe and China is clearly visible. Against this broad backdrop we turn our attention to footballer moves between clubs.

8.3   Network flows of footballers between clubs to reach the EPL

There may be one 'large network' of all flows of footballers involving many players, clubs, and nations. Our (smaller) dataset has 2302 clubs from 152 countries on footballer trails to the EPL for those playing in the EPL's first 15 seasons. However, we do not treat it as a *single* large network composed of a set of trails of players defined by their movements. By definition, our focus on players going to the EPL does imply that the network we study is one large weakly connected network. Conceptually, it is more fruitful to consider the network as being made up of disjoint networks composed of footballer trails with some 'idiosyncratic' and not systematic connections between these networks.[10] Given the design of our data pertaining to movements to, within, and from the EPL, we cannot generalize our results to the general movement of players across the globe. Our overall strategy is to start by looking at the final move into the EPL from outside the EPL and then gradually expand our focus to consider longer trails of moves.

---

[7] The seemingly black regions on the western coasts of Canada and Chile are due to their highly irregular indented coastlines. The same holds for the far north of Canada, the coasts of Greenland and the islands to the north of the European mainland.

[8] For the 1992–1996 period the percentage of players from the UK and Ireland was 46%, for 1997–2001 it was 18%, and it was 15% in 2002–2007.

[9] There are no USA players from Alaska, the shaded region to the west of Canada and detached from the lower 48 states of the USA.

[10] In a similar vein, we do not consider all phone calls through a single exchange or cell phone tower as a 'single network'.

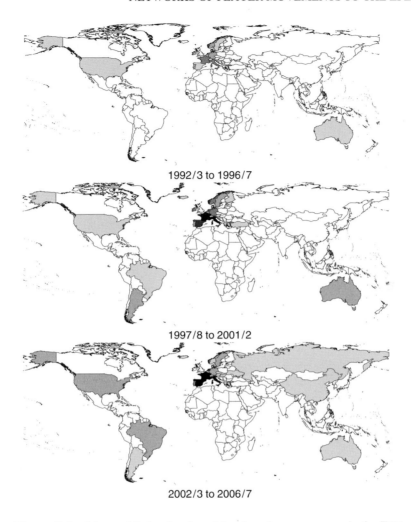

1992/3 to 1996/7

1997/8 to 2001/2

2002/3 to 2006/7

Figure 8.4   Maps with the density of foreign player presence in the EPL.
(The darker the shading for countries, the greater the presence of players from these countries in the EPL.)

8.3.1   Moving directly into the EPL from local and non-local clubs

Here, we focus on the 2625 footballers who *moved to* the EPL during the first 15 seasons of this league. This group is divided into 1301 players playing for clubs promoted to the EPL and remained on these club squads[11] and 1324 players who moved to the EPL as individuals through transfers and loans. We focus on flows from non-EPL clubs to EPL clubs for the first time entry of footballers to the EPL. Flows within the EPL are considered separately.

---

[11] One part of the movement of players between clubs is triggered by promotion. Some players who were good enough to help win the promotion of a club to the EPL are deemed not good enough to play in the EPL. They are discarded in the attempt to build a squad good enough for the club to have a chance of remaining in the EPL.

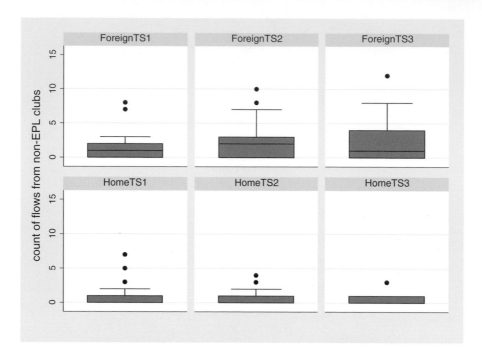

Figure 8.5   Flows from non-EPL clubs to EPL clubs.
TS1 is a label for the first time slice, TS2 for the second, and TS3 for the third time slice.

The summarized item is the count of non-EPL clubs sending footballers to the EPL. Figure 8.5 is organized by the three time slices, and whether the sending clubs are from the Home Countries or from elsewhere. The units of analysis are clubs: the distributions are skewed. They are consistent with Proposition 7.9.4 (there being decreasing proportions of English players.[12] on squads) and Proposition 7.9.5 (lower proportions of players coming from the Celtic Fringe to the EPL). They also provide some implicit support for Proposition 7.9.3 (in the early part of the EPL, the clubs tended to recruit players most resembling indigenous players).

For EPL clubs receiving players from elsewhere, the high outliers for the first two time slices were Chelsea and Arsenal in that order. For the final time slice, Chelsea is the sole high outlier. Neither of these clubs has been relegated from the EPL. Different clubs emerge as outliers when players coming from Home clubs are considered. For the first time slice, the high outliers in order (from the top) are Coventry City, Southampton, and Wimbledon. All were relegated and did not return to the EPL in the 15 years we are considering.[13] Coventry City was the high outlier in the second time slice with three other clubs in the second highest outlier that is marked (Everton, Newcastle United, and Derby County). Only the last of these was relegated in the period we study. For the final time slice, the high outlier has two clubs (Blackburn Rovers[14] and Tottenham Hotspur). Table 8.2 reported no *systematic* effect of

---

[12] English players tend not to play professional football for non-local clubs.
[13] Southampton did return for the 2012/3 season.
[14] Blackburn Rovers was relegated in 1999 but returned to the EPL in 2001.

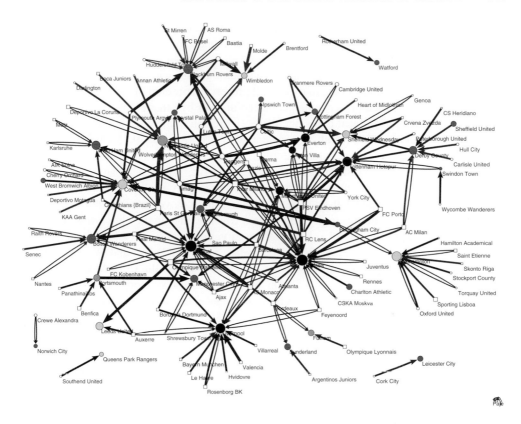

Figure 8.6   Simple moves to the EPL from outside the EPL in three time slices.
(Only flows involving at least two transfers are shown.)

foreign player presence on success. The results in Figure 8.5 suggest an effect but only at the extremes regarding failure.

8.3.2   Direct moves of players to the EPL from non-EPL clubs

Figure 8.6 is rather complex even while showing only *direct* player moves to the EPL. Clubs are the vertices. Circles represent local clubs. Black circles represent the never-relegated EPL clubs – Arsenal, Aston Villa, Chelsea, Everton, Liverpool, and Manchester United. White circles represent local clubs never in the EPL. Between these extremes, light gray circles repesent clubs that have been relegated from the EPL. Dark gray circles denote clubs that have been both promoted to and relegated from the EPL during the period that we study. White squares represent clubs located elsewhere. Vertex sizes reflect the number of moves in and out of clubs. The network ties in Figure 8.6 are black for time slice 1, red for time slice 2, and blue for time slice 3. Clubs never playing in the EPL have only outgoing ties. No moves *within* the EPL are considered here.[15]

---

[15] The flows from clubs relegated from the EPL to EPL clubs all came after they were relegated from the EPL. Flows from clubs promoted to the EPL are for before their promotion.

We constructed the device of an 'imaginary club' to deal with gaps in the playing careers of footballers. This is described more fully in the Appendix describing our data and the construction of our dataset. When players have been discarded by their clubs, they do not always find another club quickly. For the time when they are without a club, and have not left the game, we label this an imaginary club. It preserves continuity in the data to cover the gaps. It contains discarded players who get picked up later by other clubs.

Confining attention to local clubs, there are 21 English clubs outside the EPL that sent players to EPL clubs. This is fully consistent with Proposition 7.9.15 (that EPL player recruitment is predatory). These smaller clubs, as noted in Chapter 7, are unable to keep their best players – often, they must sell them in order to survive. While the positive aspect, from a player's viewpoint, is that they have the opportunity to play in the highest league in the country, this does not diminish the predatory aspect of these movements for clubs. Their role in the English league system is to develop players who are outside the elite levels. There are eight clubs that did play in the EPL – Birmingham City, Bolton Wanderers, Charlton Athletic, Portsmouth, Sheffield United, Sheffield Wednesday, Swindon Town, and West Bromwich Albion – and sent players to EPL clubs when they were outside the EPL. Most often these moves came after a club was relegated[16] from the EPL. This also supports Proposition 7.9.15, perhaps even more strongly. Good players play on relegated teams, have had EPL experience, and are attractive targets for recruitment.[17]

Seven Scottish clubs sent players to the EPL during its first 15 seasons. The Scottish Premier League (SPL) has long been imbalanced and dominated by 'the Old Firm' (Celtic and Rangers). Rangers[18] has won 54 national titles while Celtic has won 43. (It added another in 2013.) These two clubs are the main Scottish sources for player moves to the EPL. Both are big clubs implying some ambiguity about these moves resulting from predation. However, SPL's standing in world football is much lower than the EPL's standing.[19] Player movements from the other (much smaller) five Scottish clubs are predatory. Two Irish clubs sent players to the EPL. As the Irish league is low-ranked, this is consistent with Proposition 7.9.15. One, Cherry Orchard, has long sent Irish players to top-level clubs in England.

France had 13 clubs sending players to the EPL in its first 15 seasons. It was the primary external source of players for EPL clubs. As Ligue 1 has a lower general standing than the EPL (despite having two founding members of the G-14 and one additional member), such moves were predatory. The evidence is less clear for Italy, Spain, and Germany. Eight Italian clubs sent players to the EPL. However, three were AC Milan, Inter Milano, and Juventus. Given their stature in world football, as indicated by their founding memberships in the former G14 and their UEFA successes, moves from these three clubs to the EPL cannot be viewed as predatory. However, the moves from the other five Italian clubs can be viewed as resulting from predation. Five Spanish clubs sent footballers to the EPL in the same 15 years.

---

[16] Bolton Wanderers and Portsmouth sent players to the EPL before they were promoted.

[17] Sometimes relegated clubs can no longer afford to pay the salaries of their top players given the sharp drop in revenue that comes with playing in the second tier. Players can also have clauses in their contracts that free them from their contracts if the club is relegated.

[18] Rangers went bankrupt in the 2011/2 season and, after failed negotiations with creditors, its assets were bought by a new owner. It was relaunched but its application for readmission to the SPL was denied. In 2012/3, they played in the third tier of Scottish football.

[19] Its UEFA ranking compared to England has always been lower. Further, while the top three EPL clubs qualify directly for the next season's Champions League, only the top SPL club does. The fourth-place EPL club and the second-place SPL club enter the qualifying rounds.

Two were Barcelona and Real Madrid, both founding members of the G14. As both have repeatedly won the European Cup, moves from them to the EPL cannot be viewed as resulting from predation. Three German clubs sent players to the EPL. Two were Bayern München and Borussia Dortmund, both founding members of the G14 with stellar records. This suggests that *the elite clubs of Europe form a set of clubs preying on all other clubs while trading players between each other in a non-predatory fashion.*

Three clubs from the Netherlands sent players to the EPL between 1992 and 2007. All three – Ajax, PSV Eindhoven and Feyenoord – have long dominated the Eredivisie, their top league, with Ajax and PSV Eindhoven also being founding members of the G14. However, the Eredivisie is a far less prominent league. Most of their best players, sooner or later, move to the top European leagues. Such moves are the result of predation. The same holds for all of the other European and South American clubs sending players to the EPL.

Proposition 7.9.13 regarding clubs preferring repeated transactions is not easy to assess. Even so, there is modest provisional support for it in Figure 8.6, albeit with some counter-examples. Arcs in this figure vary in thickness depending on the flow volumes between pairs of clubs. We focus first on the heavy flows *within* time slices. Three of them are in the first: Rangers → Everton; Birmingham City → Coventry City; and Portsmouth → Manchester City. All of these are local flows. There are five heavy flows in the second time slice: Celtic → Sheffield Wednesday; Paris Saint-Germain → Newcastle United; RC Lens → Middlesbrough; Barcelona → Chelsea; and Molde → Wimbledon. Only the first is a local flow. The third time slice has five heavy flows,[20] none of which are local: PSV Eindhoven → Chelsea; PSV Eindhoven → Tottenham Hotspur; FC Porto → Chelsea; Real Madrid → Bolton Wanderers; and Olympique de Marseille → Leeds United. These flows support there being a preference for repeated transactions *within* 5-year time slices, suggesting that such preferences are shorter-term.

There are many other less thick arcs, indicating smaller flows, in Figure 8.6, revealing more repeated transactions within time slices. Yet there are many multiple arcs between pairs of clubs that reveal transactions between them in multiple time slices. This also provides provisional support of the 'repeat transaction' preference of McGovern. This evidence is the strongest for pairs of adjacent time slices. However, there are only three flows between pairs of clubs in all three time slices: AS Monaco → Arsenal; Parma → Chelsea; and PSV Eindhoven → Aston Villa. It may be important to distinguish the reputation of a club for being reliable as a source of talent and repeat flows being dependent on particular people involved in these transactions. Arsene Wenger managed AS Monaco from 1987 through 1994 and became Arsenal's manager in 1996, a position he has held through 2013. The repeated flows of AS Monaco → Arsenal appear to depend primarily on one individual.[21] Two of the flows from the Italian club Parma → Chelsea may be partially explained by having Italians (Gianluca Vialli, 1998–2000, and Claudio Ranieri, 2000–4, both with deep knowledge of the Italian football system) as managers of Chelsea. Incidently, the heavy flow FC Porto → Chelsea followed the move of FC Porto's then manager José Mourinho (2002–2004) to Chelsea (2004–2007) and suggests player movements based on the nationality of Mourinho,

---

[20] There is a heavy flow, 'imaginary club' → Birmingham City, which states that the club picked up players without a club in this time slice. Only four other clubs, all based in London, recruited players without clubs in the three time slices. The imaginary club is far more prominent in flows away from the EPL.

[21] No doubt his deep knowledge of the French football scene is a part of other flows of players from other French clubs to Arsenal.

the relations he had with players at FC Porto, and his knowledge of the Portuguese football league. These flows provide modest support to Proposition 7.9.13. They suggest that one mechanism for repeated transactions is the involvement of trusted individuals.

The repeated flows of PSV Eindhoven → Aston Villa may reflect that the source club is trusted as a reliable developer of footballing talent rather than resting on specific personal ties. If there is a general, or even a partial, preference for repeated transactions between pairs of clubs, identifying repeated transactions permits the exploration of detailed reasons for them. This remains an item for future work.

8.4   Moves from EPL clubs

We now consider moves from EPL clubs either to other EPL clubs or to clubs outside the EPL for each of the time slices. Only players in the EPL's first 15 seasons are involved. Even so, these networks are very complicated. We do not study all these moves for two reasons. One is that the resulting network diagrams are visually uninterpretable. The second, more important, reason is to focus on systematic patterns. Occasional moves are idiosyncratic to the players and clubs involved: they are highly unlikely to cumulate into anything systemic. When players will become available is uncertain, and the needs of clubs can change quickly. Matching the two depends on who is available when. We first restrict attention to pairs of clubs where with at least four moves from one to the other in a time slice before expanding to include pairs of clubs with at least three moves. We report our results for each time slice. These flows are mostly from EPL clubs to non-EPL clubs plus non-EPL clubs to other non-EPL clubs. Only a small number are moves between EPL clubs (for players remaining in the EPL) or from non-EPL clubs to EPL clubs (for playes returning to the EPL). Vertex sizes in Figures 8.7–8.9 represent numbers of players moving to or from clubs.

8.4.1   The 1992–1996 time slice flows with at least three moves

Figure 8.7 (upper panel) shows the between-club flows with at least four moves in this time slice. The largest flow shown in this figure is West Ham United → AFC Bournemouth, a team which played in the third tier of English football during this time slice. There was a less heavy flow of Queens Park Rangers → Bristol Rovers, a team that played in the second tier at the beginning of this time slice and in the third tier at the end of it. Clearly, both flows involved players whose EPL playing careers had come to an end. This discarding of players from the EPL is the reverse of predation. Yet the two are part of a coherent system: younger players seen to have potential are recruited upwards to stronger clubs while older players whose potential is waning are discarded and move to weaker clubs if they do not quit the game. Another flow was Tottenham Hotspur → Ipswich Town. As this receiving club spent the first three years of this period in the EPL before being relegated, these moves are from a stronger club to a weaker club. The same holds for the flow Manchester United → Middlesbrough, a club that bounced between the EPL and the second tier in this period. The flow Everton → Portsmouth featured players that were in the EPL before moving to a club in the second tier.

Crystal Palace bounced between the EPL and the second tier in this time slice: flows from this club were to weaker clubs: Brentford spent one season in the second tier before dropping to the third; and both Charlton Athletic and Wolverhampton Wanderers played in the second tier. Leicester City played just one season in the EPL in this time slice and sent players

Flows with at least four transfers between clubs

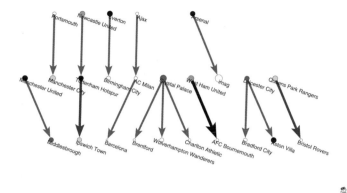

Flows with at least three transfers between clubs

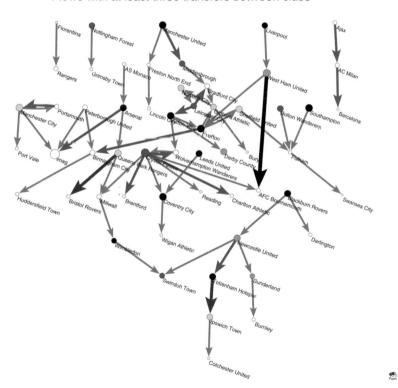

Figure 8.7   Within and from EPL club flows: 1992–1996.

The different sizes of the vertices reflect differences in club participation in transfers. The different thicknesses of the arcs depict different flow volumes involving multiple players.

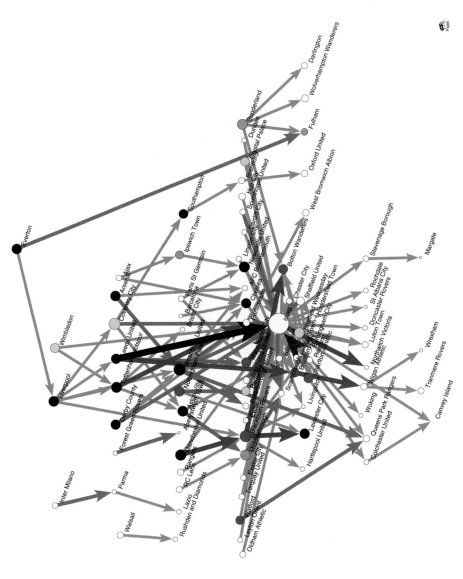

Figure 8.8 Time slice 1997–2001 featuring at least three transfers between clubs. The different sizes of the vertices reflect differences in club participation in transfers. The different thicknesses of the arcs depict different flow volumes between clubs.

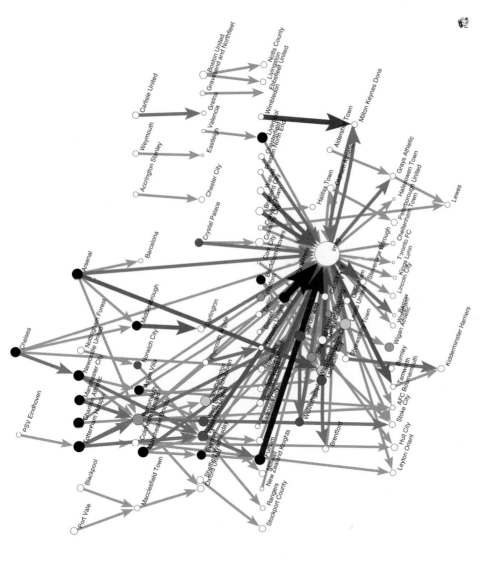

Figure 8.9 Time slice 2002–2006 with at least three transfers between clubs.

The different sizes of the vertices reflect differences in club participation in transfers. The different thicknesses of the arcs depict different flow volumes between clubs.

to Aston Villa, a much stronger club in this period, following their relegations. It also sent players to Bradford City in the third tier of English football. The Portsmouth → Manchester City flow was before Portsmouth reached the EPL (in time slice 3). The Newcastle United → Tottenham Hotspur flow was within the EPL. Arsenal → 'imaginary club' was Arsenal discarding players who were unable to find another club within three months.

The lower panel of Figure 8.7 expands the upper panel by including flows between clubs with three players in this time slice. Manchester United sent three players to Preston North End, another Lancashire club playing in the third and fourth tiers during these five seasons. These involved players who never again played in the EPL. There was a further flow from Preston North End to Lincoln City, a club playing in the fourth tier. Not all of the same players who arrived from Manchester United were in this flow. Instead, they came directly, or indirectly, from other EPL clubs in tiny flows to Preston North End before moving to Lincoln City. Southampton and Bolton Wanderers had flows to Fulham before Fulham reached the EPL (in 2001). Again, these are players moving to a club in a lower tier. Also, the Fulham → Swansea City flow sent players to a club in the third tier.

There is a chain of moves linking AS Monaco with Swindon Town. We have discussed the AS Monaco → Arsenal flow. Arsenal sent players to Queens Park Rangers (QPR), a much weaker team which played in the EPL before being relegated in 1996. QPR sent players to Millwall, a club in the second tier. The next link is interesting because it represents moves from a second-tier club to an EPL club, but not all of the upwardly mobile players in the arc had *prior* EPL experience. Players moving from Wimbledon to Swindon Town did so after the latter team was relegated in 1994 after just one season in the EPL. It is important to note that the arcs between clubs in a path made up of flows linking more than two clubs need not involve the same players.

While there are some apparent anomalies in Figure 8.7, all are legitimate flows. The chain Ajax (Holland) → AC Milan (Italy) → Barcelona (Spain) are all flows involving players previously in the EPL. They left different EPL clubs so there is no large flow from an EPL club to Ajax.

8.4.2   The 1997–2001 time slice flows with at least three moves

Figure 8.8 shows the flows within and from the EPL for the second time slice. This is a larger and denser network of moves between clubs. It corresponds to the expansion of EPL club squads shown in Figure 7.6. Another clear feature of Figure 8.8 is the growth of the imaginary club capturing players without a club. The largest flow to this destination was from Leeds United. When attention is confined to flows with at least four players, there are are 12 clubs discarding players, of which three are EPL clubs. Expanding the focus by including flows with three moves, the number of clubs discarding players increases to 22. All of these discarded players were once playing in the EPL. Their moves can be directly to the imaginary club or via clubs playing in lower league levels. This points to yet another risk for EPL footballers because none of these playing careers were over.

Clubs receiving players from the imaginary club all picked up discarded players. In the main, these receiving clubs are low-ranking clubs in the league system or, more likely, outside the Football League. Among the latter, Northwich Victoria and Scarborough are prominent as receiving clubs for former EPL players near the end of their playing careers. The sequence: imaginary club → Stevenage Borough → Margate represents movements completely outside the Football League for former EPL players. In terms of the alleged 'glamour' of the EPL, these are moves into oblivion. Yet they reveal also the resilience and determination of some footballers to continue playing outside the 'big time' limelight. Proposition 7.9.19 is supported strongly.

Leeds United sent at least four players to Bradford City (another West Yorkshire club) that played just one season in the EPL before relegation. It also sent players to Wigan Athletic (well before Wigan was promoted to the EPL at the end of the third time slice). These moves are all 'downwardly mobile' transitions for the players involved in them. The flow Everton → Fulham also involved players leaving the EPL. Similarly, all the flows from Sunderland to Fulham, Wolverhampton Wanderers, and Darlington (a club in the northeast of England and close to Sunderland) feature downward mobility. There are many such moves between clubs in Figure 8.8 which form a regular, if unglamorous, feature of professional football. Moves such as Walsall (that spent the 2000/1 season in the second tier) → Rushden and Diamonds (a club outside the Football League in this time slice) involved further delayed downward mobility as the players leaving Walsall had gotten there at earlier time points after exiting the EPL.

There are flows within the EPL including Everton → Liverpool, Arsenal → West Ham United; and Blackburn Rovers → Leicester City. However, these are in the minority among the moves depicted in Figure 8.8. Also, some of the flows between clubs that were in the EPL feature moves from clubs once in the EPL to EPL clubs or to clubs seeking good players to gain promotion to the EPL.

The chain Inter Milano → Parma → Lazio features flows between Italian clubs. Yet they involve players previously in the EPL. Again, with players leaving different EPL clubs, there is no large flow from an EPL club to Inter Milano. As these clubs were in Serie A, the players involved continued to play at a high level. The flow RC Lens → Middlesbrough features flows of players who were once in the EPL and are returning. The flows Ajax → Chelsea, Ajax → Ipswich Town, and Barcelona → Chelsea also feature players returning to the EPL. However, in general, once players depart the EPL, very few return, consistent with Proposition 7.9.19.

8.4.3   The 2002–2006 time slice flows with at least three moves

Figure 8.9 shows the flows for the final time slice with at least three moves between clubs. This network remains large and dense. There was a further expansion of the size of the imaginary club. The largest direct contributor to the number of discarded players is Bolton Wanderers, a club promoted to the EPL in 2001 (after two earlier one-season stays in England's top league). This club was discarding players to make room for the new arrivals as shown in Figure 8.6. Confining attention to between-club moves involving at least four players, there are 15 clubs 'sending' players to the imaginary club. As in the previous time slice, the primary beneficiaries of recruiting discarded players are found in the lower reaches of the Football League (e.g. Leyton Orient, AFC Bournemouth, Barrow, Cheltenham Town, and Lincoln City) and clubs outside this League (e.g. Tamworth, Stevenage Borough, Grays Athletic, Woking, and St Albans City).

The large flow involving at least four players from Wimbledon to Milton Keynes Dons is not a real flow. As noted in Appendix A.4 documenting our data, this is primarily a flow of players 'moving' as a result of the demise of Wimbledon and the formation of Milton Keynes Dons from its remains: they simply stayed with the new club when it moved to a new location. It is of some interest that in the aftermath of the creation of MK Dons that it 'exchanged' players with the imaginary club. Put differently, they discarded players but found other acceptable players previously discarded by other clubs. The flows in the chain Chelsea → Newcastle United → West Ham United → Fulham all involve EPL clubs following Fulham's promotion to the EPL in 2001. Fulham discarded the second largest number of players to the imaginary club following its promotion. This is consistent with players being good enough to help a club win promotion but are discarded as a promoted club seeks to strengthen its squad so as to survive, or even thrive, in a higher league. Fulham has remained in the EPL since its promotion. However, it was relegated in 2014.

Again, there are indirect downwardly mobile moves. This includes the moves of Manchester United → Sunderland, a club bouncing between the first and second levels of English football in this time slice. The moves Sunderland → Wolverhampton Wanderers (a.k.a Wolves) are also downward, save for the few players moving to Wolves for its one season in the EPL in this time slice. Wolves also discarded players to the imaginary club. Leeds United picked up players from Leicester City after Leicester City was relegated in 2001 and again in 2004. In turn, Leeds United[22] sent players to Sheffield United, another club from West Yorkshire. Middlesbrough again sent players to Darlington that discarded players. There are many such indirect downwardly mobile sequences in Figure 8.9.

The moves involved in Shrewsbury Town → Kidderminster Harriers; Reading → Brentford; Plymouth Argyle → Torquay United; and Carlisle United → Gretna are all delayed downward moves of former EPL players. In the main, once players leave the EPL, there is little chance of them returning. Former EPL players face only movement down league levels if they want to continue playing football. Again Proposition 7.9.19 receives support.

8.5 Moves solely within the EPL

We consider only moves between clubs *while they were both in the EPL*. All other clubs are ignored in this section. Players move either through a loan from one club to another or by being transferred between clubs. Figures 8.10 and 8.11 show loans and transfers respectively for the full 15-year period. In both figures, the sizes of the vertices show the outdegree (number of players leaving clubs).

8.5.1 Loans

We consider, first, the loans between EPL clubs as shown in Figure 8.10. As noted before, many players want first-team playing experience. A loan move is beneficial for peripheral players in top-level squads, especially when they are young, when they would not otherwise be playing regularly. Both of the clubs stand to benefit from loans. Loaning clubs gain when the loaned players get EPL playing experience. Receiving clubs receive good players when they lack the resources to buy players.

The clearest feature of Figure 8.10 is differential club involvement in loaning players. Vertex sizes are proportional to the numbers of players involved in loans. The core of the EPL loaning network features the 'Big Four' clubs: Arsenal; Chelsea; Liverpool, and Manchester United, as well as Tottenham Hotspur. Additional benefits to clubs with squads large enough

---

[22] The club overspent both in terms of transfer fees and player salaries in an effort to seek European glory. But the manager and players did not deliver. The club went into administration, lost points, had to sell their best players (at cut rate prices) and was relegated in 2004. It has not returned to the EPL.

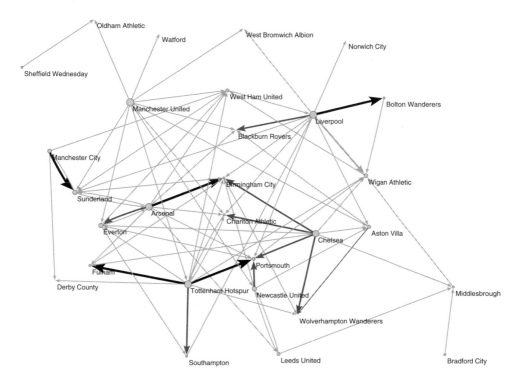

Figure 8.10   The network of loans between EPL clubs: 1992–2006.

The different sizes of the vertices reflect differences in club participation in transfers (outdegrees). The different thicknesses of the arcs depict different flow volumes between clubs.

to loan players, in addition to their loaned players get playing experience elsewhere, include: 1) they receive money from the receiving club, and 2) loaned players developing quickly as skilled players can be recalled when needed.

Loan flow volumes between EPL clubs differ greatly. With the exception of Manchester United, the big clubs are involved in large flows. The flow Liverpool → Bolton Wanderers is large. Substantial flows from Liverpool to Blackburn Rovers and Wigan Athletic were present. All four clubs are geographically close in the northwest of England with Liverpool the loaning focus. The other large flows are Arsenal → Birmingham City; Tottenham Hotspur → Portsmouth; Tottenham Hotspur → Fulham (also in London); and Manchester City → Sunderland. If we include Tottenham Hotspur → Portsmouth we have loans between southern clubs centered on Tottenham Hotspur. Chelsea is the focus of loans to Charlton Athletic (also in London), Portsmouth, Birmingham City, and Wolverhampton Wanderers.

There are very few reciprocal ties loaning between EPL clubs. Just three are present: Manchester City and Sunderland (heavily imbalanced in favor of Manchester City); Arsenal and Everton (imbalanced in favor of Arsenal); and Leeds United and Newcastle United. There are no loans between local clubs involved in deep, even bitter, historical rivalries: Manchester City and Manchester United; Everton and Liverpool (in Liverpool); Arsenal and Tottenham Hotspur (North London); Aston Villa and Birmingham City (in Birmingham); and Newcastle United and Sunderland (in northeast England).

8.5.2 Transfers

Transfer flows within the EPL are shown in Figure 8.11 for the full 15-year period. This network is highly centralized with a clear core of clubs heavily involved in transfers and a set of peripheral clubs with limited transfer activity. In part, this is a trivial result. Clubs having shorter or very short stays cannot trade players within the EPL after their relegation. However, the implications of this pattern are rather ominous for these 'lesser' clubs. As noted earlier, once relegated, they risk losing their best players to clubs remaining in the EPL. When they receive players from EPL clubs, these players are often being unloaded. This is consistent with the pattern shown in Figure 7.11: some clubs have never been relegated from the EPL; some clubs had long stays; other clubs bounce between the top two tiers like yo-yos; and some clubs have very short EPL stays. This pattern is unlikely to change because it is reinforced through transfers creating a system of cumulative advantage for a few clubs and cumulative disadvantage for the remaining clubs.

Transfer flows between clubs differ greatly in their sizes. The heaviest flow in Figure 8.11 is Liverpool → West Ham United. West Ham also received many players from Arsenal and Manchester City. Other heavy flows include Charlton Athletic → Leeds United, Liverpool → Leeds United, Manchester United → Everton and Liverpool → Tottenham Hotspur. As the details of this dense core are hard to discern because of the number of transfers, we look more closely at transfers in each of the time slices. We removed all transfer flows between clubs involving only one player to focus on *multiple* transfers between clubs. There were 120 single player links in the first time slice, 116 in the second, and 170 in the third time slice. We note that these single player flows contradict the idea that clubs preferring repeated transactions, Proposition 7.9.13, at least within these time slices. The heaviest flows in each of the time slices involve four players. For each time slice, we have kept the locations of the vertices representing clubs fixed for better comparisons between time slices. Again, vertex sizes distinguish club involvements in transfer activity.

Figure 8.12 shows the transfer network for the first time slice, Figure 8.13 contains the corresponding EPL transfer network for the second, and Figure 8.14 shows that for the third. Recall that these figures are drawn so that clubs present in successive time slices have very similar coordinates. A total of 29 clubs played in the EPL in the first time slice with 22 of them involved in the transfer moves in Figure 8.12 (involving two or more players in a flow). The heaviest incidence of within-EPL transfer activity is Liverpool → West Ham United. This holds also in the second time slice but disappears in the third. Indeed, there are no strong repetitions holding across all time slices. The reinforces the idea of preferences for repeated transactions between clubs (McGovern 2002), should they exist, not being long-term preferences. First-team club managers are highly transient. Their frequent departures from clubs, and later arrivals at other clubs, may help account for this. Other notable but less heavy flows in the first time slice were: Chelsea → Southampton; Chelsea → Everton; Blackburn Rovers → Newcastle United; Leeds United → Coventry City; Leicester City → Aston Villa; and Newcastle United → Tottenham Hotspur. Of these, the only flow to repeat in the second time slice was Chelsea → Southampton. Indeed, only three of the flows in the first time slice are repeated in the second. While some of this is due to clubs being relegated, most appear to reflect an absence of long-term preferences for repeated transactions.

In Figure 8.13 for the second time slice, the heavy flow Liverpool → West Ham United is joined by another heavy flow, Arsenal → West Ham United. Among the less heavy but large flows, two also feature Liverpool: Liverpool → Tottenham Hotspur and Everton →

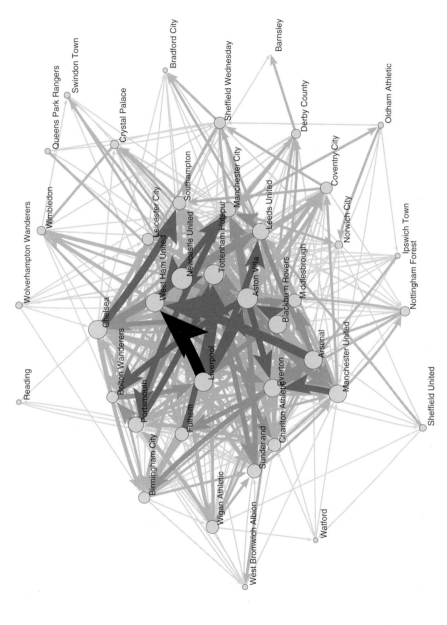

Figure 8.11 Within and from EPL transfer flows, 1992–2006.

The different sizes of the vertices reflect differences in club participation in transfers. The different thicknesses of the arcs depict different flow volumes between clubs involving at least two players.

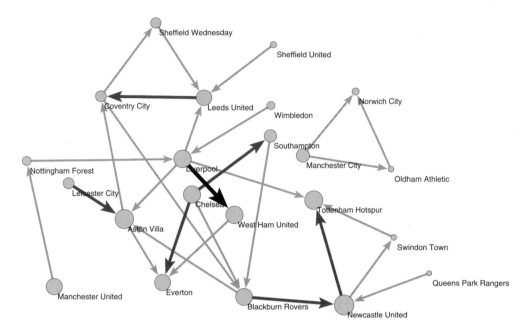

Figure 8.12   Within and from EPL transfer flows: Time slice 1.

The different sizes of the vertices reflect differences in club participation in transfers. The different thicknesses of the arcs depict different flow volumes between clubs involving at least two players.

Liverpool. This club was highly active in transfers within the EPL. Other less heavy flows are Aston Villa → Everton, Coventry City → Aston Villa, West Ham → Wimbledon, and Wimbledon → Tottenham Hotspur.

Preferences for repeated transactions need not involve flows in one direction. For example, Aston Villa → Coventry City in the first time slice was followed by a heavier reverse flow in the second time slice. The heavy Blackburn Rovers → Newcastle United flow of the first time slice was followed by a lighter reverse flow in the second.

The heaviest flows in the third time slice, see Figure 8.14, were Arsenal → Birmingham City, Chelsea → Newcastle United, and Manchester City → Portsmouth. None were present in the two previous time slices. There are more less heavy flows in this time slice. We do not list them all. Instead, we turn to consider reciprocal flows within time slices. There were none in the first time slice. However, two appear in the second (see Figure 8.13): Aston Villa ↔ Manchester United, and Everton ↔ Newcastle United. While these are not repeated in the third time slice, four others appear: Aston Villa ↔ Newcastle United; Chelsea ↔ Birmingham City; Everton ↔ Manchester United; and Charlton Athletic ↔ Manchester City. Except for the reciprocal transfer links involving Birmingham City and Charlton Athletic, all involve clubs in the EPL for all of the 15 years we consider.

In terms of changes through time, several features stand out. One concerns the heaviest flows. One existed in the first time slice, two in the second, and three in the third. Second, the number of the next highest heavy flows also increased across the time slices. This provides additional support for Proposition 7.9.13. Third, consistent with the first two features, the density of these flows increases through time. While players enter the EPL from English clubs in lower levels of the English league system, as shown in Figure 8.6, increasingly, entrances

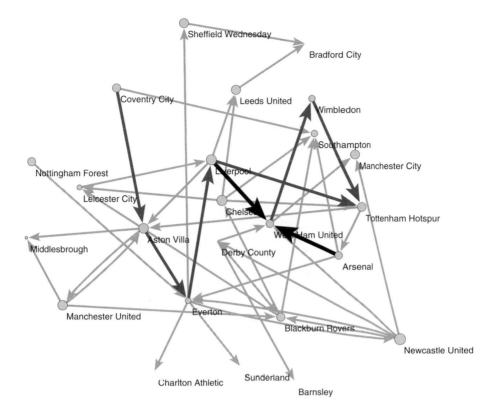

Figure 8.13   Within and from EPL transfer flows: time slice 2.
The different sizes of the vertices reflect differences in club participation in transfers. The different thicknesses of the arcs depict different flow volumes between clubs involving at least two players.

to the EPL are from non-local clubs. This provides additional support for Proposition 7.9.4 as well as Proposition 7.9.5. This result emphasizes the diminishing opportunities for British players getting to the EPL from local clubs over time for both loans and transfers. The structural implication of this closure is the EPL becoming more sealed off from the rest of the English system and more coupled to external sources with regard to player transfers.

The implication of Figures 8.12, 8.13, and 8.14, for Proposition 7.9.13, at best is mixed. The presence of sufficiently many single player flows between pairs of clubs contradicts this claim. Yet the steady growth of heavy transfer flows between pairs of EPL clubs over the three time slices, suggests that this preference is present and increased, supporting the proposition. If one can pick the (relatively few) multiple flows in a time slice as supporting the proposition or the (far more frequent) single player links between clubs in a time slice as not supporting the hypothesis, it suggests that the hypothesis is not truly testable:[23] evidence for or against depends on the eye of the beholder.

---

[23] The proposition uses the term 'preference', an inherently subjective concept, while the data are in the form of realized player flows. In the end, these flows are the ones that count. While preferences are unobserved, their realization is constrained by the conditions when transfers take place.

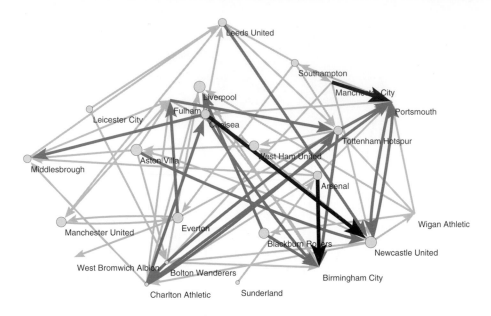

Figure 8.14   Within and from EPL transfer flows: time slice 3.

The different sizes of the vertices reflect differences in club participation in transfers. The different thicknesses of the arcs depict different flow volumes between clubs involving at least two players.

Looking at the full distribution of transfer flows over the three time periods provides a more exact test. These are shown in Table 8.3. Transfer volumes increased through the three time slices: more players were on the move between clubs over time. There appears to have been a lot of 'churning'. Each time slice has five years. In the first slice, 90% of the transfers were single transactions between clubs. For both of the second two time slices, it is slightly over 87%. Even the number of times there were just two transfers between pairs of clubs in a 5-year interval is rather small. If one and two transfers are counted as evidence against a preference for repeated transactions, the percentages for the three time slices are 98%, 97%, and 97%. This is very strong evidence against Proposition 7.9.13. Of course, a preference for repeated transactions may exist. However, such preferences have to be realized in large numbers to support Proposition 7.9.13. With few exceptions, they are not realized. Player moves result from complex (and competing) negotiations made under time constraints imposed by club budgets and player availability. These conditions make it unlikely that preferences for repeated transactions, if they exist, will be realized.

8.6   All trails of footballers to the EPL

We expand our focus to consider longer sequences (trails) of footballer moves to the EPL.

Given our focus on reaching England's top league, these trails can be viewed in different ways. As noted in the data description, the number of stays on these trails varies from 1 to 25. As the length (number of player moves) of these trails varies greatly, we seek a broad generic summary. To illustrate some of the problems involved in doing this, we show the trail for the French player Xavier Graverlaine in Figure 8.15. Ignoring loan moves, this trail

Table 8.3 Counts of transfers between pairs of clubs by time slices.

Time slice 1		Time slice 1		Time slice 1	
Count	Frequency (percent)	Count	Frequency (percent)	Count	Frequency (percent)
1	2562 (89.99)	1	3638 (87.62)	1	4157 (87.39)
2	225 (7.90)	2	396 (9.54)	2	457 (9.61)
3	45 (1.58)	3	88 (2.12)	3	102 (2.1)
4	12 (0.42)	4	20 (0.48)	4	29 (0.61)
5	2 (0.07)	5	8 (0.19)	5	9 (0.19)
6	1 (0.03)	6	1 (0.02)	6	1 (0.02)
		7	1 (0.02)	7	1 (0.02)
				8	1 (0.02)
Totals	2847		4152		4757

is the longest for players reaching the EPL in the second time slice. The size of vertices in this trail represents the total length of his stays at clubs.

Graverlaine moved through 11 clubs (or 13 clubs if Paris Saint-Germain is counted three times). He started at Nantes. After a long time there, he moved through four other French clubs before reaching Paris Saint-Germain for the first time. He then moved from Paris Saint-Germain to Racing Strasbourg before returning to Paris Saint-Germain. Next, he traveled on a cycle through three other clubs before again returning to Paris Saint-Germain. Finally he moved to Watford in the EPL. From the argument of Maguire and Stead (1998) this is a fairly long trail and it is unique. It would add to the count of the many trails over which players traveled to the EPL. Yet, it could be viewed as a simple trail as far as movement to the EPL is concerned. In terms of clubs we have Paris Saint-Germain → Watford and for countries it is France → England. We sought a way of summarizing and classifying the trails to the EPL while being attentive to their lengths and complexities but without getting bogged down in all of the details.

8.6.1   Counted features of trails to the EPL

Footballers start their careers in the country of their nationality even though this does not always hold: future players can migrate as children and, as players, have a choice about the country they can represent at the international level. Most often, this involved migrations from former colonies to former colonial powers. We use the term 'nation of origin' to refer to their nationality. In the first time slice, 361 players started their careers in the EPL of

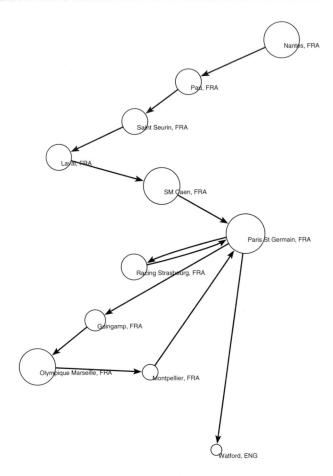

Figure 8.15 The transfer trail for Xavier Graverlaine.

The clubs in this trail are ordered from his earliest club (Nantes) to his latest club (Watford). The one exception to the convention is Paris Saint-Germain where he had multiple stays.

whom 4.2% did not start the careers in their nation of origin. In the second time slice, 516 players started their careers in the EPL with 3.7% starting their careers outside their country of origin. The number of players starting on the EPL in the third time slice dropped to 244 with 7.4% of them starting outside their country of origin. These serve as a backdrop to our primary interest of moves into the EPL. These counts are shown in Table 8.4.

In addition to the three time slices defined earlier, we include a time slice 0 for the players that were in Division One clubs that stayed in the top tier when the EPL was formed for the 1992/3 season. Included in this table are players having more than one transfer move with loans included, to get a total count of trails. The counts are for players *entering* the EPL in each of the time slices. The number of players moving to the EPL increased from the 516 arrivals in time slice 0. In this time slice the largest number of moves on a trail was 15 with 282 clubs being on these footballer trails. These clubs were located in 30 countries. The number of players entering the EPL for time slices 1, 2, and 3 were 637, 787, and 686 respectively, consistent with our findings regarding the impact of the Bosman Decision. The increase in

Table 8.4 Counts of trails to the EPL by times slices.

Time slice	Number of players	Largest move count	Club counts	Country count	Players not starting in country of origin
slice 0	516	15	282	30	95
slice 1	637	15	555	51	98
slice 2	787	21	787	72	137
slice 3	686	20	816	75	123

these numbers from time slice 0 to time slice 2 was reversed in time slice 3, consistent with the data shown on the left in Figure 7.6. The largest number of moves remained the same in time slice 1 but increased for the last two time slices. The number of clubs located on the football trails increased steadily through the time slices, as did the number of countries where these clubs were located. This is further evidence of the increased linkage of the EPL to leagues and clubs in other countries as sources for recruiting players.

A visual representation of some of the contents of Table 8.4 is shown in Figure 8.16. More precisely, the graphs represent the distributions of the number of clubs for which footballers played before reaching an EPL club. For each time slice, the modal category has the smallest number of such clubs. In terms of relative sizes within each time slice, the number increased from time slice 0 to time slice 2 before dropping in time slice 3. The distributions are more skewed right in time slices 3 and 4. Note that this ignores the identities of the previous clubs and is more fundamental than distinguishing trails using club names.

One final set of counts takes the form of loans as the first move into the EPL, a small part of the counts shown in Table 8.4. We divided players entering the EPL into two categories: those starting in 1) the Home Countries and 2) other countries. The counts are shown in Table 8.5. Loaning as a way of entering the top tier of English football was negligible in time slice 0. These loans jumped for the second time slice for both categories of players. Thereafter the two trajectories diverged sharply. Loans as a way of entering the EPL for players from the Home Countries dropped while loans from clubs elsewhere increased dramatically. Entering the EPL on loans opened up for player from elsewhere while diminishing for home-grown players, consistent with the larger trend for transfers of disadvantaging players from the Home Countries for entry into the Premiership.

Table 8.5 Number of players entering the EPL through loans.

Time slice	Players from the Home Countries and Ireland	Players from elsewhere
slice 0	1	0
slice 1	20	12
slice 2	12	63
slice 3	2	92

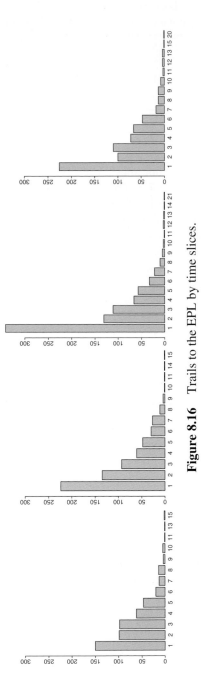

Figure 8.16 Trails to the EPL by time slices.
The time slices are ordered from the left as 0, 1, 2, and 3.

8.6.2   Clustering player trails

Player careers consist of games played, successes on the field, club successes where they played, and sequences of clubs for which they played. We focus narrowly on the last, labeling this 'player careers'. Our data have 3749 players who moved among 2301 clubs. The linked set of their trails of moves results in a large temporally distributed network. Visually, it is incomprehensible. We sought a summary of these trails. Counting types of individual trails has little value. Our response was to cluster player careers, a choice raising methodological issues. Very few players spend their entire careers at one club – Gary Neville at Manchester United and Paolo Maldini at AC Milan are rare exceptions. Such players, having no trails, were excluded.

Additional restraints were required to perform the clustering of trails with a set of players who had the same length of time playing. This implies setting both age and time limits. Concerning time, some players suffer career-ending injuries well before they would have retired. Some quit the game for various reasons, including lack of success. Such shortened player careers were excluded. Regarding age, there are players in our data whose careers have just started. Their careers are too short for inclusion. At the other extreme, some players have very long playing careers but they are the exception.

The youngest players in our data are 15. However, very few young players play for their club's first team while serving an apprenticeship. If they do play, it is likely to be for the reserve or junior squads of their clubs. Importantly, there are very few moves between different clubs for such young players. Also, many players in our data started when they are older than 15, implying a need for a higher low threshold. Considering career endings, waiting for all players in our database to complete their playing careers was impractical. An upper age limit was needed.

To set lower and upper thresholds, we examined the distribution of ages when players start and end their playing careers. These are shown in Figure 8.17. Both distributions are skewed in opposite (but obvious) directions. The left-hand display in Figure 8.17 confirms 15 as too low for classifying careers. However, it is not unusual for there to be 18–19-year-old footballers playing in first-team squads. Examining the distribution of starting ages suggests that 19 is a more appropriate lower threshold.

An additional issue arises with setting the upper limit, for which there is no obvious solution. Given our concern here with with movement *to* the EPL, we have less interest in the tail-ends of playing careers. Indeed, as noted earlier, the majority of players, once they leave the EPL, descend to lower leagues and often end up playing non-League football. This is the case in our data and we do not want to include them here.[24] Conventional wisdom claims that the majority of players reach their performance peaks in their late 20s. (There are variations by playing position with goalkeepers overall having longer career.) As players age beyond their peak, it is more likely they will be transferred to a lower-level club or discarded. As a result, we have set the upper age limit for clustering player careers at 30.

As a partial check on this line of reasoning, the number of players in the dataset for different starting and ending ages is shown in Table 8.6. For starting ages of 15–18, the number of cases seems too small for clustering. There are 1287 players with a starting age of 19 and an ending age of 30. We focus on these cases. As shown in the lowest three rows of

---

[24] A future project is one that clusters the tail-ends of careers to get a better summary of where players go after leaving the EPL as they age and continue playing.

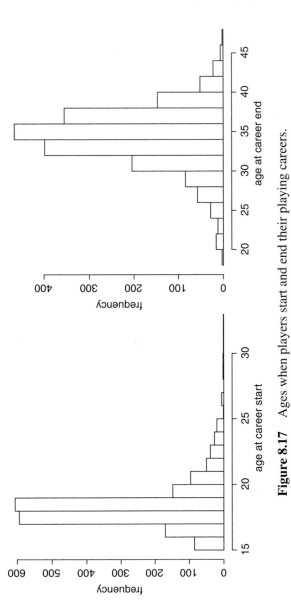

Figure 8.17 Ages when players start and end their playing careers.

Table 8.6 Number of players by starting and ending ages.

Start age	End age	Number of players
15	30	8
16	30	79
17	30	237
18	30	788
19	**30**	**1287**
20	30	1422
19	29	1328
19	28	1328

the table, some additional cases are obtained but this is counteracted by losing some years for the length of player careers. The bolded (19, 30) pairing is a reasonable trade-off regarding having enough cases with a reasonable number of career playing years.

We turn to define what is clustered. Given the many countries and clubs on footballer trails, constructing 'variables' to represent them using club names, places, and geographical distances traveled has little value.[25] As noted earlier, it is widely accepted that the top European leagues are the EPL, Ligue 1, the Bundesliga, Serie A, and La Liga. Commentators for EPL games and pundits talking about these games often claim that 'the EPL is the best league in the world.' This seems little more than hubris buoyed by the presence of many players from elsewhere. No doubt, similar claims are made about the other top European leagues. But this does raise the issue of ranking clubs rather than leagues. As described in Appendix A.4, we assembled data for use in constructing the ranks of all clubs on player trails through time. These ranks form the foundation for constructing the data used to the cluster player careers. As players move *to* top leagues, a reasonable expectation is that many move to better-ranked clubs and, in general, as they age, they start moving to lower ranked clubs as they continue their careers.

Evaluating footballer 'success' in their playing careers can be done in many ways, one of which is to read every footballer's bibliography to learn about their achievements. In principle, this provides the most complete information. However, sifting through player narratives for 1287 players – never mind doing so for over 3700 players – is impractical for three reasons. First, the volume of information is huge. More importantly, second, these narratives differ greatly in their completeness from multiple books to nothing. Third, combining information about items such as the number of times a player appears in games, the number of wins for the club when he plays, the number of goals scored, the number of assists provided, and the number of times a player represents his country is close to impossible, even assuming such data could be assembled. As a result, we define and operationalize, for this section, the 'success' of a player by using the ranks of clubs at the time for which he was at them as an approximate measure of success.[26]

---

[25] Also, there being no systematic and complete data on club characteristics such as size, financial health, and ownership precludes constructing indices from them.

[26] Club rank is not always a valid indicator of a player quality. Peripheral members of a club's squad seldom play. Yet squad membership implies that they have enough value for the club to retain their services. This suggests that our operationalization of success has merit.

Ideally, an unambiguous temporal ranking of all clubs across the globe that takes into account the changing ranks of clubs over time exists. Alas, there is no such official ordering of all clubs. Indeed, it is doubtful that one could be established. So an approximation to this ideal ranking was sought. Ideally, this would use the ranks of leagues at each point in time together with the ranks of clubs within leagues. To this end, UEFA has produced orderings of European football leagues since 1980 and the rankings of football clubs that play in the UEFA-sponsored cup competitions since 1990. Necessarily, these are incomplete rankings. The end-of-season league tables provide rankings of clubs within European and non-European football leagues often going back to the end of WWII. To get an approximation to an absolute ordering of clubs, decisions were needed about combining these partial and diverse fragments. Since the dataset consists of EPL players from the season 1992/3 to 2006/7, many of these players came from Europe or were playing in Europe. Therefore, UEFA rankings are useful despite being incomplete.

Attention was paid to the EPL rankings, as well as to the rankings of the other four top European football leagues, because these were complete. Even so, this was merely a first step. The scheme in Figure 8.18 shows the criteria used to implement our algorithm for approximating absolute club ranks over time. The top rank was set at 1. Constructing lower ranks was especially difficult because UEFA's rankings system excludes all lower leagues in national league systems. Further, ranking the myriad lower-level clubs has little practical value – all that matters is that they are ranked very low. Somewhat arbitrarily, we set the lowest possible club rank at 100. In England, this includes non-league clubs. All steps of Figure 8.18 were done in relation to UEFA rankings.

The criteria listed in Figure 8.18 were implemented in an algorithm to assign ranks to all clubs on player trails to the EPL. Reading down this figure, the steps are as follows:

1. The country in which each club plays was identified. There were three possible outcomes: 1) the country is one of the top five UEFA leagues (England, France, Germany, Italy, and Spain); 2) it is another UEFA country, and 3) it is a non-UEFA country.
2. This step has several separate substeps within it:

    (a) Club ranks in the top five UEFA leagues were the minimum (i.e best rank) of their direct seasonal ranks and the UEFA league seasonal ranks.
    (b) Clubs playing in very low leagues, e.g. non-league football in England, were assigned 100 as their ranks.[27]
    (c) If a club did not play in one of the top five UEFA leagues but has a UEFA rank, it was assigned the rank of UEFA league it played in.
    (d) For non-UEFA clubs, we distinguished clubs from Argentina, Brazil, and Uruguay.[28] Clubs from these three countries were treated differently given these nations' prominence in world football.
    (e) Clubs in all other countries were placed in a separate category. Note that the # symbol in Figure 8.18 represents 'not in'.

---

[27] The same was done for obscure clubs outside any of the national league systems, regardless of where they were located.

[28] National teams from these three countries have lifted the World Cup trophy nine times out of the 19 times it has been played. Further, clubs from these countries have consistently supplied footballers for the European leagues.

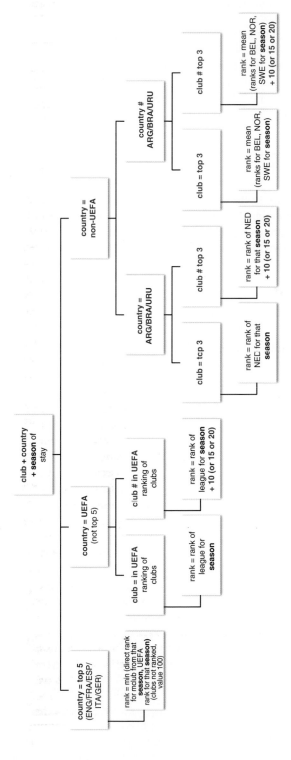

Figure 8.18 Outline of criteria for constructing absolute ranks of clubs through time.

3. There are substeps in this overall step:

   (a) For European clubs not in one of the top five UEFA leagues, their ranks were constructed in one of two ways: 1) if the club did play in a UEFA sponsored cup competition, it received the rank of its league as computed by UEFA; and 2) if it did not play in such a competition, then 10 was added to the UEFA rank for its league, to make sure that it is ranked below the better clubs of its league.[29]

   (b) From historical records, we established the top three club in Argentina, Brazil, and Uruguay and separated them from the rest of the clubs in their countries.[30]

   (c) For each of the other league systems we also distinguished the top three clubs from the rest of the clubs in every country.

4. This step applies only to the non-UEFA clubs, and ranks were made in comparison to the known UEFA rankings. There is a second tier of leagues in Europe that includes the Netherlands, Belgium, and leagues from Scandinavia. Of these, the Netherlands is more prominent (having two of the founding clubs of the G14). There are four steps:

   (a) All of the top three clubs in Argentina, Brazil, and Uruguay were assigned the UEFA rank for the Netherlands;

   (b) The remaining clubs from Argentina, Brazil, and Uruguay were assigned the rank of the Netherlands' top league +10;

   (c) The top three clubs from the other non-UEFA leagues were assigned the mean rank of the leagues from Belgium, Norway, and Sweden;

   (d) The remaining non-UEFA clubs were assigned the mean rank of the leagues from Belgium, Norway, and Sweden + 10.

Given the decision to use data on the 1287 players in the bolded line of Table 8.6, there are 11 1-year time intervals denoted by $1 \leq i \leq n$ with $n = 11$. For each player, the mean rank of the clubs for which he played in the ith time interval was used. We defined x_i as a vector composed of all the player ranks for the ith time interval. The set of player careers is represented by $X_p = [x_1, x_2, \ldots, x_n]$. The lowest value of the mean club rank variable is 1 (for players in the best ranked clubs). The largest value is 100 (for players playing for non-league clubs in any country). For each player career, a row of X_p is a symbolic object. These objects were clustered. This symbolic variable has a fully comparable metric for players, one that does *not* have to be normalized.

The clustering of symbolic objects was performed in the R package used in Chapter 5 for the US patent citation network.[31] Given the relatively small number of units, the *adapted*

---

[29] We experimented with placing them 10, 15, or 20 ranks lower in the absence of a known rank. There is some arbitrariness regarding these choices of lower ranks. Even though a conventional sensitivity analysis cannot be done, we checked the impact of these choices on the sum of squared errors (residuals), SSE, when doing survival analyses. For adding 10, 15, and 20 for such lower ranked clubs, the corresponding values for SSE were 109 million, 110 million, and 110.9 million. The differences seem minor and the results we report came from adding 10 to a league rank.

[30] In general, this is an easy step because most national league systems have been dominated by a tiny handful of clubs, another indicator of the marked inequality within league systems.

[31] The procedure is described in Chapter 3.

hierarchical procedure was used. As with all clustering methods, a dissimilarity measure must be chosen. The results from using the generalized Euclidean distance, δ_1, were less than satisfactory. It produced clusters whose internal variabilities were too large. A better dissimilarity measure for these data is the generalized measure, δ_3, introduced in Chapter 3. The results of the symbolic clustering using δ_3 are shown in Figure 8.19. Below the dendrogram are the career plots (gray trajectories) with the patterns of the four obtained leaders (black lines) superimposed. These plots are shown in the order of the clusters in the dendrogram. Reading from the left of Figure 8.19, we label these clusters of player careers C_1, C_2, C_3, and C_4. The largest of these clusters is C_4, with C_3 coming second in size, consistent with the EPL being an elite league attracting talented players. The smallest cluster is C_2 and is made up of the players who never made it at the highest level in England.

In interpreting these trajectories it is important to recall that rank and success are reverse coded: a low rank means a high success level for the players in terms of the ranks of the clubs for which they played and high ranks correspond to less success. Players in C_1 start, on average, at clubs with quite good success levels. Their success increases slightly for the first year or so but, over the remaining years, their club level success gradually drops. Players in C_2 fare far less well. On average, they start at clubs with a slightly lower success level than players in C_1. After this start, the success level of the clubs for which they played drops dramatically until about age 22. Thereafter, their declines in success levels are less precipitous. Even so, the players in this cluster are downwardly mobile throughout their careers. Together, the footballers in C_1 and C_2 were the less successful players who once played in the EPL. Of course, there are exceptions, as shown by the gray trajectories in the boxes at the bottom of Figure 8.19.

In contrast, the footballers in C_3 and C_4 had careers marked by upward mobility and greater success. Players in C_3, on average, started at low-level clubs (with high ranks) and dropped a little further in the next year. Thereafter, they had a steady rise in success until about age 27. Their overall success level dropped slightly until age 30. However, within this group, as shown by the gray trajectories, there was a sizeable number who remained at the top levels after age 22. Players in C_4 were even more successful. They started their careers at successful clubs, gradually moving to higher ranking clubs through their careers until about age 28. There was a slight decline in success thereafter. Player trajectories in C_4 were the most homogeneous: many played at the highest level throughout their careers. These were the elite footballers who played in the EPL during its first 15 seasons.

The EPL is a glamorous and highly competitive league. Players with enough talent and ambition from the UK and Ireland may seek to play in the EPL at the outset of their careers. Others come to the notice of clubs and are recruited to football academies or show promise at lower league levels. Through TV coverage, EPL games are watched in many countries. Top clubs have marketing strategies to improve their brands' visibility. This also raises awareness of the EPL. As a result, young players elsewhere may harbor ambitions to play in England's top league. Even so, players in other countries are far less likely to have such early ambitions. However, as they gain prominence in their own leagues, they become known to EPL clubs. Some are sought by them. Whatever their origins, many players reach the EPL. However, Figure 8.19 depicts a fundamental divide for players who reached and played in the EPL regarding overall success.

Footballers in clusters C_3 and C_4, in the main achieved the success they sought while those in C_1 and C_2 were less successful. When a league season starts, all clubs have ambitions and zero points. When the season ends there is an exact ordering with ranks from 1 to 20.

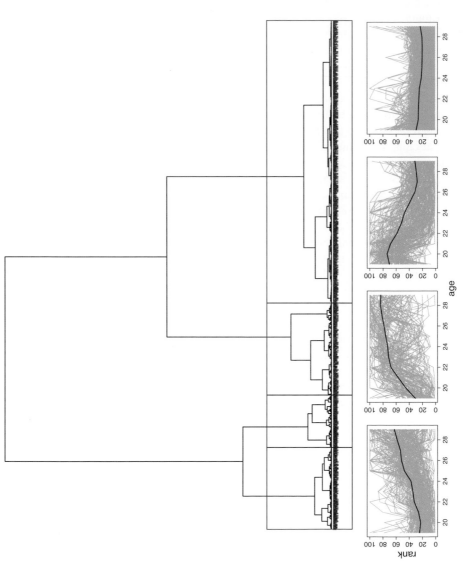

Figure 8.19 Dendrogram for hierarchical clustering of player careers.

The boxes below the dendrogram contain the player trajectories (in gray) and the leader trajectory (in black). They are ordered from left to right by C_1, C_2, C_3, and C_4.

As noted in Chapter 7, club success is rewarded handsomely while club failure is punished severely. Players at failing (relegated) clubs suffer consequences for club failures. In terms of talent, equally talented players could have been 'successful' or not depending on where they arrived in the EPL. Clubs vary in squad sizes, implying that not all players have equal playing opportunities. Players can fall out of favor with their managers for a variety of reasons. Changes in a club's perceived needs add further uncertainties for players. Some of the 'success' and 'failure' may be due to players not making the grade or being replaced at consistently successful clubs. Some of these differences regarding success is a system property: there must be groups of successful players and groups of less successful players regardless of talent or performance. Consistent with this system property, Proposition 7.9.12 (claiming the existence of fundamental player routes to the EPL) is supported. Of course, any clustering effort produces clusters. However, the four fundamental player migration profiles are genuinely different, interpretable, and meaningful.

We next provide some specific player career profiles for those who appeared in the EPL in two sections: 1) Home Country players, and 2) players from elsewhere.

8.6.2.1   Some Home Country profiles

Figure 8.20 shows player trajectories for four players from the Home Countries, one for each of the identified clusters. They are ordered in the same way as in Figure 8.19. Graham Kavanagh's (Ireland) trajectory is in C_1, Mike Basham's (England) is in C_2, Martin Ling's (England) is in C_3 and Jamie Scowcroft's (England) is in C_4. These players do not have a lot of name recognition in the wider football world except, perhaps, Kavanagh.[32] All four players had short spells in the EPL for the ages shown in Figure 8.20.

Kavanagh (first panel) started at Home County, a well-known Irish source for players in England before moving to Middlesbrough when the club was in the second tier. It was promoted to the EPL for the 1993/4 season (when there were 22 clubs in the league) but was immediately relegated. He and the club played two seasons in the second tier before again being promoted to the EPL. He then moved to Stoke City and Cardiff City, both clubs at lower levels.[33]

Along with all players in the C_2 cluster, Basham (second panel) represents the typical non-glamorous and downwardly mobile football career. He started playing at West Ham United, a club relegated from the EPL in his first season. Although the club was promoted one year later, Basham does not appear to have played a role in this success having been loaned to Colchester United, a club then in the fourth tier. He played for other lower tier clubs such as Swansea City (also recently promoted to the EPL after a remarkable rise from the fourth tier over a decade), Peterborough United, Barnet, and York City. Not shown in Figure 8.20 is a career ending in non-league football as he continued playing well into his 30s.

Ling (third panel) started his career with lowly Exeter City, then in the third tier. He then played most for Southend United and Swindon Town. With the latter club he enjoyed

---

[32] He played five times for Ireland's under-21 team before 16 full international appearances for Ireland. Scowcroft made five appearances for England's under-21 team.

[33] There is movement of some clubs between levels as both Stoke and Cardiff are now in the Premiership with Cardiff promoted to this league for the 2013/4 season. However, consistent with our argument concerning the closure at the top level, all of these clubs had played at the top level earlier.

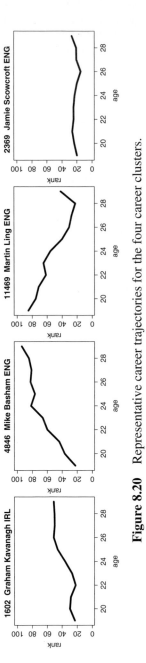

Figure 8.20 Representative career trajectories for the four career clusters.

promotion to the EPL but the club was immediately relegated. He finished with Leyton Orient, a fourth-tier club. Scowcroft was in the EPL with Ipswich Town before the club was relegated and spent two seasons in the EPL with Leicester City before it was relegated at the end of the 2002/3 season. He then also played for Leyton Orient before finishing his career in non-league football with his home-town club.

We note that the cluster analysis used players in the 19–30 age range. Many careers last longer than this and, as we have noted, there were salient features for the players we have discussed occurring after age 29 when playing careers for footballers generally go into decline.[34] While Scowcroft played in the EPL, his was a less successful career[35] than other players in C_4.

8.6.2.2   Some profiles of players from elsewhere

We pursue this further because of the considerable variation of the trajectories within each of these four clusters by considering some of the players within them, especially in C_4. Figure 8.21 shows the trajectories for eight *non-English* players clustered into C_4. The top panel has the trajectories of four players who had great success in all senses of the term, while the second panel shows the trajectories of another four top players.

The trajectory on the top left of the first panel is for Marc Overmars from the Netherlands. The start of his career includes his time at Ajax, when this club won both the Intercontinental Cup and the Champions League in 1995. (They lost the 1996 Champions League final only on penalties after a drawn game.) He moved to Arsenal in the EPL before moving on to FC Barcelona. His career was played primarily at the very top level of European club football before he returned to the Go Ahead Eagles, his youth club. He also played for his country 86 times. The top right panel has the trajectory for Juan Sebastián Verón who started his career with Estudiantes de la Plata before moving to Boca Juniors. Both were, and remain, top Argentinean clubs. He then played for Sampdoria, Parma, and Lazio, all clubs in the Italian Serie A. Manchester United recruited him from Lazio. Not shown in the panel is his move to Chelsea before he returned to his youth club in Argentina. He played for Argentina 73 times.

The bottom left panel has the trajectory for Peter Schmeichel who played in Denmark for most of his 20s before Manchester United recruited him at age 28. He played in goal for the Danish national team when it won the Euro1992 competition for national teams.[36] During his time at Brøndby the club won the UEFA Cup in 1991 and were the Danish top level champions for the previous four seasons. Not shown in the trajectory is the rest of his career at Manchester United where he played eight years, winning five EPL titles, three FA Cups, one League Cup, and the UEFA Champions League.[37] He played for Denmark 129 times. In part, his longevity at the top level was due to his playing in goal.

---

[34] We plan further analyses of the ending sequence of playing careers because very few end their careers at the top level.

[35] However, he played in over 500 games within League Football, a noteworthy achievement in itself.

[36] He was voted as 'The World's Best Goalkeeper' in 1992 and 1993. Source: http://www.iffhs.de/, accessed 13 September 2013.

[37] Manchester United won the 'Treble' (the EPL title, the FA Cup, and the UEFA Champions League) in his final season with the club.

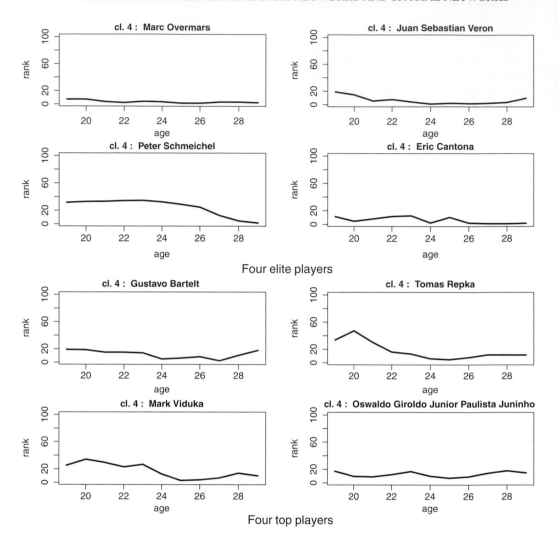

Figure 8.21   Trajectories to the EPL for players in cluster C_4.

The final trajectory in the top panel of Figure 8.21 is for Eric Cantona who played for top-level French clubs, three of which won Ligue 1 titles, in his early career before moving to Leeds United in 1992. Leeds was the last First Division champion prior to the formation of the EPL with Cantona being very instrumental in this club's success. He then moved to Manchester United which won four Premier League titles in five years. The club also did the League and FA Cup double twice during his stay. He played for France 45 times.

Also grouped in C_4 is David Beckham with a 'boring' flat profile given his entry into the EPL with Manchester United and long stay with the club in the top flight. He then moved to Real Madrid in Spain's La Liga,[38] before dropping levels to join the LA Galaxy for five

---

[38] Real Madrid and FC Barcelona have won far more La Liga titles than all other Spanish clubs.

years (although he had two loan spells with AC Milan). He played for England in 115 games. Phil Neville[39] had another flat profile with Manchester United before moving to Everton, a lesser club in the EPL. His tally of England caps is 59. Having such players clustered in C_4 adds to our confidence in the classification method we used.

We note the presence of Ajax, Arsenal, FC Barcelona, Inter Milano, Manchester United, Olympique de Marseille (one of Cantona's clubs), and Real Madrid in the club histories of the elite players we have just considered. All clubs belonged to the former G-14 described briefly in Chapter 7. The moves between them are consistent with the circulation of elite players among the elite European clubs.

Four additional players in C_4 merit some attention, even though they were less 'successful' in terms of the clubs for which they played and trophies won. The first trajectory in the second panel of Figure 8.21 is for Gustavo Bartelt who played for Lanús, a club in the top Argentine league, before moving to AS Roma at age 24. During his stay there, the club was the Serie A champion in 2001. However, he did not play many games in his five-year stay and was loaned twice. One loan was to Aston Villa where he made no first-team appearances.[40] From AS Roma he returned to play in Argentina but for lower-level clubs, especially after the period shown in the trajectory. He never played for Argentina.

Tomáš Řepka played in his native Czechoslovakia until age 24, mainly for Sparta Praha, a powerful club based in Prague. He moved to Fiorentina in Serie A before moving to West Ham United. He finished his career playing in the Czech Republic. He was capped once by Czechoslovakia and played 46 times for the Czech Republic. Mark Viduka, an Australian of Croatian descent, played for Dinamo Zagreb before moving to the Scottish club Celtic and then to Leeds United. Not shown in Figure 8.21 are subsequent stays in the EPL with Middlesbrough and Newcastle United. He played for the Australian national team 43 times (after a combined 38 appearances for the under 20 and under 23 national teams).

Juniñho started with the Brazilian club São Paulo at the end of one its golden eras before moving, surprisingly, to Middlesbrough in the EPL for a two-year stay. After the club was relegated, he moved on to Atlético Madrid in La Liga. During his time there he was loaned back to Middlesbrough for second EPL stay. He played 49 times for Brazil.

The age distribution of the arrivals of these eight players is: Juniñho (22); Overmars (24); Viduka (25); Bartelt (26); Cantona (26); Řepka (27); Schmeichel (28); and Verón (29). All but one (Bartelt) were successful in the EPL. The youngest arrival Juniñho is contrary to Proposition 7.9.10, a reminder that the systematic (and supportive) test of this hypothesis provided in Section 7.11 is the *only* sound way of assessing the proposition. Harris (2006) makes it clear that Middlesbrough went out of its way to settle their recruited player *and his family* in a heavily industrial and gloomy city, surely an alien place compared to Brazil. This created a deep bond between the player, the club, and the fans, one that seems alien to the current mercenary environment. We suspect that no such efforts were made for Bartelt at Aston Villa, given that the move was simply a loan.

These career trajectories suggest some issues including: 1) Are loans as a route to the EPL productive for foreign players and the receiving clubs? 2) Is player success due solely

---

[39] His brother, Gary Neville (capped 85 times for England), played his entire career at Manchester United and is not in our dataset which was defined to include only players who moved. So did Ryan Giggs with 64 caps for Wales. Both players would have been in C_4 given their career club trajectories.

[40] This cannot be viewed as a successful stay, regardless of the reasons for him never playing for the club. He was on loan to a Spanish club, Rayo Vallecano, when AS Roma won the Serie A title.

to playing performance on the field, or do clubs need to invest in more than the transfer fees and salaries expended in recruiting players in order to settle players? and 3) How are the uncertainties of recruiting young players resolved? Kuper and Szymanski (2009) suggest some answers based on the experience of Olympique Lyonnais, better known as Lyon. Some of their suggestions include: 1) 'Help your foreign recruits locate' (p. 71), consistent with Middlesbrough's treatment of Juniñho, 2) 'The best time to buy a player is when he is in his early twenties' (p. 69) something not tested directly in Section 7.11 but may be worthy of further attention, and 3) 'Sell a player if another club offers more than he is worth'[41] (p. 72).

Figure 8.22 shows the trajectories for another two players included for additional interest value. The one on the right is the trajectory for Dennis Bergkamp (also clustered in C_4), who also started his career at Ajax and was a part of the squad winning the 1986/7 European Cup Winners Cup.[42] He moved to Inter Milano in the Serie A, at face value, a good move. His time there was miserable during the so-called 'dark times' for this club, in a period of instability. His trajectory shows this drop to lower ranks with Inter's poor performances. While his quality as a player remained, turmoil at the club affected his form. Playing for a top club need not be beneficial if the club performs poorly. His move to Arsenal brought a dramatic increase in his footballing fortunes under Arsene Wenger. He was instrumental in helping Arsenal complete a domestic league and cup double in the 1996/7 season, a feat repeated in the 2001/2 season. The drop shown in Figure 8.22 for this player (the upwards spike in his trajectory) is a reminder of the importance of the standing of the clubs for which a footballer plays in terms of playing careers.

The second player trajectory, for another British Commonwealth player, was classified in C_1 and belongs to the Australian player Richard Johnson. It tells a very different story to the elite and top-level players of Figure 8.21 but similar to the career of Graham Kavanagh. After playing for Australian clubs, he joined Watford at age 17. The peak of his career came at age 26 when Watford was promoted to the EPL. Its stay in England's top league was short: it was immediately relegated. Thereafter, not shown in the trajectory, he played for clubs in the lower-level English leagues before returning to Australia for the inaugural season of the A-League. He finished his playing career with two New Zealand clubs[43] having even lower standing.

Every player appearing for clubs in national football leagues has a career history. All have some interest value as indicated by the career trajectories shown in Figures 8.20, 8.21, and 8.22. Indeed, these careers have been recorded in club documents, player websites, and general websites, especially Wikipedia. However, organized football is a *system organized around clubs*. In this sense, players are secondary even those with star status. We have focused on playing careers as a sequence of clubs for which players have played, despite

---

[41] This seems remarkably prescient, given the transfers that occurred at the end of the 2013 summer transfer window: the transfer of Gareth Bale for Tottenham Hotspur to Real Madrid for £85.3 million (setting a new world record transfer fee); the move of Mesut Özil from Real Madrid to Arsenal for £42.3 million, shattering Arsenal's transfer fee record and the move of Marouane Fellaini from Everton to Manchester United for £27.5 million.

[42] He was the top scorer in the Eredivisie (the top league in the Netherlands) from 1991 through 1993. As a result, he was voted Dutch Footballer of the Year in 1992 and 1993.

[43] Maintaining professional football clubs in New Zealand has been difficult. Johnson played for the New Zealand Knights and the Wellington Phoenix. The former club replaced the defunct Football Kingz, based in Auckland, before it also became defunct and was replaced by the Wellington Phoenix in the Hyundai A-League of Australia.

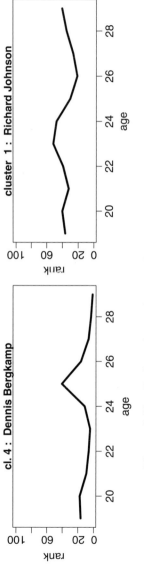

Figure 8.22 Trajectories to the EPL for players in different clusters.

the stellar careers of some individual players. Stars such as Peter Schmeichel, Eric Cantona, David Beckham, Ryan Giggs, Phil Neville, Mark Overmars, and Thierry Henry no doubt had an impact on the success of their clubs in the EPL. But clubs have impacts on players as the case of Dennis Bergkamp at Inter Milano shows.

8.6.3   Interpreting the clusters of player careers

The clustering of player careers was done *solely* in terms of the club's ranks (success) of players. We now consider position played and national origin with regard to the four clusters. See Figure 8.23. The top panel shows the distribution of positions played in each cluster. Reading from the left, the ordering is football playing positions: defenders, midfielders, forwards, and goalkeepers. The distributions show the same general pattern across the four clusters but with an increased presence of midfielders, joining forwards, in the most successful cluster, C_4. They play in the most noticed (glamor) positions because they contain most of the goal scorers.[44]

The lower panel of Figure 8.23 shows the distribution of players born in the Home Countries (on the right) and players born elsewhere (on the left) players in each of the four clusters of playing careers. In all four clusters, there are more local-born players than foreign-born players. Of greater interest is the higher number of foreign-born players in the two clusters of more successful players. Much of the alleged superiority of the EPL as a league is driven by the recruitment of foreign-born players.

We did examine the departures of foreign-born players from the EPL. In doing so, all loans were ignored because we were more interested in transfers. In the main, these players did not return. Of the 1182 foreign players only 138 came back after leaving England's top league. About 90% once they left stayed away. Even this number is slightly overstated because there were two type of returnees. Some left the EPL when their clubs were relegated and returned when their clubs were later promoted or went to other EPL clubs. Others went to clubs and leagues in other countries. When they returned to the EPL they did so from top clubs in these other leagues. When players departed England for lesser clubs and leagues, their chances of returning bordered on zero. Once foreign players returned to their homelands, as most did, the chance of returning was zero, in large part because their overall careers were coming to an end. Proposition 7.9.19 was confirmed.

8.7   Summary and conclusions

A partial summary of results were reported in Chapter 7 and will not be repeated here. However, additional support for Proposition 7.9.4 is in Section 8.3.1 and in Section 8.4, with further partial support for Proposition 7.9.3 in Section 8.3.1. Decisive support for Proposition 7.9.11 on the irrelevance of the proportion of non-national players for club success is in Section 8.1. By far, the dominant determinant of EPL club success is club salary levels. Of course, this argument is not new (Kuper and Szymanski 2009) but there is some evidence that the link between salary levels and success is getting tighter over time.

Proposition 7.9.12 claiming the presence of fundamental routes is confirmed in Section 8.6. There are four fundamental clusters of player careers. The largest cluster,

---

[44] In many databases there is the fetish for reporting goals scored by all positions. It reaches the absurdity of reporting goals by all goalkeepers, though few such players have scored.

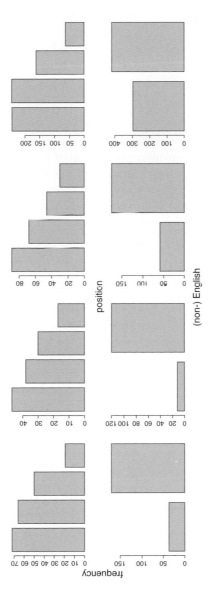

Figure 8.23 Bar plots for positions and origins of players by success cluster.

For positions the ordering from left to right is defenders, midfielders, forwards, and goalkeepers. The ordering from left for national origin is foreign-born and native-born.

C_4 on the right of Figure 8.19, has the elite and most successful players who start near the top, move to even higher clubs, and remain at high levels. Another group of successful players, C_3 in Figure 8.19, starts at lower levels but climbs the hierarchy of clubs without reaching the same levels of success. The remaining two clusters of player careers, C_1 and C_2 on the left of Figure 8.19 *are the smallest and feature the less successful players*. The minority of EPL players have been less successful and downwardly mobile through most of their careers. As the sizes of C_3 and C_4 are larger than the sizes of C_1 and C_2, these results provide some indirect support for Proposition 7.9.19 regarding departures from the EPL being permanent. Strong direct support for this proposition is in Section 8.4.

Section 8.3.2 provided some provisional support for Proposition 7.9.13 regarding club preferences for repeated transactions. Section 8.5 appeared to provide considerable support for this proposition with many multiple flows of players between clubs. However, the figures depicting the flows in this section were restricted to higher volumes. When the many single player flows were counted, they provided compelling evidence against Proposition 7.9.13. Section 8.3.2 provides considerable evidence in support of Proposition 7.9.15 on predation being the primary relation between clubs regarding player movements between clubs. We noted another structural feature in Section 8.5 with the sealing off of the EPL from the other English leagues, including the second level. These two features – predation by the top clubs and the overall sealing off of the EPL within England – were reinforcing mechanisms.

9

Mapping spatial diversity in the United States of America

We turn our attention to mapping *spatial* distributions for large diverse social systems. We focus on the continental United States (excluding Alaska and Hawaii) as an example having 3111 units (counties[1]) that differ greatly and are located in a vast ecologically diverse geographical space. The social diversity across these units can be viewed as being driven by the presence of different peoples, living in them over long periods of time and carving out places to live in different ecosystems. These areas also have quite different cultures, life styles and preferences contained within them. Attempting to understand the spatial distribution of the social, economic, political, and physical features of such a large system is a daunting task.

One way of doing this is to put a primary emphasis on *contiguous* areas and study these directly linked areas. Examples of this approach include 1) the work of the US Census Bureau with its broad regions, 2) a provocative book, *The Nine Nations of America*, by Garreau (1981), and 3) *American Nations: A History of the Eleven Rival Regional Cultures of North America*, an in-depth historical examination of the formation of the United States and its change over time by Woodard (2011). A second approach to understanding the spatial distribution of America's diversity is to focus on carefully selected indicators, summarize them, and describe areal units in terms of these summaries. Chinni and Gimpel (2010) did this in *Our Patchwork Nation: The Surprising Truth about the 'Real' America*. Together, these three books form the departure points for our depiction of US spatial diversity.

Geography plays two very different methodological roles in these two broad approaches to studying spatial distributions. Garreau (1981) and Woodard (2011) as well as the Census Bureau adopt an approach where delineating broad contiguous regions takes center stage as an organizing principle. In contrast, Chinni and Gimpel (2010) focus on the attributes of places while ignoring geography completely until they plot the summarized statistical data

---

[1] In Louisiana, parishes are, in essence, counties.

Understanding Large Temporal Networks and Spatial Networks: Exploration, Pattern Searching, Visualization and Network Evolution, First Edition. Vladimir Batagelj, Patrick Doreian, Anuška Ferligoj and Nataša Kejžar.
© 2014 John Wiley & Sons, Ltd. Published 2014 by John Wiley & Sons, Ltd.

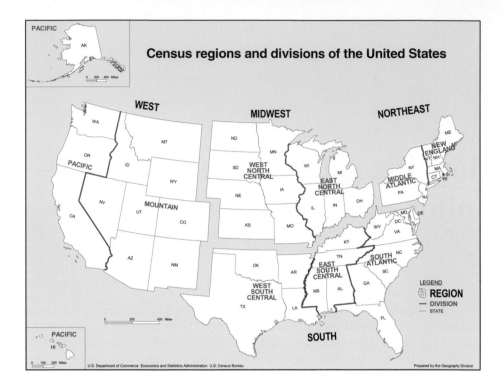

Figure 9.1   The standard census map of regions. *Source:* U.S. Government.

in geographic space. The former approach appears to emphasize spatial contiguity too much while the latter emphasizes it too little. We steer a middle course between these approaches. Our goal is clustering spatial units in terms of their attributes while, at the same time, taking into account the contiguity (adjacency) of the units. The methodology used builds on the constrained clustering approach developed by Ferligoj and Batagelj (1982, 1983).

9.1   Mapping nations as spatial units of the United States

Woodard (2011) makes a distinction between nations and states. A *state* is a sovereign political entity and, as such, it is eligible for membership in the United Nations (UN). Clearly, the United States of America is a state in this sense. In contrast, 'a *nation* is a group of people who share – or think they share – a common culture, ethnic origin, language, historical experience, artifacts and symbols,' Woodard (2011): 3. This conception allows him to claim the existence of eleven nations in North America. Earlier, Garreau (1981) suggested that there are nine such nations. Both claim 'The United States of America' is composed of nations and fragments of larger nations.[2] Both authors assert also that our understanding of the USA is too limited when we use only the Census Bureau's classification of units shown in Figure 9.1.

---

[2] Despite Woodard's distinction between nation and state, we henceforth use 'state' in the sense of the political units represented in the US Senate. The reason is simple: our attention is on mapping the spatial diversity of the USA, where the term 'state' has a long-accepted usage.

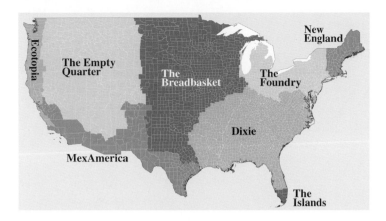

Figure 9.2   The Garreau's 'nine nations' of North America.

Garreau (1981) threw down a gauntlet when he wrote 'forget about the borders dividing the United States, Canada, and Mexico, those pale barriers so thoroughly porous to money, immigrants and ideas. Forget the bilge you were taught in sixth-grade geography about East and West, North and South, faint echoes of glorious pasts that never really existed save in sanitized textbooks,' (1981: 1). From his perspective, the map shown in Figure 9.1 is of little value for describing the spatial diversity of the United States. In its place, he defined nine distinct nations covering Canada, the contiguous 48 states of the USA, Northern Mexico, and some Caribbean islands.

These nine nations[3] are: New England; The Foundry; Dixie; The Islands; MexAmerica; Ecotopia; The Empty Quarter; The Breadbasket, and Quebec.[4] Drawing upon partial histories, local interviews, regional folk wisdom, and a wide variety of documents, he lays out the map shown in Figure 9.2.

Woodard (2011) presents an alternative set of eleven nations, for which we have drawn the schematic map shown in Figure 9.3. While he locates these nations in the same overall geographical space as Garreau, he also delves into the long histories of how these nations were formed and, equally important, the identities of their founders. The eleven nations he presents are: Yankeedom; New Netherland; The Midlands;[5] Tidewater; Greater Appalachia; The Deep South; New France (most of which is in Canada); El Norte; The Left Coast; The Far West, and The First Nation (also in Canada).

There are considerable overlaps between the two sets of nations. El Norte is essentially the same as MexAmerica. The Left Coast is the same as Ecotopia. Woodard's New Netherlands is very close to New York City (with a few more counties included), one of Garreau's aberrations. In Canada, New France includes Garreau's Quebec. However, Woodard separates

---

[3] He allows some exceptions which he calls 'aberrations': New York City, Washington DC, Hawaii, and Alaska. The first two are unique cities because of their economic and political roles respectively. We ignore Alaska and Hawaii because they are physically detached from the other states and separated by large distances.

[4] With Quebec in Canada this reduces the Nine Nations to eight in the USA. Even so, we use his title, Nine Nations, throughout this chapter.

[5] His Midlands extends into Canada. We ignore this part of the Midlands, given our focus on the continental USA.

Figure 9.3   The Woodard's 'eleven nations' of North America.

the Cajun parishes of Louisiana to put them into (a separated part of) New France. This was reasonable given the historical migration from the Canadian part of New France to the Cajun area in southern Louisiana. Also, Louisiana's legal code is based primarily on the French legal code rather than on English common law. Confining attention to the USA, The Far West is approximately The Empty Quarter.

There are clear differences in their two regional narratives. Woodard separates Greater Appalachia from Garreau's Dixie as a separate nation. He then expanded this nation to include areas in a southwesterly direction. Tidewater has also been removed from Dixie as a separate nation. Yankeedom includes the US part of Garreau's New England but is expanded to include large areas of Garreau's The Foundry. Neither The Foundry nor The Breadbasket is preserved within Woodard's scheme and Garreau does not include The First Nation.

Woodard's account is attentive to the founders[6] of nations and is more overtly political in the sense of looking at conflicts between pairs of nations and the alliances they formed over time. He sees Yankeedom and The Deep South as locked into a permanent struggle for the control of the US federal government. Our partial summary (in Chapter 6) of the political factions at the Philadelphia meeting where the US Constitution was written is fully consistent with this argument. Greater Appalachia and Tidewater were uneasy allies of the Deep South while The Left Coast and New Netherland sided with Yankeedom in this struggle, one that has lasted for centuries. There are more recent sharp differences. The Left Coast is seen as the political womb of the environmental movement while The Deep South is an integral part of the core of resistance to environmental regulation along with Texas. These conflicts point to deep divides across geographic space regarding culture, values, and attitudes towards government in these nations of the USA.

---

[6] For him this is particularly important for New Netherland, even though the early Dutch settlers were quickly driven out. This is important also for Tidewater, a nation he sees as 'founded by younger sons of Southern English gentry who aimed to reproduce the semi-Feudal manorial society of the English countryside where economic, political, and social affairs were run by and for the landed aristocrats,' Woodard (2011): 7. Part of his broader claim is that such early features of his nations still affect them. If he is correct, these nations will still have distinctive characteristics.

While depicting broad regions sharing some important characteristics makes considerable sense, the nations in both Woodard's and Garreau's accounts do not map neatly in the divisions used by the Census Bureau. At face value, both the Nine Nation and Eleven Nation partitions seem more persuasive than the somewhat arbitrary and simplified administrative definition of regions used by the Census. Garreau included maps of each of his nine nations containing boundaries, some of which were within the US states rather than simply between them. This allowed us to construct the map shown in Figure 9.2, which represents those nations (or parts of nations) wholly or partially contained in the Continental USA.

Woodard (2011) provides a county level map which we used to construct another regional map.[7] This is shown in Figure 9.3. Both Figures 9.2 and 9.3 are drawn in a format we use hereafter. Note also that there is again a distinction within Florida with a southern tip. However, this seems more like a blank within Woodard's narrative than part of a 'nation'.

We will attempt to 'reconstruct' these maps by clustering quantitative descriptions of areal units while, as noted above, constraining the clustering of these units by their contiguity. We note that large cities, and the counties containing them, are often very different from the adjacent counties surrounding them.

9.1.1   The counties of the United States

Thus far, we have been silent about the detailed (spatial) 'units' of the USA. At face value, choosing areal units for the USA is quite straightforward. Ideally, these units would be fairly homogenous. Local units such as neighborhoods come the closest to homogeneity. However, they tend to be unstable over time. Also, systematic comprehensive long-term data for these units over space do not exist. US states are very heterogeneous and, while systematic data are available for states over long historical periods, such data ignore many of the details informing Garreau's (1981) and Woodard's (2011) narratives. Chinni and Gimpel (2010) propose sound reasons for using counties to map the spatial diversity of the USA. Most importantly, county boundaries are stable and continuous through time, especially for the period after WWII that we consider. Second, data have been collected at the county level by organizations other than the Census Bureau. This broadens the range of available data. One explicit feature of the accounts of Garreau and of Woodard is that social life is *conditioned*, but not determined, by the contexts within which human lives are led. Another reason for using counties is that they each have a governing authority that sets broad common conditions for their inhabitants. While counties can be diverse in terms of the data about them, they are a reasonable compromise between homogeneous areas and using highly diverse US states as the areal units.

Chinni and Gimpel (2010) used factor analytic methods to develop a classification of the counties into 12 types. Their thumbnail descriptions of these type include: 1) Boom Towns (384 counties, 59.3 million people) having relatively wealthy locals, are growing rapidly and have increasing minority populations; 2) Campus and Careers (71 counties, 13.1 million people) have a younger population with many students and many post-grads; 3) Emptying Nests (250 counties, 12.1 million people) have older populations with plenty of boomers and retirees, but are less diverse; 4) Evangelical Epicenters (468 counties, 14.1 million people)

---

[7] This eliminated all guesswork about county membership in nations, in contrast to Garreau's maps where some boundaries between nations were located within states.

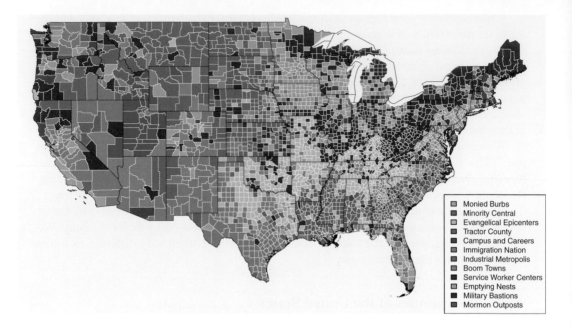

Figure 9.4   Chinni and Gimpel's patchwork nation map of the USA.

are full of young evangelicals and their families having overall lower median household incomes; 5) Immigration Nation (204 counties, 20.7 million people) is located primarily in the Southwest, is largely Hispanic, has lower income levels and higher than average poverty levels; 6) Industrial Metropolis (41 counties, 53.9 million people) is home to big industrial cities, is more densely packed, younger, and more diverse; 7) Military Bastions (55 counties, 8.4 million people) are middle-income locations near military bases that are full of soldiers and their families; 8) Minority Central (364 counties, 13.5 million people) has large populations of African Americans and Native Americans, has lower incomes and higher poverty levels; 9) Monied Burbs (286 counties, 69.1 million people) have higher household incomes, higher numbers of educated people, and are closely split in elections; 10) Mormon Outposts (44 counties, 1.7 million people) are located in the Mormon West and are often rural and sparsely populated; 11) Service Worker Centers (663 counties, 31 million people) includes tourist places, with midsized towns where many people live without employee benefits and live on the margins; and 12) Tractor Country (311 counties, 2.3 million people) are places that are mainly white, rural, and remote with sparse populations where farming and agribusiness form the economic foundations.

A reconstructed version of their 'patchwork' map of the continental USA is shown in Figure 9.4. While some counties of the same type are adjacent, joint membership in a type did not depend on adjacency. Clusters of a particular type can be distributed widely across the USA although there is some evidence of spatial clustering for some of the clusters. The result is a patchy map consistent with their well-chosen book title, 'Our Patchwork Nation'. The approach we take attempts to 'strengthen' the use of adjacency in space while 'weakening' the regionalism of Garreau, Woodard, and the Census. We use data on counties, focusing on broad demographic features including: age and population density; education levels; poverty levels; income distributions; labor force participation; employment indices; housing stocks; crime levels; land types; water availability; and political participation. The detailed descriptions of these variables are provided in Appendix A.5, with a listing of them provided in Table 9.2, on page 369. In terms of the categories used, these variables overlap considerably with those used by Chinni and Gimpel. However, the clustering method we use differs sharply from their use of factor analytic tools, which ignored geographic information when constructing their clusters of counties. We constrain our clustering by taking into account the spatial contiguity of counties both within and between states. As noted above, this places us between a strictly regional approach and classifying counties based solely on their attributes. In essence, we attempt to couple these two approaches by using a new efficient algorithm for clustering with relational constraints when network datasets are large.

9.2   Representing networks in space

The counties neighboring relation (network) is described by two subrelations: **Rook** – a pair of neighboring counties has a common border, and **Queen** – a pair of neighboring counties meet only in a common point (for example (Mason, Newaygo) and (Lake, Oceana)). To make the network connected we added the subrelation **Sea** that links counties on islands to the closest counties on the land.

Pajek allows the inclusion of coordinates for vertices (units). For spatially distributed units linked by adjaceny, the coordinates can be their latitude and longitude. An example of doing this is shown in Figure 9.5 where the communes of Slovenia are represented as vertices. Additionally, this figure includes an attribute in the form of the degree of the vertices which ranges from 1 to 13. Of some interest is the presence of cut-vertices and a bridge. The Pajek

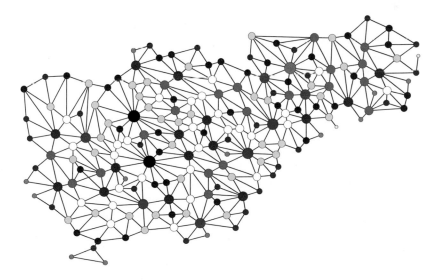

Figure 9.5   The network of Slovenian communes linked by adjacency in geographic space.
Colors represent degree sizes. Vertex colors and sizes represent degrees.

commands for obtaining this figure, after reading the network files with the coordinates, are:

```
Network/Create Partition/ Degree [Input]
Partition/Copy to Vector
Draw/Network + First Partition + First Vector
```

Figure 9.6 shows a similar outline map for the USA. There are bridges and cut-vertices in this map also. The range of degrees is 0 to 14. There is an isolate, San Juan Islands County, off the west coast in Washington State. Despite being almost contiguous, there are no bridges to the San Juan Islands. Instead of representing vertex degrees, the sizes of the vertices in Figure 9.6 represent the area sizes of counties. The layout on the left is suggestive of Garreau's use of the term Empty Quarter where county sizes are larger and populations are lower (except on the coast).

Drawing such networks in space with one variable is too simple when the areal units are characterized by many variables. A way of using these many variables, while considering the constraint of adjacency is described in the next section.

9.3   Clustering with a relational constraint

We denote the set of units (in our case US counties) forming a geographic area by $\mathcal{U}$. These units are described by attribute data (as a set of variables). Further, two units $u_i, u_j \in \mathcal{U}$ are adjacent if u_i and u_j share a *common* boundary. Adjacency defines geographical contiguity, a *binary relation* denoted by R. Our goal is to cluster these units in terms of their similarity across the variables while considering also R. This relation imposes constraints on the set of feasible clusterings $\mathbf{\Phi}$ where $\mathbf{\Phi} = \{\mathbf{C}$: each cluster $C \in \mathbf{C}$ induces a subgraph in the graph $(\mathcal{U}, R)$ of a required type of connectedness.$\}$

In general, different types of feasible clusterings can be defined for a binary relation R for directed networks. Some examples of types of relational constraints, Φ^i, are shown in Table 9.1. These constraints specify the type of connectivity that must be satisfied by the units within each of the clusters.

A set of units, $L \subseteq C$, is a *center* of cluster C in the clustering of type $\Phi^i(R)$ iff the subgraph induced by L is strongly connected and $R(L) \cap (C \setminus L) = \emptyset$.

The sets of feasible clusterings $\Phi^i(R)$ in Table 9.1 are linked as follows:

$\Phi^4(R) \subseteq \Phi^3(R) \subseteq \Phi^2(R) \subseteq \Phi^1(R)$ and $\Phi^4(R) \subseteq \Phi^5(R) \subseteq \Phi^2(R)$

Further, if the relation, R, is symmetric, then $\Phi^3(R) = \Phi^1(R)$. Also, if the relation, R, is an equivalence relation, then $\Phi^4(R) = \Phi^1(R)$.

Figure 9.7 shows some graphs illustrating the connectedness ideas in Table 9.1.

Given the contiguity relation for the US counties that we consider is symmetric, this means that Φ^1, Φ^2 and Φ^3 all imply the same constraint. Units within each cluster must be connected within the cluster. We present a general treatment of clustering with relational constraints applicable for all of the connectedness types in Table 9.1.

In general, we can use both hierarchical and local optimization methods for solving some types of problems with a relational constraint (Ferligoj and Batagelj 1983). The broad scheme for the *adjusted hierarchical algorithm* used here is presented in the following sections.

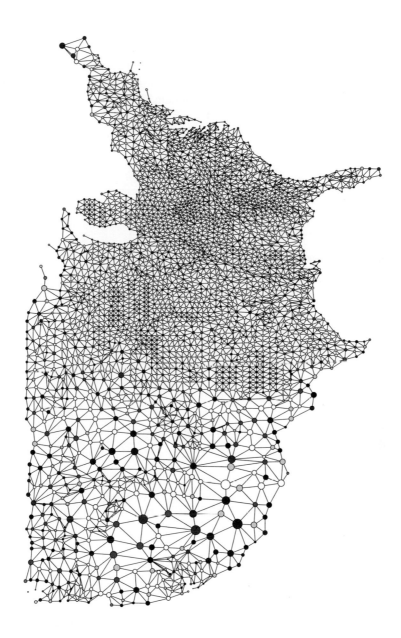

Figure 9.6   The network of US counties linked by adjacency in geographic space.
Vertex sizes represent areas of counties.

9.3.1   Conditions for hierarchical clustering methods

The set of all partitions is denoted by $\Pi(\mathcal{U})$ with the set of feasible clusterings denoted by $\boldsymbol{\Phi}$: $\boldsymbol{\Phi} \subseteq \Pi(\mathcal{U})$. $\boldsymbol{\Phi}$ determines the *feasibility predicate* $\Phi(\mathbf{C}) \equiv \mathbf{C} \in \boldsymbol{\Phi}$ defined on $\mathcal{P}(\mathcal{P}(\mathcal{U}) \setminus \{\emptyset\})$; and, conversely, $\boldsymbol{\Phi} \equiv \{\mathbf{C} \in \mathcal{P}(\mathcal{P}(\mathcal{U}) \setminus \{\emptyset\}) : \Phi(\mathbf{C})\}$.

Table 9.1 Some types of relational constraints.

Constraint	Type of connectedness
Φ^1	Weakly connected units
Φ^2	Weakly connected units that contain at most one center
Φ^3	Strongly connected units
Φ^4	Clique
Φ^5	The existence of a trail containing all the units of the cluster

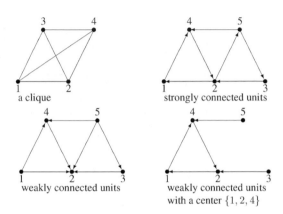

Figure 9.7 Examples of different types of connectedness.

In the set, $\mathbf{\Phi}$, the relation of *clustering inclusion*, $\sqsubseteq$, can be introduced by

$$\mathbf{C}_1 \sqsubseteq \mathbf{C}_2 \equiv \forall C_1 \in \mathbf{C}_1, C_2 \in \mathbf{C}_2 : C_1 \cap C_2 \in \{\emptyset, C_1\}.$$

When there is a clustering inclusion, the clustering $\mathbf{C}_1$ is a *refinement* of the clustering $\mathbf{C}_2$.

It is well known that $(\Pi(\mathcal{U}), \sqsubseteq)$ is a partially ordered set.[8] Because any subset of a partially ordered set is also partially ordered, from $\mathbf{\Phi} \subseteq \Pi(\mathcal{U})$, it follows that $(\mathbf{\Phi}, \sqsubseteq)$ is a partially ordered set.

The clustering inclusion determines two further relations on $\mathbf{\Phi}$:

1. $\mathbf{C}_1 \sqsubset \mathbf{C}_2 \equiv \mathbf{C}_1 \sqsubseteq \mathbf{C}_2 \wedge \mathbf{C}_1 \neq \mathbf{C}_2$ (this implies $\sqsubset$ is a strict inclusion)

2. $\mathbf{C}_1 \sqsubset\!\!\!\cdot\ \mathbf{C}_2 \equiv \mathbf{C}_1 \sqsubset \mathbf{C}_2 \wedge \neg \exists \mathbf{C} \in \mathbf{\Phi} : (\mathbf{C}_1 \sqsubset \mathbf{C} \wedge \mathbf{C} \sqsubset \mathbf{C}_2)$ (this defines $\sqsubset\!\!\!\cdot$ as a *predecessor* relation).

Henceforth, we assume that the set of feasible clusterings, $\mathbf{\Phi} \subseteq \Pi(\mathcal{U})$, satisfies all of the following conditions:

F1. $\mathbf{O} \equiv \{\{X\} : X \in \mathcal{U}\} \in \mathbf{\Phi}$. Intuitively, the partition where every element of $\mathcal{U}$ is a singleton in its own cluster is a feasible clustering.

---

[8] A stronger claim states that $(\Pi(\mathcal{U}), \sqsubseteq)$ is a semimodular lattice.

F2. The feasibility predicate, Φ, is *local* by having the form of $\Phi(\mathbf{C}) = \bigwedge_{C \in \mathbf{C}} \varphi(C)$ where $\varphi(C)$ is a predicate defined on $\mathcal{P}(\mathcal{U}) \setminus \{\emptyset\}$ (clusters).

The intuitive meaning of $\varphi(C)$ is: $\varphi(C) \equiv$ the cluster C is good.[9] Therefore, the locality condition can be read as stating that a good clustering $\mathbf{C} \in \Phi$ consists of good clusters.

F3. The predicate, Φ, has the property of *binary heredity* with respect to the *fusibility* predicate $\psi(C_1, C_2)$, that is,

$$C_1 \cap C_2 = \emptyset \wedge \varphi(C_1) \wedge \varphi(C_2) \wedge \psi(C_1, C_2) \Rightarrow \varphi(C_1 \cup C_2).$$

This condition means that in a good clustering, a fusion of two fusible clusters produces a good clustering.

F4. The predicate ψ is *compatible* with clustering inclusion $\sqsubseteq$, that is,

$$\forall \mathbf{C}_1, \mathbf{C}_2 \in \Phi : (\mathbf{C}_1 \sqsubset \mathbf{C}_2 \wedge \mathbf{C}_1 \setminus \mathbf{C}_2 = \{C_1, C_2\} \Rightarrow \psi(C_1, C_2) \vee \psi(C_2, C_1)).$$

F5. The *interpolation* property holds in Φ, that is, $\forall \mathbf{C}_1, \mathbf{C}_2 \in \Phi :$

$$(\mathbf{C}_1 \sqsubset \mathbf{C}_2 \wedge \mathrm{card}(\mathbf{C}_1) > \mathrm{card}(\mathbf{C}_2) + 1 \Rightarrow \exists \mathbf{C} \in \Phi : (\mathbf{C}_1 \sqsubset \mathbf{C} \wedge \mathbf{C} \sqsubset \mathbf{C}_2))$$

These conditions provide a framework in which the hierarchical methods can be applied also for constrained clustering problems. The number of clusters in a partition is denoted by k, the value of which is fixed ahead of time or selected during the clustering analysis. Partitions with higher values of k have lower values of the criterion function but may be too detailed for useful summaries. There is always a compromise between the quality of a solution and having an efficient and more useful summary. The set of feasible partitions with k clusters is a subset of all partitions with k clusters: $\Phi_k(\mathcal{U}) \subset \Pi_k(\mathcal{U})$.

In the case of ordinary clustering problems both the predicates $\varphi(C)$ and $\psi(C_p, C_q)$ are always true: all of the conditions F1–F5 are satisfied.

9.3.2   Clustering with a relational constraint

We return to the motivating problem of clustering spatial units in terms of their properties, while taking into account the spatial adjacency relation, R. In general, the units are described by attribute data and each unit X is mapped by a mapping, a, to its data values – description $[X]$. Formally: $a \colon \mathcal{U} \to [\mathcal{U}]$. Further, the units are related by a binary *relation* $R \subseteq \mathcal{U} \times \mathcal{U}$. Together, these determine the *relational data* $(\mathcal{U}, R, a)$ as defined by $\mathcal{U}$, R, and a.

Clustering the units according to the similarity of their descriptions while also considering the relation, R, imposes *constraints* on the set of feasible clusterings as defined in Section 9.3.1. This relational constraint can be stated formally in a general fashion:

$$\Phi(R) = \{\mathbf{C} \in P(\mathcal{U}) : \text{each cluster } C \in \mathbf{C} \text{ is a subgraph } (C, R \cap C \times C)$$
$$\text{in the graph, } (\mathcal{U}, R), \text{ of the required type of connectedness}\}$$

---

[9] The property 'good' is described formally in the next section by a predicate describing the specific constraint.

In the case of regional units, while they are clustered in terms of their attributes, this clustering is constrained by having the units also clustered into regions forming contiguous parts of the territory covered by all of the units in a cluster.

Table 9.1 listed five types of relational constraints. For each of them, the corresponding fusibility predicates can be stated. They are

$$\psi^1(C_1, C_2) \equiv \exists X \in C_1 \exists Y \in C_2 : (XRY \vee YRX)$$
$$\psi^2(C_1, C_2) \equiv (\exists X \in L_1 \exists Y \in C_2 : XRY) \vee (\exists X \in C_1 \exists Y \in L_2 : YRX)$$
$$\psi^3(C_1, C_2) \equiv (\exists X \in C_1 \exists Y \in C_2 : XRY) \wedge (\exists X \in C_1 \exists Y \in C_2 : YRX)$$
$$\psi^4(C_1, C_2) \equiv \forall X \in C_1 \forall Y \in C_2 : (XRY \wedge YRX)$$
$$\psi^5(C_1, C_2) \equiv (\exists X \in T_1 \exists Y \in I_2 : XRY) \vee (\exists X \in I_1 \exists Y \in T_2 : YRX).$$

For ψ^3 the property F5 does not hold if R is non-symmetric.

It is necessary to consider also the details of the clustering method used for clustering attributes in relation to fusing clusters. The fusibility condition $\psi(C_i, C_j)$ is equivalent to $C_i R C_j$ for the tolerant, leader, and strict clustering methods. Also, $\psi(C_i, C_j)$ is equivalent to $C_i R C_j \wedge C_j R C_i$ for the two-way clustering method.

In the original approach of Ferligoj and Batagelj (1983), a complete dissimilarity matrix (with dissimilarities for all pairs of units) was required. This worked well for small problems, but for a system of the size considered here, a much faster algorithm is necessary. One way of obtaining one is to consider *only* the dissimilarities between *adjacent* units in R. Let $(\mathcal{U}, R)$, where $R \subseteq \mathcal{U} \times \mathcal{U}$, be a graph and $\emptyset \subset S, T \subset \mathcal{U}$, and $S \cap T = \emptyset$ where S and T are clusters of elements in R. A *block* (c.f. the set of ties between units in a pair of positions in generalized blockmodeling Doreian et al. (2005) of the relation R for S and T is $R(S, T) = R \cap S \times T$. We denote the *symmetric closure* of the relation R by $\hat{R}$. It follows that $\hat{R} = R \cup R^{-1}$. Further, $\hat{R}(S, T) = \hat{R}(T, S)$. For all dissimilarities, d, between units, $s \in S$ and $t \in T$, we define a dissimilarity between clusters, $D(S, T)$, as

$$D(\{s\}, \{t\}) = \begin{cases} d(s, t) & s\hat{R}t \\ \infty & \text{otherwise.} \end{cases}$$

Here, we consider three options for extending D to clusters: the minumim, the maximum, and the average values which are defined as follows with the updating formula following the definition.

Minimum

$$D_{\min}(S, T) = \min_{(s,t) \in \hat{R}(S,T)} d(s, t)$$

$$D_{\min}(S, T_1 \cup T_2) = \min(D_{\min}(S, T_1), D_{\min}(S, T_2))$$

Maximum

$$D_{\max}(S, T) = \max_{(s,t) \in \hat{R}(S,T)} d(s, t)$$

$$D_{\max}(S, T_1 \cup T_2) = \max(D_{\max}(S, T_1), D_{\max}(S, T_2))$$

Average

Let $w : \mathcal{U} \to \mathbb{R}$ be a weight on units; for example $w(v) = 1$, for all $v \in \mathcal{U}$,

$$D_a(S,T) = \frac{1}{w(\hat{R}(S,T))} \sum_{(s,t)\in\hat{R}(S,T)} d(s,t)$$

$$w(\hat{R}(S,T_1 \cup T_2)) = w(\hat{R}(S,T_1)) + w(\hat{R}(S,T_2))$$

$$D_a(S,T_1 \cup T_2) = \frac{w(\hat{R}(S,T_1))}{w(\hat{R}(S,T_1 \cup T_2))} D_a(S,T_1) + \frac{w(\hat{R}(S,T_2))}{w(\hat{R}(S,T_1 \cup T_2))} D_a(S,T_2)$$

Let C_p, C_q and C_s be three clusters where $C_p, C_q, C_s \subset \mathcal{U}$ and let t be a threshold. The dissimilarity D has the *reducibility* property (Bruynooghe 1977) iff

$$D(C_p, C_q) \leq \min(D(C_p, C_s), D(C_q, C_s)) \Rightarrow$$

$$\min(D(C_p, C_s), D(C_q, C_s)) \leq D(C_p \cup C_q, C_s)$$

or equivalently

$$D(C_p, C_q) \leq t, \ D(C_p, C_s) \geq t, \ D(C_q, C_s) \geq t \Rightarrow D(C_p \cup C_q, C_s) \geq t.$$

The minimum, maximum, and average dissimilarities have the reducibility property.

9.3.3   An agglomerative method for relational constraints

An agglomerative clustering method proceeds by joining (fusing) clusters successively. Finer-grained clusters (and partitions) are contained within coarser-grained clusters (and partitions) to form a hierarchy of partitions. Both hierarchical and local optimization methods can be used for solving some types of problems with a relational constraint (Ferligoj and Batagelj 1983).

The pseudo-code of an *agglomerative algorithm* for doing this can be expressed in the following form:

```
1.      k := n; C(k) := {{X} : X ∈ 𝒰}; h_D({X}) = 0, X ∈ 𝒰;
2.      while ∃C_i, C_j ∈ C(k): (i ≠ j ∧ ψ(C_i, C_j)) repeat
2.1.        (C_p, C_q) := argmin{ D(C_i, C_j): i ≠ j ∧ ψ(C_i, C_j)};
2.2.        C := C_p ∪ C_q; h_D(C) = D(C_p, C_q); k := k − 1;
2.3.        C(k) := C(k + 1) \ {C_p, C_q} ∪ {C};
2.4.        determine D(C, C_s) for all C_s ∈ C(k)
2.5.        adjust the relation R as required by the clustering type
3.      m := k
```

For a selected strategy, in step 2.5, it is necessary to specify how to *adjust the relation after joining* C_p and C_q to form $C = C_p \cup C_q$, where C_s is any other cluster from $\mathbf{C}(k)$. The rules for the strict, leader, and tolerant strategies are presented in Figure 9.8.

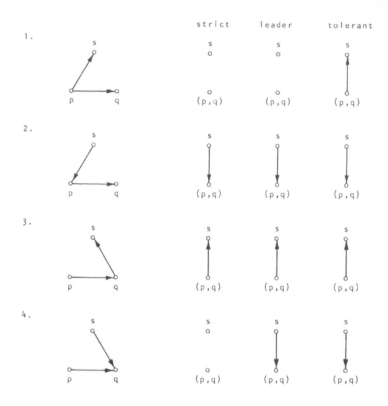

Figure 9.8   Relation adjustment rules for three clustering strategies.

For $p \in C_p$, $q \in C_q$, and $s \in C_s$, there are four configurations to consider when there is only an unreciprocated arc, $p \to q$, when p and q are both placed in C. These configurations are shown on the left in Figure 9.8. The arc (p, q) is always present with the four configurations distinguished by having only one of arcs (p, s), (s, p), (p, s), or (s, q). The adjusted ties between s and (p, q) for the three methods are shown on the right in three columns with each set of adjustments in a column.

The adjustment rules are much simpler for the two-way strategy with $p \leftrightarrow q$: the cluster C_s is linked with C only in the two cases presented in Figure 9.9.

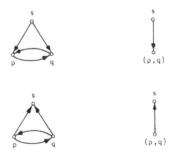

Figure 9.9   Relation adjustment rules for the two-way strategy.

The adjustment rules shown in Figures 9.8 and 9.9 work in the matched pairs: 1)Φ^1 and the tolerant method; 2) Φ^2 and the leader method; 3) Φ^4 and the two-way method; and 4) Φ^5 with the strict method.

The condition $\psi(C_i, C_j)$ is equivalent to $C_i RC_j$ for the tolerant, leader, and strict methods; and to $C_i RC_j \wedge C_j RC_i$ for two-way method.

Also, in step 2.4, it is necessary to update the *dissimilarities between clusters*. The foundations for doing this are described in Section 9.3.1. For the minimum, maximum, and average measures, in ordinary clustering cases, it is straightforward to derive the update formulas allowing us to determine the dissimilarity $D(C, C_s)$, where $C = C_p \cup C_q$, from the dissimilarities $D(C_p, C_s)$ and $D(C_q, C_s)$. Note that $T_1 \cap T_2 = \emptyset$.

9.3.4 Hierarchies

The agglomerative clustering procedure produces a series of feasible clusterings $\mathbf{C}(n)$, $\mathbf{C}(n - 1), \ldots, \mathbf{C}(m)$ with $\mathbf{C}(m) \in \mathrm{Max}\,\mathbf{\Phi}$ (the maximal elements for $\sqsubseteq$).

Their union, $\mathcal{T} = \bigcup_{k=m}^{n} \mathbf{C}(k)$, is called a *hierarchy* and has the property

$$\forall C_p, C_q \in \mathcal{T}: C_p \cap C_q \in \{\emptyset, C_p, C_q\}.$$

The set inclusion, $\subseteq$, is a *tree* or *hierarchical* order on $\mathcal{T}$. The hierarchy, $\mathcal{T}$, is *complete* iff $\mathcal{U} \in \mathcal{T}$.

For $W \subseteq \mathcal{U}$ we define the *smallest cluster* $C_\mathcal{T}(W)$ from $\mathcal{T}$ containing W by two conditions:

C1. $W \subseteq C_\mathcal{T}(W)$ and
C2. $\forall C \in \mathcal{T}: (W \subseteq C \Rightarrow C_\mathcal{T}(W) \subseteq C)$
        $C_\mathcal{T}$ is a *closure* on $\mathcal{T}$ with a special property

$$Z \notin C_\mathcal{T}(\{X, Y\}) \Rightarrow C_\mathcal{T}(\{X, Y\}) \subset C_\mathcal{T}(\{X, Y, Z\}) = C_\mathcal{T}(\{X, Z\}) = C_\mathcal{T}(\{Y, Z\})$$

We can assign the levels to the clusters of a hierarchy. A mapping $h : \mathcal{T} \to \mathbb{R}_0^+$ is a *level function* on $\mathcal{T}$ iff

L1. $\forall X \in \mathcal{U} : h(\{X\}) = 0$ and
L2. $C_p \subseteq C_q \Rightarrow h(C_p) \leq h(C_q)$.
A simple example of level function is $h(C) = \mathrm{card}(C) - 1$.

Every hierarchy/level function determines an ultrametric dissimilarity on $\mathcal{U}$:

$$\delta(X, Y) = h(C_\mathcal{T}(\{X, Y\}))$$

The converse is also true (see Dieudonné (1960)): Let d be an ultrametric on $\mathcal{U}$. Denote $\bar{B}(X, r) = \{Y \in \mathcal{U} : d(X, Y) \leq r\}$. Then for any given set $A \subset \mathbb{R}^+$ the set

$$\mathbf{C}(A) = \{\bar{B}(X, r) : X \in \mathcal{U}, r \in A\} \cup \{\{\mathcal{U}\}\} \cup \{\{X\} : X \in \mathcal{U}\}$$

is a complete hierarchy, and $h(C) = \mathrm{diam}(C)$ is a level function.

The pair $(\mathcal{T}, h)$ is called a *dendrogram* or a *clustering tree* because it can be visualized as a tree.

Theorem 9.1. (Monotonicity) *If a dissimilarity, D, has the reducibility property, then h_D is a level function.*

The minimum, maximum and average dissimilarities have the reducible property.

9.3.5   Fast agglomerative clustering algorithms

When the reducibility property is satisfied, the nearest neighbors network for a given network is preserved after joining the nearest clusters. This permits the development of very fast agglomerative hierarchical clustering procedures (Batagelj et al. 2008).

9.3.5.1   Nearest neighbors graphs

For a given dissimilarity, d, on the set of units, $\mathcal{U}$, and relational constraint, R, the *k nearest neighbors graph* $\mathbf{G}_{NN} = (\mathcal{U}, \mathcal{A})$ is defined as

$$(X, Y) \in \mathcal{A} \Leftrightarrow Y \text{ is selected among the } k \text{ nearest neighbors of X, and } X\hat{R}Y.$$

For $(X, Y) \in \mathcal{A}$ its value can be set to $w((X, Y)) = d(X, Y)$ to obtain a nearest neighbor network $\mathcal{N}_{NN} = (\mathcal{U}, \mathcal{A}, w)$.

When there are equidistant pairs of units, a choice is required about selecting from them. There are two options. One is to include all equidistant pairs of units in a nearest neighbor graph. The other option is to specify an additional selection rule for including only some of them. We denote the graph where all equidistant pairs are included by $\mathbf{G}_{NN}^*$. When the choice is to include only a single nearest neighbor, we denote this graph by $\mathbf{G}_{NN}$.

9.3.5.2   The structure and some properties of nearest neighbor graphs

Let $\mathcal{N}_{NN} = (\mathcal{U}, \mathcal{A}, w)$ be a nearest neighbor network. A pair of units $X, Y \in \mathcal{U}$ are *reciprocal nearest neighbors*, or RNNs, iff $(X, Y) \in \mathcal{A}$ and $(Y, X) \in \mathcal{A}$.

Suppose card($\mathcal{U}$) > 1 and R has no isolated units. Then in $\mathcal{N}$ (Murtagh 1985):

- every unit/vertex $X \in \mathcal{U}$ has outdeg(X) $\geq$ 1 (there are no isolated units), and

- along every walk the values of w are not increasing.

Using these two observations, some additional properties in $\mathcal{N}_{NN}^*$ are:

- all the values of w on a closed walk are the same and all its arcs are reciprocal (all arcs between units in a non-trivial strong component are reciprocal);

- every maximal (cannot be extended) elementary (no arc is repeated) walk ends in an RNNs pair;

- there exists at least one RNNs pair – corresponding to $\min_{X,Y \in \mathcal{U}, X \neq Y} d(X, Y)$.

9.3.5.3   The algorithm

Any network $\mathcal{N}_{NN}$ is a subnetwork of $\mathcal{N}_{NN}^*$. Its connected components are directed (acyclic) trees with a single RNNs pair in the root. It can be shown that if the clustering method has the reducibility property then the NN-chain remains a NN-chain also after the agglomeration of an RNNs pair.

Therefore, based on the nearest neighbor network, very efficient algorithms for agglomerative clustering for methods with the reducibility property can be built. In pseudo-code:

```
chain := [ ]; W := 𝒰;
while card(W) > 1 do begin
    if chain = [ ] then select an arbitrary unit X ∈ W else X := last(chain);
    grow a NN-chain from X until a pair (Y, Z) of RNNs are obtained;
    agglomerate Y and Z:
        T := Y ∪ Z; W := W \ {Y, Z} ∪ {T}; compute D(T, W), W ∈ W
end;
```

Its implementation for the methods minimum, maximum, and average is available in Pajek using the command

    Network/Create Hierarchy/Clustering with Relational Constraint/

When Clustering with Relational Constraint has been reached there are three options:

    Run
    Make Partition
    Extract Subtree as a Hierarchy

9.4 Data for constrained spatial clustering

As described in Appendix A.5, the data for the analyses presented here come from a variety of sources: 1) a spatial contiguity network file; 2) county shape files from the Global Administrative Areas website (http://gadm.org); 3) statistical data about US counties from the US Census Bureau (http://www.census.gov/support/USACdataDownloads.html); and 4) additional county data from the National Oceanic and Atmospheric Administration (http://www.noaa.gov). Combining information from these databases was complicated (see Appendix A.5). We selected county-level variables congruent with *Nine Nations of North America* and available[10] for 2000. Many variables were excluded due to collinearity. We selected 42 variables covering the broad areas listed in Section 9.1.1. These variables are listed in Table 9.2.

9.4.1 Discriminant analysis for Garreau's nations

Starting with the Nine Nations set of nations implies having a categorical variable (nations are the categories) into which counties are placed. It is reasonable to check whether these 42 variables can predict this categorical variable. If they cannot do so, then using them for the clustering with a relational constraint becomes suspect. But if they do have predictive value, this cluster analysis is justified. There are eight nations in the USA according to Garreau. These groups are known (see Figure 9.2) so no clustering (partitioning) using these variables was sought. Instead, discriminant function analysis was used to 'predict' membership in the categorical variable. In short, we sought to see if these variables can discriminate the categories (nations).

---

[10] There were some exceptions when data from nearby years were used.

Table 9.2 Final county variables used in constrained clustering analyses.

Variable	Variable name
1	Median Age
2	Civilian labor force unemployment rate
3	Percent of population with at least a Bachelor's (for people 25 and older)
4	Percentage change of housing units between 1990 and 2000
5	Median household income
6	Percentage of population living in poverty
7	Per capita personal income
8	Percentage change of population between 1990 and 2000
9	Population density (population per square mile)
10	Percentage of population female
11	Percentage of population Black or African American
12	Percentage of population Native American
13	Percentage of population Asian
14	Percentage of population Hispanic or Latino
15	Overall birth rate
16	Overall death rate
17	Infant mortality rate
18	Per capita water use
19	Percentage of population under 18
20	Percentage of population over 85
21	Percentage of land area in farms
22	Percentage of labor force employed in the construction industry
23	Percentage of labor force employed in manufacturing
24	Percentage of labor force employed in transportation and warehousing
25	Percentage of labor force employed in finance and insurance
26	Percentage of labor force employed in professions, science, and technology
27	Percentage of labor force employed in education and health
28	Percentage of population over 25 with less than a ninth grade education
29	Percentage of labor force employed in farming
30	Percentage of labor force employed in government (federal, state, and local)
31	Percentage of housing units owner occupied
32	Percentage of occupied housing units lacking indoor plumbing
33	Percentage of population that is rural
34	Percentage change of the urban population between 1990 and 2000
35	Change in per capita income between 1989 and 1999
36	Per capita ground water use
37	Percentage net domestic migration
38	Percentage of native population born in state of residence
39	Ratio in labor force: male to female
40	Percentage voting difference of Democrats over Republicans
41	Percentage public high school enrollment
42	Percentage change of poverty between 1995 and 2000

Table 9.3 Centroids for the first two discriminant functions.

Nation	First DF centroids	Nation	Second DF centroids
MexAmerica	−2.946	MexAmerica	−4.268
The Bread basket	−1.547	The Islands	−2.252
The Empty Quarter	−1.378	Ecotopia	−1.032
Ecotopia	−1.208	The Empty Quarter	−0.459
The Islands	−0.522	Dixie	−0.327
New England	−0.487	The Foundry	0.750
The Foundry	0.000	The Bread basket	0.927
Dixie	1.930	New England	1.040

Nations are ordered from the lowest to the highest centroid scores. DF stands for discriminant function.

A total of seven discriminant functions were established. Of these, the first four discriminant functions discriminate nations well. However, the last three discriminant functions fare less well.[11] All are interpretable. For an interpretation we use the loadings (coefficients for the linear combination defining a discriminant function) of the discriminant functions and centroids (means of the discriminant functions for each nation). The first discriminant function is defined primarily by the following variables (loadings with absolute values higher than 0.5 are shown in parentheses):

- race (percent Hispanic or Latino population (−0.850) and percent Black or African American population (−0.775))

- age (percent of the population under 18 (−0.775) and the percent of the population over 85 (−0.775))

- percentage of the population living in poverty (0.531)

- median household income

- education (percentage of population over 25 with less than a ninth grade education (0.553))

- percentage of land area in farms

- percentage of native population born in the state of residence

- civilian labor force unemployment rate

Table 9.3 contains the centroids for each of Garreau's nations in the USA. For the first discriminant function centroids (left panel), higher positive values occur for lower percentages of Hispanic or Latino populations, lower percentages of young people (below 18) in the population, higher percentages of Black or African Americans in the population, lower percentages of old people (above 85) in the population, higher percentages of people with lower education levels in the population, and higher percentages of the population living in poverty. MexAmerica typically has: the highest percentage of Hispanics or Latinos and the

---

[11] The eigenvalues for each discriminant function are: 35.33; 24.44; 19.04; 17.47; 8.66; 6.58; and 4.48.

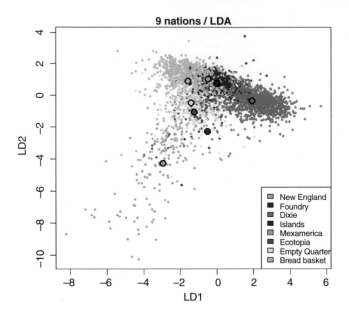

Figure 9.10   Plot of the first two discriminant functions with counties marked by nation colors.

The eight nations in the US are distinguished by colors. The black circles mark the locations of the centroids.

lowest percentage of Blacks or African Americans in the population; the highest percentage of the young people in the population; the highest percentage of older people in the population; the highest percentage of people with lower educational levels and the highest rate of poverty. The Breadbasket, The Empty Quarter, and Ecotopia follow. Dixie is at the opposite extreme compared to MexAmerica.

The second discriminant function is defined mostly by percentages of Hispanics or Latinos in the population (coefficient is -1.223) followed weakly by median household income (coefficient is 0.458).[12] High values on the second discriminant function are defined by lower values (higher percentages) of Hispanics or Latinos in the population and higher values (lower levels) for household income. The centroids for this function are in the right panel of Table 9.3. Without surprise, MexAmerica has by far the largest presence of the Hispanic or Latino population. The Islands and Ecotopia follow. New England is at the opposite extreme. Both The Foundary and The Breadbasket have a lower percentage of Hispanics and Latinos.

The values of the first two discriminant functions are plotted in Figure 9.10. Dixie (green points) has the most tightly clustered set of counties in this (arbitrary) two-dimensional space. The Breadbasket (gray points) also has tightly clustered counties. The Foundry (red points) comes next with overlaps in the spaces defined by Dixie and The Breadbasket. MexAmerica's counties, while not so tightly clustered in space, are located entirely in the lower left part of Figure 9.10. Most of The Empty Quarter counties (yellow points) occupy one region in the two-dimensional space. Ecotopia's counties (magenta points) are more scattered, overlapping the regions for The Empty Quarter and MexAmerica. Although New England (orange points)

---

[12] We included this for interpretational purposes even though the coefficient is below 0.5.

Table 9.4 Centroids for the second two discriminant functions.

Nation	Third DF centroids	Nation	Fourth DF centroids
New England	−2.109	New England	−1.559
Ecotopia	−2.048	The Foundry	−1.516
The Empty Quarter	−1.949	MexAmerica	−1.004
The Foundry	−0.937	Ecotopia	−0.134
The Islands	−0.645	Dixie	0.177
Dixie	0.239	The Bread basket	0.210
The Bread basket	0.753	The Islands	0.626
MexAmerica	0.771	Empty Quarter	1.817

Nations are ordered from the lowest to the highest centroid scores. DF stands for discriminant function.

and The Islands (purple points) are less clearly occupying a space, Figure 9.10 suggests that the eight nations in the USA are discriminated well in the two-dimensional space defined by the first two discriminant functions.

We consider next the third and fourth discriminant functions. High values on the third discriminant function are driven mainly by two variables (with coefficients in parentheses) high values for the percentage of land that is in farms (0.722) and high percentages of Blacks or African Americans in the population (0.560). The ordered centroids for this function are in the left panel of Table 9.4. Here, the most typical nations with the lowest percentage of farm land and the lowest percentage of Black or African American population are New England, Ecotopia, The Empty Quarter, and The Foundry. Dixie, The Bread Basket, and MexAmerica have the opposite characteristics but they are less extreme.

The fourth discriminant function is mostly defined by the following variables (coefficients in parentheses): median household income (−1.042); percent Hispanic or Latino population (−0.711); percent native population born in the state of residence (−0.615); percent people living in poverty (−0.583); percent of labor force employed in manufacturing (−0.553); and percent of the population 25 years old and over having a Bachelor's degree or higher (0.553). The centroids for this discriminant function are in the right panel of Table 9.4. The Empty Quarter has lower household income and also lower percentages of Hispanic or Latino populations, lower percentages of native-born people in the state of residence, lower percentages of the labor force employed in manufacturing and lower levels of higher education attainment. Both New England and The Foundry, overall, have the opposite characteristics. The remaining discriminant functions fare less well in discriminating nations, consistent with their low eigenvalues. Even so, the results of using them are interpretable. The centroids for these dimensions are shown in Table 9.5. The fifth discriminant function is defined by the following variables (with their coefficients in parentheses): percent of the population 25 years or over with less than a ninth grade education (0.709); percent of the population at least 25 years old with Bachelor's degree or higher (0.672); difference of voting Democratic over Republican (0.533); the civilian labor force unemployment rate (−0.519); percent of the native-born population in their state of residence (−0.515). The corresponding centroids are in the left panel of Table 9.5. Typically, New England and also The Islands (here, the southern tip of Florida) have higher percentages of both people with very low educational attainment and with very high educational attainment, higher levels of voting Democratic over Republican, lower unemployment rates, and lower percentages of native-born people

Table 9.5 Centroids for the last three discriminant functions.

Nation	Fifth centroids	Nation	Sixth centroids	Nation	Seventh centroids
Foundry	−0.5228	Ecotopia	−1.98706	Ecotopia	−0.36186
Empty Quarter	−0.2202	Islands	−1.38047	New England	−0.22428
MexAmerica	0.0208	Breadbasket	−0.04461	MexAmerica	−0.04802
Dixie	0.0373	Dixie	−0.00244	Dixie	−0.01740
Breadbasket	0.0435	Foundry	0.09901	Breadbasket	−0.00498
Ecotopia	0.1092	MexAmerica	0.20593	Empty Quarter	0.01749
Islands	1.9537	Empty Quarter	0.30408	Foundry	0.08922
New England	2.4103	New England	0.45916	Islands	4.23150

Nations are ordered from the lowest to the highest centroid scores.

living in the state where they were born. At the opposite end of this distribution is The Foundry and The Empty Quarter, but they are far less extreme in these characteristics.

The sixth discriminant function is mostly defined by the percentage of Asians in the population (-0.756) and the civilian unemployment rate (-0.508). The corresponding centroids are shown in the middle panel of Table 9.5. Ecotopia's counties, in general, are extreme in having the highest percentage of Asians in the population as well as the highest unemployment rate. The counties of the southern tip of Florida follow. The last discriminant function is defined by the following variables (coefficients in parentheses): percentages of the population 25 years old and over with Bachelor's degree or higher (-0.858); percentages of Asians in the population (-0.682); percentages of the population under 18 years (-0.645); and percentages of the workforce employed in the professions, sciences, and technology (0.537). Their centroids are shown in the right panel of Table 9.5. The Islands (southern tip of Florida) stands out as extreme having a high presence of people with low education attainment, lower percentages of Asians in the population, lower percentages of young people (under 18), and a higher percentage of people employed in the professions, sciences, and technology.

The discriminant analysis results using attributes of counties are consistent with much of the descriptive narrative of Garreau even though he described conditions in late 1970s and early 1980s while the data we use come mostly from 2000. This is consistent with the idea advanced by Woodard about change in these large regions being relatively slow.

Because, Woodard's account is based on historical records and contexts, we did not think it justifiable to construct a set of quantitative variables as we did for Garreau's account. This implied that we could not perform a discriminant analysis for his set of nations.

The analysis of this section did not utilize the adjacency of counties in geographic space. We turn next to consider clustering counties in terms of their attributes but constrained by their adjacency relations. We provide clustering results for different values of k as Garreau and Woodard have different numbers of nations in their characterizations of the USA.

9.5 Clustering the US counties with a spatial relational constraint

The selected 42 variables, discriminating the nations in terms of the counties placed in them, sets a foundation for the clustering with relational constraints. Some decisions were needed to obtain a clustering according to the selected variables and contiguity relation. This was done

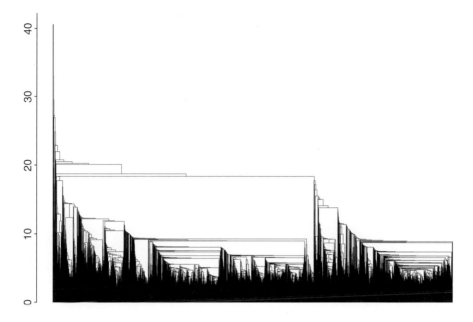

Figure 9.11   The first (height) dendrogram for clustering with a relational constraint.

in a straightforward fashion: 1) the variables were standardized; 2) the Euclidean distances between *directly linked* counties were computed; 3) the *tolerant* strategy (where each cluster induces a connected subgraph) was adopted; and 4) the *maximum* dissimilarity described in Section 9.3.2 was used for clustering. The first dendrogram for the constrained clustering is shown in Figure 9.11. It shows a very clear partition into two large clusters. This partition is shown in Figure 9.12. It is considered first in Section 9.5.1. We consider this partition first as an illustration before considering a partition with $k = 8$, the number of Garreau's nations in the USA. In drawing these images of partitioned counties, we fill areas for each of the counties to match the layouts of Figures 9.2 and 9.3.

9.5.1   The eight Garreau nations in the USA

Without surprise, the clear partition of the US counties into two clusters does not match fully Garreau's image of the USA. However, these two areas do have some correspondence with his eight nations in the USA. The area shaded in light gray includes New England completely, and virtually all of The Foundry. It includes most of Dixie, the southern tip of Florida, and most of MexAmerica. However, it includes some contiguous areas within The Breadbasket and some of The Empty Quarter. The area marked in dark gray contains all of Ecotopia, virtually all of The Empty Quarter, and most of The Breadbasket. There are contiguous parts of the dark gray region in two areas within Garreau's Dixie. The dark gray region is mainly in the northwest and the west coast of the USA. In brief, the light gray area contains the east coast, most of the south, and a swath of counties stretching to the west along the border.[13]

---

[13] There are some counties belonging to neither cluster because they are too dissimilar to their immediate neighbors. They are represented by the white shaded areas. Black areas represent the set of 'too small' clusters.

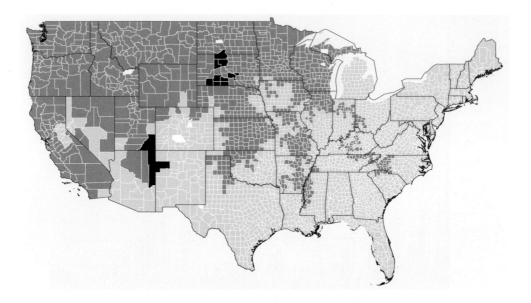

Figure 9.12   The spatial clustering with two clusters of counties.

In order to have a clearer image of a more detailed partition of counties than is possible to discern from Figure 9.11, ranks can be used instead as clustering levels to magnify differences between clusters. The resulting dendrogram is shown in Figure 9.13. There are choices regarding the number of clusters to select when reading a dendrogram. We opted to consider two in detail. One has eight clusters while the other has 15 clusters. The clustering into the largest clusters (with smaller clusters all shaded in plum) is shown in Figure 9.14. There are also several clusters with only one county (singletons) that are shaded in white. This partition was chosen to facilitate a better comparison with Garreau's eight nations within the USA.

While there are some correspondences between the partitions shown in this figure and Garreau's image of the USA, the match is not a very good one. There are eight clusters shown in colors. Again, there are some counties, shown in white, not belonging to any cluster because they are highly distinctive *in their immediate spatial context* of adjacent tables. This introduces some patchiness consistent with Chinni and Gimpel's analysis but inconsistent with Garreau's claimed homogeneous regions.

The large green area on the right (east) of the map does contain New England and the Foundry, which is only partially consistent with Garreau. The Foundry is more diverse than Garreau's narrative suggests. Further, the area marked in green also contains a large area to the south and west, incorporating much of Dixie and part of MexAmerica. It also stretches further west into The Breadbasket. The pink area on the right contains parts of the Foundry and Dixie and has a subarea more akin to the Breadbasket region.

While the large orange area contains much of the Breadbasket, it stretches west into The Empty Quarter and south into Dixie. The dark blue area in Alabama and Georgia is contained within Dixie. The disconnected region(s) marked in purple have parts in The Breadbasket, Dixie, and MexAmerica. Clearly, the monolithic Breadbasket and Dixie regions depicted by Garreau are much more diverse than his account suggests.

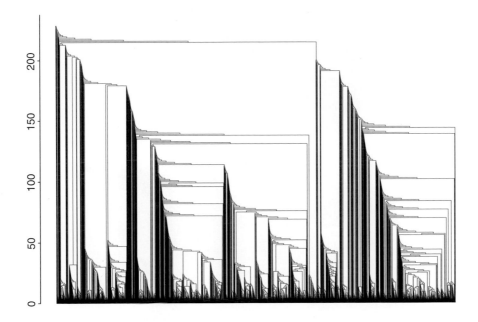

Figure 9.13   A second (rank) dendrogram for clustering with a relational constraint.

There is a light blue area depicted in the south western area of the USA. While it corresponds to the MexAmerica of Garreau's account, it does not have the green and purple areas in Texas but stretches further north into The Empty Quarter. Only part of Nevada is shown as being in the light blue area. The scattered counties marked in white were not classified into any of the eight clusters. They can be viewed as residuals.

Ecotopia is largely intact – the exception is the area around San Francisco – inside the area shown in yellow in Figure 9.14. However, the counties marked with yellow stretch east into The Empty Quarter and also into the northwestern part of MexAmerica. At a minimum, the area called the Empty Quarter is better seen as having at least two distinct parts. Much of MexAmerica is intact. Of Garreau's nations, only New England in the northeast of the USA and Ecotopia (with the exception of the area around San Francisco) on the west coast of the USA remain intact when clustering using relational constraints is used.

These results suggest two things: 1) with the exception of New England and Ecotopia, the monolithic nations in Garreau's are far more heterogeneous than his simple classification into nations suggests, and 2) combining systematic county attribute data with attention to geographic adjacency makes a huge difference in the classification of counties. It is highly unlikely that the changes in the USA from 1980 to 2000 are sufficient to account for these differences. The results argue against both the large aggregations of counties into the nations of Garreau and the patchwork account of Chinni and Gimpel, by returning a partition between the two. Of course, this is not surprising. Garreau's account is rich in anecdotal detail and items of local history. However, it is unlikely that Garreau visited all 3111 counties of the USA. The result appears to be the inclusion of some counties into nations using details of other counties in roughly the same area. One surprise with the discriminant analyses was

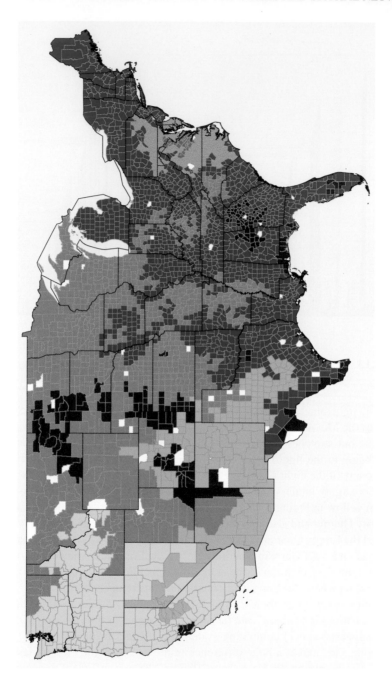

Figure 9.14   The spatial clustering with eight clusters of counties.

how few of the 42 variables drove the seven discriminant functions. It seems that including geographic contiguity, in addition to it playing a direct role in the partitioning, brought some of the other county attributes into play. In contrast to Chinni and Gimpel's analysis, the areas delimited here have greater geographic coherence. The result is fewer regions rather than more patches.

9.5.2   The ten Woodard nations in the USA

We consider a finer-grained partition than the one shown in Figure 9.14. From the dendrogram in Figure 9.13 we identified the largest 15 clusters. The dendrogram was obtained from the dendrogram for maximal method using the rank transformation. The clusters are shown in Figure 9.15 with the primary goal being a closer look at Woodard's set of nations. In doing so, we also take another look at Garreau's nations. Again, there are counties, colored white, sufficiently distinctive that they are not clustered. These include major metropolitan areas (always very distinct from the areas around them) and university towns. Overall, there is only partial confirmation of the two sets of nations depicted by Garreau and by Woodard – along with some surprises.

In contrast to Figure 9.14, Garreau's New England nation is identified with some counties of The Foundry joining it in the red region in the northeast of Figure 9.15. This is contained within Woodard's Yankeedom. Much of The Foundry is found together in the yellow region of Figure 9.15. This seems more consistent with Garreau than with Woodard, especially with the latter's nation of The Midlands having 'disappeared'. In the east, the alleged nation is completely within the yellow region and inside The Foundry, which is better captured in this partition with 15 clusters than in the partition with eight clusters. However, the yellow region extends south to almost the northern border of Florida, consistent with the earlier finding of a heterogeneous Dixie.

The orange area on the East Coast seems very consistent with Woodard's Tidewater, something missed completely in Garreau's depiction of the USA. Surrounding Tidewater is a larger region shown in dark green, suggesting that Tidewater extends further inland and north than Woodard suggests. However, he identified a seemingly coherent nation missed by Garreau even though it may have two distinct parts. Woodard's Greater Appalachia does not survive either. Neither does Garreau's Dixie nor Woodard's Deep South as single monolithic entities. Also not present is Woodard's New France. These Cajun parishes are a part of the jade green region, one that stretches to the north but is interrupted by other regions. This connected region (having the appearance some of the gerrymandered[14] voting districts of the USA!) includes a widely dispersed set of counties. There are two small regions within either Garreau's Dixie or Woodard's Deep South and Greater Appalachia. One is shown in dark blue with the other being brick red. The latter includes parts of the Ozarks in the western part of Greater Appalachia.

The straggly region marked in purple seems rather bizarre. It covers parts of the South (be it Dixie or Greater Appalachia), stretches into the Breadbasket and then west into The

---

[14] Gerrymandering is the drawing of the boundaries of electoral districts in a way that gives one party an unfair advantage over its rivals. It has a long history in the USA. Indeed, the term was coined in 1812 because the governor of Massachusetts, whose surname was Gerry, headed an administration that constructed senatorial boundaries to favor the Democratic-Republicans over the Federalists. (The terms Democrats and Republicans have very different meanings today.)

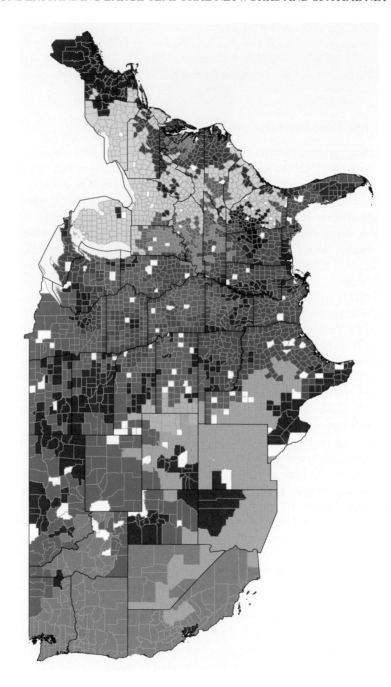

Figure 9.15   The spatial clustering with the 15 largest clusters of counties.

Empty Quarter. There is a cornflower blue region that is a part of Garreau's Breadbasket or Woodard's Yankeedom, suggesting that neither of the monolithic areas in these author's characterization of the USA is appropriate for this part of the USA. The patches in light blue and light brown, as well as the white area, reinforce this impression.

Looking at the west coast of the USA, both Garreau's Ecotopia and Woodard's The Left Coast are in the violet region. However, this region extends well into Garreau's Empty Quarter and Woodard's Far West (which overlap heavily). Neither of these areas is anywhere near as homogeneous as these authors suggest. Parts of Garreau's MexAmerica and Woodard's El Norte are in the pink region of Figure 9.15 but not as cleanly as their depictions of nations imply. In short, while there is great appeal in the notion of mapping out separate nations comprising the USA, the country does not conform to these tidy depictions.

9.6 Summary

This chapter accomplished three things. One was a sustained examination of the spatial distribution of administratively defined units, for which data can be assembled, in terms of their 'types' in geographical space. This issue was couched in a substantive context defined by two distinctive approaches to mapping spatial diversity. Garreau (1981) and Woodard (2011) exemplify the approach of delineating large regions of national systems as homogeneous entities regarding their long-term histories and their social and political composition. The other approach is exemplified by Chinni and Gimpel's (2010) examination of the statistical attributes but ignoring geographical space during the statistical analyses. Having completed their factor analyses to create types of units, they did plot their 12 identified types of spatial units in space to reveal a patchwork nation. Neither approach, by itself, suffices because each approach ignores a *crucial* feature regarding spatial distributions of areal units. We identified a middle ground between these approaches.

The second contribution was the creation, and programming, of a fast algorithm to cluster spatial units in terms of their statistical attributes while constraining the resulting partitions by a spatial relation defined by adjacency of spatial units. This general method, termed clustering with a relational constraint, combines statistical data on units with relational data for them.

The third contribution was to combine the substantive concern with the new methodological development to produce new partitions of spatial units (of the USA in this analysis) and to characterize their distribution in geographic space. Our results, without great surprise, are located between the results of the two broad approaches motivating our study. While the spatial distribution of county types shares *some* similarities with the nations identified by Garreau and Woodard, there are considerable differences. There is far greater variation within the allegedly homogeneous nations than the approach of mapping large swaths of terrain. This variation within 'nations' merits more attention, some of which is available in the detailed narratives of Garreau and Woodard. Our results also show *some* consistencies with the patchwork nation of Chinni and Gimpel but with far less patchiness. In general, this lays the foundations for a more nuanced interpretation of the spatial distribution of diversity.

Finally, clustering with a relational constraint is a method applicable to many social network situations and is not confined solely to networks distributed in geographic space. While the formal developments outlined in this chapter were motivated by studying a large network that happened to be distributed in space, they have great potential utility for studying social networks, larger or smaller.

10

On studying large networks

Our interest in preparing this book was driven by a commitment to coupling substance and methods in a quest to *understand* large networks and the network processes operating both in, and on, them. This required, for us, the analysis of sound empirical data, an essential feature of scientific research.

Substance without methods amounts to armchair philosophy when the systematic analysis of relevant data plays little or no role. While ideas 'in the air' can have a useful place, we are aware of 'ideas being a dime a dozen', especially when they are not critically examined by considering systematic empirical information. Similarly, using methods without substance seems an empty endeavor when understanding network phenomena is the central concern. While it may be pointless to compare these two overly limited approaches, we think that methods have potentially greater value for understanding the dynamics of temporal networks and spatial networks than speculations not based a sound scrutiny of empirical evidence. Yes, we are biased regarding this because we are more interested in doing social science than participating in social studies. As we argue in Chapter 1, integrating substance, technique, and sound data is critical for understanding social phenomena. This implies a need to choose very carefully with regard to all components of this triumvirate.

10.1 Substance

Substance, for most researchers, is driven by a deep interest in understanding certain phenomena. As such, it can be regarded as an individual choice. This choice can stem from a burning desire to understand something specific, or interacting with persuasive others who have such an interest, or by being employed in an organization where such an interest is present. While such interests can be instrumental, for example earning a living or making a profit, our concern was driven by, and created by, a passion that was both general and substantively specific. Parenthetically, it is fair to claim that we (the authors of this volume) can get very interested in many social network questions: an abiding interest in network

Understanding Large Temporal Networks and Spatial Networks: Exploration, Pattern Searching, Visualization and Network Evolution, First Edition. Vladimir Batagelj, Patrick Doreian, Anuška Ferligoj and Nataša Kejžar.

phenomena is a part of our collective soul. Even so, specific choices regarding substantive issues were required.

Given that we have all contributed to the SNA literature, we were drawn to the citation literature of our field. As a result, we were curious about 'our' literature. Studying it was natural for us. However, it meant looking at a much wider network literature beyond the patches with which we were familiar. We wanted to look at the field as a whole, as a substantive concern, beyond a narrow conception based on a small set of specific topics.

At face value, patents could be viewed as utterly boring and narrowly focused on technical documents belonging to an arcane and remote world. Yet most social change within societies can be seen as driven by technological change, at least in the physical features forming the basic infrastructure of societies. This societal infrastructure is made up of invented things, some of which have great value. Patents, and the protections they provide to inventors, are crucial for providing coherence, and are integral to this kind of change. As a result, the patent citation network was added to the substantive domains interesting us.

The US Supreme Court, by design, even though the protections it provides are less than obvious to most people, is a core US institution: the decisions made by this court affect the economic, political, and social life of this particular nation. No doubt this is true for all other countries around the globe with such a court. Citations between Supreme Court decisions, and the role of such a court in national affairs, were added to the substantive interests pursued in this book.

These three examples are all citation networks in which documents cite other documents. To broaden the scope of temporal networks that we considered, the movement of football (soccer) players around the goal to play their game was included as a substantive problem area. While the players are the individuals about whom many discussions about football are centered, they play football within organized social contexts. From a substantive perspective, these institutionalized and global contexts of football are more interesting than the game itself.

The primary units of these contexts are the clubs for which footballers play. Our focus, again driven by a passionate interest, was on the organization of football. The networks formed by players moving across the globe link clubs (which negotiate the huge majority of player moves) and, secondarily, the countries between which they migrate. These temporal networks became our focus in the hope of using them to better understand national football league systems, the role of national concerns of key actors, and the international organization of 'the beautiful game.'

We included spatial networks for substantive reasons also. As outlined in Chapters 1 and 9, there are rival approaches to studying the spatial distribution of social attributes in space for areal units. For a country the size of the USA, there are many very different ecosystems occupied by people having different cultures with distinctive histories, values, and political orientations. It is natural to try to understand these spatial distributions in order to establish a coherent account of them. Two books, *The Nine Nations of America* and *American Nations: A History of the Eleven Rival Regional Cultures of North America*, along with the bureaucratic approach of the US Census Bureau, emphasize the delineation of large *contiguous* areas within a vast space. Another approach, exemplified by *Our Patchwork Nation: The Surprising Truth about the 'Real' America*, completely ignores geographic space to focus on analyzing the statistical properties of areal units so as to cluster them. Each approach had immense appeal for us, especially the grouping of contiguous areas not conforming to the arbitrary dictates of the Census Bureau. Yet we were troubled by each of them ignoring something crucial about the phenomena studied. It was, for us, an obvious quest to think about combining the two approaches in a systematic fashion.

For both the organization of football and the geographic distribution of social attributes in space, there is a considerable amount of conventional wisdom, about which we are naturally skeptical, especially in the context of English football. This became an additional source of motivation, along with a deep desire to see if studying networks could add anything of value to our understanding of social phenomena when most discussions of them are devoid of network analytic ideas.

We are neutral about whether social science is best done in an exploratory fashion or in ways considered as deductive. More accurately, we think such contrasts are ill-posed and have little value. As a result, both induction and deduction were used in this book, but with a bias more in favor of induction. We view the value of each broad approach as being conditioned by the state of our knowledge. The more we know, or think we know, the better the chances of taking a deductive approach – provided the goal is to critically test this knowledge. Of the substantive topics we considered, only one permitted a deductive approach at the outset. However, when we look further into the other substantive domains that we have examined in this book, we believe that we have created the foundations for being more deductive when considering them in the future.

10.2   Methods, techniques, and algorithms

The formal core of the book is contained in Chapters 2 and 3, where we presented two broad sets of ideas: 1) foundations with network analytic concepts, and 2) a coherent framework for analyzing temporal networks. Methods used in multiple chapters were presented there. Additional formal tools and methods were presented in Chapter 9 for the analysis of spatial networks.

We do have a particular fondness for a set of methods, a fondness that may have shown throughout the book. We mention them here. Given the historical origins of main path analysis, we like this method because it answers a particular set of questions well. However, our results showed it to be quite limited. Another very useful method is the establishment of islands, particularly line islands, a method used for studying all of the citation networks. While it is useful for delineating coherent parts of networks, we show that there are other methods that can be used for similar purposes. Clustering symbolic data had great value for studying both the patent citation network and a quite different problem relating to the football data. Finally, clustering with a relational constraint was crucial for the analysis of the large spatial network we considered.

While network analytic techniques are considered often as highly portable, and therefore are assumed to be relevant for all networks, we have considerable skepticism on that front as well. In part, this is driven by having to think about substance, substantive issues, and the appropriateness (or not) of specific techniques for seeking substantive understandings of empirical phenomena. In short, we do not think there are cookie cutter methods applicable to every network problem. Not every problem can be solved by using a hammer!

We think there are two potentially troublesome – at least for us – trends in contemporary network analysis. One is the presumption of universal portability, with attempts to use any technique simply because it happens to be in fashion. One example that we encountered took the form of trying to fit a scale-free distribution, developed usefully for a class of large networks, to the degree distributions of very small networks, despite such a usage making no sense. The second is the presumption that 'my method' is the best of all methods. While this is a choice each researcher can make, it often comes with attempts to enforce an orthodoxy regarding how networks 'should' be analyzed. These two strands of thought strike us as unproductive, and so we resist them.

In large part, this resistance stems also from realizing the importance of matching substance with methods in the pursuit of understanding. The tools we used were coupled in ways we thought appropriate for the substantive issues engaging our interest. While they worked well for us, we are not claiming they are the only useful tools. Nor do we claim that we could have obtained our results only be using them. Other tools could be equally useful, perhaps even better for our purposes, and our hope is that different tools provide complementary results and insights for understanding social phenomena. If so, it seems best to couple them rather than confine attention to one, and only one, set of preferred tools.

Another relevant motivation for the book stemmed from the need for fast, efficient algorithms that are practical for analyzing large networks. Without such new algorithms, none of what we have done would have been possible. Most of the algorithms supporting the methods for which we have great fondness were programmed with efficiency as a primary design criterion.

During our partial survey of other approaches to large networks in totally different areas, we were struck by how they were designed with specific substantive problems in mind. No doubt, they are also portable, but it does not follow that they must be used to study the kinds of networks we considered, nor that the methods we have developed and used are appropriate for other substantive domains. Portability is fine and we greatly like to see ideas flowing between diverse areas. However, neither the blind importation of methods nor the attempt to enforce an unthinking orthodoxy have any appeal. Matching of methods to substantive problems is crucial. Given this, we do think that some of the methods we surveyed from other domains will have great value when adapted for studying social network analytic problems and we plan to include them in future work.

The methods we used throughout this book for network analyses were programmed as integral parts of Pajek. As we noted earlier, Pajek served our purposes very well, especially for large networks. Of course, we hope that our efforts were persuasive regarding this. Despite our commitment to Pajek, we have no interest in claiming that it is the only relevant program for studying large networks. Again, combining results from using different methods implemented in diverse programs has great merit.

10.3 Network data

Currently, there are many large network datasets available. Indeed, they have become abundant – perhaps too abundant. While potentially useful, as many are, the value of such datasets rests upon 1) how they were conceptualized, 2) how key concepts were operationalized, and 3) how the data were collected. If any part of the conceptualization, the operationalization, or the data collection is fatally flawed then the resulting data are likely to have little or no value when understanding network phenomena is the goal. The risk seems particularly acute for data collected electronically and remotely. We add a fourth crucial requirement, that the boundary specification problem be solved either completely or very well. These four criteria imply, at a minimum, that data have to be collected and selected with great care.

Faced with this general selection problem we used two different approaches for obtaining the data for our empirical analyses. One was to use already collected data appearing to pass muster on all four of the data adequacy criteria that we think are essential for making such choices. The other was to collect our own data.

10.3.1.1   The network datasets we used

The patent data we used, covering 30 years, were the cleanest network data among the citation networks we studied. While this seemed likely at the outset, as we studied the patent application process, we realized how clean these data really were. Similarly, the Supreme Court data appeared to be well defined and were extracted from the complete record of all Supreme Court decisions handed down in a period lasting more than 200 years. The SNA citation data were the least clean of the citation datasets as we noted in our discussion of cleaning them. For both the Supreme Court data and the SNA citation data, cleaning them moved these datasets to a form satisfying all four selection criteria.

At face value, the football network data are quite bizarre, given that the initial motivation for collecting them was whimsy and idle curiosity. The initial, seemingly naive, questions were: 1) Where did the players come from? 2) How did they get here (the English Premier league)? and 3) Where did they go? The effort of collecting these data was monumental, as described in Appendix A.4, because there was no one reliable data source for extant data. Cleaning these data, once they had been assembled from a huge number of data sources, bordered on being a nightmare.

By focusing on England's top football league, the sampling design was rather restrictive. Yet the idiosyncratic characteristics of these data, coupled to our reading about football as it has been organized in England, are instructive for considering the interrelation of substance, method, and data. Always, the questions asked have to be questions answerable when using the data at hand. Our questions centered on the organization of football in England, especially regarding the importation of players from elsewhere. As a result, the complete data we assembled for a specific 15-year period were ideal for pursuing these questions after they were expressed in the form of explicit hypotheses.

The areal unit adjacency network data for the counties of the United States were available from a single source. They were the only 'easy' data to obtain and came from a highly trusted source.

There is a limitation inherent in considering separate large networks representing very different substantive domains regarding the generalizability of our results.

- Both the centrality and SNA literatures are only parts of a much larger literature. Some of the processes operative in this segment of the overall scientific literature may operate elsewhere, at least to some extent. Even for the narrowly defined centrality literature, we learned that our initial concern with SNA was overly restrictive. There are markedly different co-authorship cultures, differing levels of institutional control of publication by professional associations, and a heavier dominance by particular research centers in other scientific domains. These differences will limit simple generalizations from our results but the idea of including these 'external' influences has general relevance for understanding the multiple structures of scientific citation networks.

- The procedures of the USPTO seem uniform and broadly consistent with similar organizations assessing patent applications elsewhere. While it is likely that our results will generalize beyond the USA, this has to be checked.

- All nations have unique histories, implying that the creation and operation of the US Supreme Court may have been unique. Certainly, this court changed, sometimes dramatically, over the years, depending on economic, political, and social conditions plus the composition of the court. This, most likely, will limit simple generalizations to other such courts. However, the idea of paying attention to the historical contexts of national court systems will generalize.

- While our results regarding the network dynamics within the organization of football in England do have implications for the global organization of football, they cannot be generalized very far beyond England. Caution will be in order also in attempts to generalize beyond the period defined by the first 15 EPL seasons, as the dynamics have changed to some extent.

- Our results concerning the delineation of the spatial diversity of the USA may not generalize beyond the USA. But, again, the applicability of the methods we used to map this diversity will generalize.

Our broad contention is that the approaches we have taken to studying large networks in a variety of substantive contexts was valuable and can serve as a (partial) template for other large networks and substantive problems. In particular, our results show the general utility of incorporating the study of large networks into the study of substantive issues. The latter are enriched by including network analytic ideas and the former become more fruitful when informed by broader substantive concerns.

10.3.1.2   The supplementary datasets we used

We were compelled to assemble and assess attribute data for the units in the large networks that we considered, and data for the contexts within which the networks were located. The amount of such information varied across these networks.

- For the scientific citation networks, most of the attribute data on, for example, authors, journals, institutional locations, and keywords, were located in the original data source. However, as outlined in the appendix, there were major problems in obtaining clean versions for these items. Such data were important for constructing a set of two-mode networks. Studying the one-mode citation network is greatly enhanced by also considering the two-mode networks constructed from this attribute data.

- Constructing two-mode network data from attributes was important for the patent citation. Although the inherent data problems were far less severe than for the scientific citation network, the same generic issues were present. Also, the mixture of very general keywords and highly specific keywords demanded attention. We accepted the rather limited attribute data for patents in the original source. Beyond reconceptualizing originality as heterogeneity for patents, we did not expand the number of attributes.

- The Supreme Court data did not come with attributes for the decisions, especially in the form of keywords, so we could not extract them. Instead, we looked at the opinions of decisions to discern the substantive content of decisions and the Constitutional principles informing them.

- Understanding the football network required the assembly of a considerable amount of supplementary information as described in the data appendix. Principal among them was the construction of historical memberships of clubs in the top league levels in England. We did the same for the top leagues in France, Italy, Germany, and Spain to assemble historical information on club participation rates and the numbers of titles won by clubs. To obtain a ranking of clubs over time, a variety of UEFA rankings of clubs and leagues were consulted. Additionally information was assembled on the flows of players from around the world into these top leagues.

- To perform both the clustering with a relational constraint and the discriminant analyses for the spatial data, statistical data were extracted from the US Census Bureau's data on counties, along with such data from other sources. Matching these data with counties, even from the Census data, was not a straightforward task, as described in the data appendix.

10.4 Surprises and issues triggered by them

Life without surprises has the potential for great boredom. Consistent with this, we believe in learning more from 'brave failures' than from 'safe successes.' While we do not think that failure characterizes the results of our efforts here, we did not play it safe while venturing outside our usual comfort zones. We learned far more about the substantive areas we tackled, the methods we used, and the quirks in the data that we analyzed than we thought possible at the outset of this adventure.

Of course, while being surprised can be taken as an indicator of a knowledge deficit, the world is complex, and learning is a prerequisite for progress. While we do *not* summarize our results here – there are summaries in each of the chapters already – we do point to *some* of the surprises enriching our efforts by prodding us to think further.

10.4.1.1 SNA citation networks

Among the surprises with these networks was the amount of effort having to go into converting them into a form we could use. The takeaway message is simple: while data on, or for, large networks may be readily available, a lot of work may be required to make them usable. While this is generally true for all datasets, the effort seemed truly large with these data. Nothing substitutes for knowing one's data well. Datasets seldom can be taken off the shelf and used blindly in an 'as is' form.

Our motivation for looking at these data concerned an interest in the impact of physicists entering the social networks realm and 'their' impact on 'our' field. As we have noted, 'the invasion of the physicists' has been a concern in the SNA community. While real, this seems remarkably parochial in retrospect. When we identified a main path in the centrality literature starting with works by {social network analysts} → {physicists}, we thought 'Aha, we were right, the invasion led to a takeover.' But that path continued to {social network analysts} → {physicists} → {neuroscientists}. We were not observing a takeover but transitions.

Moreover, these were more than simply transitions in the sense of sequential dominances. Rather, they signaled a far more complex intellectual world where similar concepts were mobilized in multiple realms. Parts of 'the centrality literature' had nothing to do with social networks: they were separate and independent developments. Who knew that SNA, atmospheric physics, and neuroscience could be part of what seemed a narrowly defined literature? This was fascinating, interesting, and sobering for us.

10.4.1.2   The patent citation network

One surprise here was just how clean these data are and, as we note, this is due to the rigorous review process for patent applications. Maybe these data could have been taken off the shelf, but it is always safer to look closely. And some processing of the data was needed anyway.

We knew that the era in which we have lived – our age range is quite wide – was and remains one dominated by computers and communication systems. Our analyses of the patent citation literature revealed, far more clearly than we expected, just how prominent these industries, together with their increased links to some of the other technological areas, have been in the modern world.

10.4.1.3   The Supreme Court citation network

As we noted in the Supreme Court chapter, we placed a bet on line islands being a useful way of identifying coherent parts of this citation network. It was nice to learn that this was a well-placed bet. In Chapter 6, we considered only the 'Native American' and 'Threats to social order' islands. While we noted that we did look at another line island devoted to railways and one for maritime law, we did not look at the remaining line islands. On tracing the Supreme Court decisions that we used when setting up our consideration of the patent citation literature, we found some of them in one of the 48 detected line islands. This connection between these two chapters came as a complete, and welcome, surprise, one leading us to include a Supreme Court line island in the patent chapter.

Delineating line islands was merely the first step in attempting to understand the workings of the court. We sought to understand these coherent patches by linking the decisions of the court to its prior history plus the social, economic, and political decisions of the times. This was far more complex than we anticipated, so much so that analyzing this as a stand-alone citation would have been a far too narrow enterprise.

This network was constructed with its links being the citations for decisions to earlier decisions. As such, the underlying presumption was that a citation is a citation. While we knew that there are negative citations in this network we thought this only took the form of some earlier decisions being overturned by later decisions. Our examination of the Dred Scott case revealed that this was only partially correct. The *text* of the opinions for cases can refer to some earlier decisions negatively. This was and remains a troublesome surprise regarding the nature of the data.

10.4.1.4   The football network

The biggest surprise with these data was the great difficulty in assembling them. At the outset, getting them seemed a straightforward, but tedious, enterprise. This illusion was shattered with the realization that there were no completely reliable data sources out there.

Initially, we thought we were collecting data for *one* large network. Indeed, the initial write-up for the data appendix included a statement to that effect. We soon realized that we were dealing with a *collection* of networks, some of which were occasionally joined. Yet these data do pertain to a coherent general system of moves. We were stunned by the large number of clubs across the globe involved in player moves for those reaching the EPL.

The glory of having hypotheses is that they can be tested. It was no surprise that some would be unsupported while others would be supported. After all, that is the nature of the game. However, we were convinced that our hypothesis regarding the impact of the Bosman Decision being minimal was correct. It failed spectacularly. Yet we could find some evidence allegedly supporting it. This served as a reminder that the preponderance of the evidence matters most, and finding *some* evidence consistent with a hypothesis does not amount to a test of that hypothesis.

Part of the current conventional wisdom considers the money flowing into football at the top levels as creating great unfairness by enhancing greatly the fortunes of wealthy clubs and diminishing the life chances of other clubs. Our surprise is the survival of this view, given that the historical levels of inequality have not changed. All that has changed are the mechanisms for creating and maintaining inequality.

10.4.1.5   The spatial network

At the outset, we were most persuaded by the narratives of *The Nine Nations of America* and *American Nations: A History of the Eleven Rival Regional Cultures of North America*. We thought our analyses of the county data for the USA would, in the main, conform to most of their narratives. Of course, we shared the presumption that the statistical order imposed by Census Bureau regions was little more than a chimera. Our surprise came in the form of how little of the 'regional' narratives remained. The spatial layout of diversity across the USA at the county level is far more complex than these authors suggested. It was not a surprise that there was less patchiness and more regional coherence than the analysis of *Our Patchwork Nation: The Surprising Truth about the 'Real' America* implied.

10.5   Future work

As shown in Chapters 4–9, we have discovered a lot of patterns with intriguing results. Yet, we are acutely aware of having barely scratched the surface of the data we have for the large networks we have studied. Among the juicy, and sometimes daunting, items for our future agenda are the following, organized by substantive chapters.

10.5.1.1   The network of scientific citations

1. Within the network we studied, there are other line islands, with different internal structures, that we have not looked at, and they demand attention.

2. We have done relatively little with the two-mode networks we have constructed. In many ways, they have the potential to be more interesting than the one-mode network because they have the potential to help us understand the structure of the one-mode citation network and, more importantly, the forces shaping it.

3. There is a wide range of initial search terms beyond centrality and SNA. Even within the social network literature, network topics such as structural holes, cohesion, block-modeling, and exponential random graph models can be used to determine citation

networks to study. Combinations of terms could be used to define citation networks. Given our concern with the idea of researchers adopting orthodoxies, this could be a way of examining the extent to which this is reflected in the literature. The idea of using multiple search terms can be expanded to include multiple fields. One example could be defined by the terms clustering networks, blockmodeling, and community detection to better grasp the flow of ideas between fields.

4. We have written very little about interdisciplinary or multidisciplinary research. It seems useful to select some disciplines with the intent of examining the flow of ideas between fields and how they are adopted by fields receiving new (to them) ideas.

10.5.1.2   The patent network

The additional issues to pursue include the following:

1. We showed the flow of citations between six broad technological areas. One (mild) hesitation that we have concerns the heterogeneity of these areas: the flows we discerned could be bundles of flows between parts of the broad technological areas. This could be examined by using the two-digit classification of technological areas.

2. We showed only the flow networks for a small set of windows (time slices). A far more detailed analysis is to examine all of the time slices, to get a clearer sense of evolutionary patterns.

3. The establishment of distinct timescales for the value of patents for subsequent patterns is intriguing and demands further exploration. The clusters of temporal profiles could be used to shed light on the dynamics of influence between technological areas.

4. We focused on citation flows between the broad technological areas while ignoring flows within these areas. The latter flows have interest value and can be explored in the same fashion by considering finer-grained partitions within these areas, using the two-digit coding.

10.5.1.3   The Supreme Court network

1. We identified 48 line islands but, as we explored a very small number of them, the remaining islands have to be treated in the same fashion. Doing so will provide a more general understanding of how islands of decisions are held together in terms of content.

2. While we have focused on exploring coherent parts of these decision separately, this is only a first step. These islands, despite their name, were linked to some other islands. The links between islands need to be explored to see how *combinations* of substantive interests and constitutional principles link *sets* of decisions.

3. One question, asked at a conference, about what we are doing with these data is quite pointed. 'Can we learn anything different from what you are doing compared to reading books on Constitutional law?' One way of addressing this is to assemble decisions grouped together in such books and compare the sets of decisions obtained by the two contrasting selection principles.

4. Two implications of our examination of the Dred Scott case are daunting. One is having to include citations from all of the opinions in a decision. The second is to look closely at the citations between decisions to get an understanding of what a citation really is. Distilling positive and negative ties is but one part of doing this.

10.5.1.4   The football network(s)

1. Our analyses focused mainly on club-to-club networks. Our focus can be broadened to look more closely at country-to-country networks. This can be done both with regard to the nationalities of players and the countries of the clubs between which players move.

2. We paid little attention to the ends of playing careers to focus on how players reached the EPL. Yet, the structure of the downward side of playing careers, once playing at the top level is over, also has interest value.

3. More can be done to account for the reasons why the lengths of overall player careers are conditioned by where players start these careers.

4. Given our caveat about generalizing our results to a later time frame, the database could be expanded to include all of the seasons of the top league in England. However, our preference is that *we* do *not* do this!

10.5.1.5   The spatial networks

1. It is possible to try to construct a set of variables so that a discriminant analysis can be done for the second set of nations defined by Woodard. Also, a closer look at places and their characteristics is merited.

2. We may have done all that we can with regard to this particular network by having pushed the analysis as far as we can. But applying the methods we used to other countries has immense appeal. We did draw a spatial map for Slovenia as a simple illustration and think a full analysis using also the statistical data available for communes would be useful.

3. The analyses we have done for the spatial network were cross-sectional. Having a temporal sequence of such analyses could shed light on the dynamics of change.

4. We know that we have not paid much attention to temporal-spatial networks studied by geographers and see this as something to which we could turn our attention to great benefit, both substantively and technically.

10.5.1.6   Other networks

Even a few of the above avenues of inquiry are enough for a sustained research effort. Yet there are other temporal networks that beckon us like sirens. Some of them are:

1. The mathematical citation networks we touched upon briefly as illustrations in the formal chapters.

2. Collaboration networks among scientists, especially for the network of all scientists in Slovenia.

10.6 Two final comments

We know that we have a lot more to learn about the substantive areas we have tackled and much to learn about methods developed in other literatures. We will seek them actively. This concluding chapter can be interpreted as an invitation extended to others to join in this effort. For us to attempt to do everything on the above agenda would be insane. Life is just too short. So we invite others to put their shoulders to this wheel, or rather, this set of wheels.

Appendix

Data documentation

Here, we document our data sources, our data processing steps, and the construction of the network datasets analyzed in this book. This documentation is organized by the chapter sequence where these datasets are analyzed. The longest description is for the football data because they were constructed from scratch from many sources.

A.1 Bibliographic networks

We start our description with some general comments on data sources for bibliographic networks before dealing specifically with the scientific networks considered in Chapter 4.

From special bibliographies (BibTeX) and bibliographic services including Web of Science/Knowledge, Scopus, CiteSeer, Zentralblatt Math, Google Scholar, DBLP, Math Sci and others, it is possible to obtain data about *works* (papers, books, reports, etc.) on selected topics. A typical description of a work contains the following data: authors; title; publisher/journal; publication year; and pages. In some sources, there is additional data including languages, classification of documents, keywords, authors' institution/country affiliation, lists of citations, and abstracts. These data can be transformed into a collection of compatible two-mode networks on selected topics: works × authors; works × keywords; works × countries; and other pairs of characteristics describing works. Besides these networks, we can also get partitions of works by their publication years, partitions of works by journals, vector of number of pages, and, in some cases, (one-mode) citation networks.

When constructing any of these networks, the first task is to specify the vertices, and which relations are linking them. In short, the *network boundary problem* (Marsden 1990) has to be solved. This includes deciding whether a network is one-mode or two-mode, and which node properties are important for the intended analyses. For specifying lines, this amounts to answering a series of questions:

1. Are the lines directed?
2. Are there different types of lines (relations) to include?

Understanding Large Temporal Networks and Spatial Networks: Exploration, Pattern Searching, Visualization and Network Evolution, First Edition. Vladimir Batagelj, Patrick Doreian, Anuška Ferligoj and Nataša Kejžar.
© 2014 John Wiley & Sons, Ltd. Published 2014 by John Wiley & Sons, Ltd.

3. Can a pair of nodes be linked with multiple lines?
4. What are the weights on the lines?
5. Is the network static, or is it changing through time?

Another problem occurring often when defining the set of vertices is the *identification* of nodes. The unit corresponding to a vertex can have different names (*synonymy*), or the same name can denote different units (*homonymy* or *ambiguity*). For example in the BibTeX bibliography from the Computational Geometry Database (Jones 2002) the same author appears under seven different names: R.S. Drysdale, Robert L. Drysdale, Robert L. Scot Drysdale, R.L. Drysdale, S. Drysdale, R. Drysdale, and R.L.S. Drysdale. Insider information is needed to decide that Otfried Schwarzkopf and Otfried Cheong are the same person. At the other extreme, there are at least 57 different mathematicians with the name Wang, Li in the MathSciNet Database (TePaske-King and Richert 2001). Its editors have tried hard, from 1985, to resolve the identification of authors problem during the data entry phase. In the future, the problem could be eliminated by general adoption of initiatives such as using ResearcherID or ORCID.

Similarly in the WoS work references we find the following journal names: NUCLEIC ACIDS RES, NUCL ACIDS RES, NUCLEIC ACIDS RES S, NUCLEIC ACIDS RES S2, NUCL ACID RES, NUCL ACIDS RES S2, NUCL ACIDS S SER, NUCL ACIDS RES S, NUCL AC RES, NUCLEIC ACIDS RES S1, Nucleic Acids Res, NUCL ACIDS RES S1 or Q J R MET SOC, Q J R METEOROL SOC, Q J ROY METEOR SO S1, Q J ROY METEOR SOC, Q J ROY METEOR SOC B, QUART J ROY METEOR S, QUART J ROY METEOROL, QUART J ROY METEOROL SOC, QUART J ROYAL METEOR. The immediate issue with all of these names is whether they denote the same journal or a small set of journals.

There exists International Standard Serial Number (ISSN 2013), an international system for the identification of serial publications and other continuing resources. The problem is that the convention is not considered in WoS in the list of work references. In resolving the journal identification problems, it is possible to use the Global Serials Directory (Ulrichsweb 2013) and Journal Abbreviation Sources (JAS 2013), and many other services and data sources.

The identification problem appears also when the units are extracted from plain text parts of documents. In producing keywords from the title or abstract of a work, the unimportant 'stopwords' must be eliminated first. The remaining (real) terms (words or phrases) are usually standardized by replacing them by a 'canonical' representative. For example, terms 'function', 'map', 'mapping', and 'transformation' in the mathematics literature can be considered as equivalent terms. A similar problem is having equivalent terms from multilingual sources. To resolve this problem it is necessary to provide lists of equivalent terms or dictionaries.

Yet another source of identification problems stem from the grammar rules of the language used in a specific document. For example the action, 'go' can appear in the text in a variety of different forms including 'go', 'goes', 'gone', 'going', and 'went'. Resolving these grammar problems requires the use of stemming or lemmatization procedures from natural language processing toolkits such as NLTK (Bird et al. 2009; Perkins 2010) or MontyLingua (Liu 2004).

The data used in the examples in this book were obtained from the Web of Science. An example of a WoS work description record is provided in Figure A.1. Such records were

```
PT J
AU Dipple, H
   Evans, B
TI The Leicestershire Huntington's disease support group: a social network
   analysis
SO HEALTH & SOCIAL CARE IN THE COMMUNITY
LA English
DT Article
C1 Rehabil Serv, Troon Way Business Ctr, Leicester LE4 9HA, Leics, England.
RP Dipple, H, Rehabil Serv, Troon Way Business Ctr, Sandringham
   Suite,Humberstone Lane, Leicester LE4 9HA, Leics, England.
CR BORGATTI SP, 1992, UCINET 4 VERSION 1 0
   FOLSTEIN S, 1989, HUNTINGTONS DIS DISO
   SCOTT J, 1991, SOCIAL NETWORK ANAL
NR 3
TC 3
PU BLACKWELL SCIENCE LTD
PI OXFORD
PA P O BOX 88, OSNEY MEAD, OXFORD OX2 0NE, OXON, ENGLAND
SN 0966-0410
J9 HEALTH SOC CARE COMMUNITY
JI Health Soc. Care Community
PD JUL
PY 1998
VL 6
IS 4
BP 286
EP 289
PG 4
SC Public, Environmental & Occupational Health; Social Work
GA 105UP
UT ISI:000075092200008
ER
```

Figure A.1   A record from the Web of Science.

transformed into Pajek's format using the program WoS2Pajek (Batagelj 2007). A set of similar programs was developed also by Leydesdorff (2013).

In the following subsection we describe the collection procedure for the centrality literature networks.

A.1.1   Centrality literature networks

The data we use were extracted from the Web of Science (WoS) by using its Advanced Search (see above) and entering the following query:

```
TS=(centrali* AND (network* OR graph))
```

Note that 'centrali*' covers all words beginning with 'centrali', including centrality and centralization. Similarly, 'network*' covers words such as network, networks, and networking. Including also 'graph' extends the search to the graph theoretical literature. Because WoS allows only up to 100,000 hits in a query, time bounds have to be set. After a query, information about the number of hits is provided. Getting the list of hits requires clicking on their number. In WoS, hits can be saved to a file. However, WoS allows only 500 hits to be saved at once, so longer lists have to be saved in parts. Saving selected hits is done as follows:

- step 1: determine the range of hits to be saved (1–500, 501–1000, 1001–1500, …);

- step 2: select **Full Record** and **plus Cited Reference** and

- step 3: select **Save to Plain Text**.

Next, when the **Save** button is clicked, a new page with **Processing Records** appears. There is a time lapse until the selected records have been processed and written to the file. When a window appears, the option **Save to Disk** appears and **OK** is clicked. It is necessary to select the directory and enter the name of the file to which the selected hits are saved. Clicking on **Save** saves them. To continue saving the selected hits, **Back to Results** is clicked. This procedure is repeated until all the identified hits are saved to the named files.

Also available is a list of articles citing the hits following a query. The top of the page with the list of hits has the option **Create Citation Report**. Clicked this brings up a new page with histograms. For our purposes, obtaining the list of citing articles was critical. This is done by clicking **View Citing Articles**. Saving them uses the same procedure as for saving the selected hits to file. Finally, all collected files can be merged into a single file.

From such data obtained from the Web of Science (Knowledge), the corresponding networks are obtained by using the program `WoS2Pajek`:

- A citation network, **Ci**: works × works

- An authorship network, **WA**: works × authors[1]

- A keywords network, **WK**: works × keywords – but only for works with a complete description

- A journals network, **WJ**: works × journals[2]

- A partition of works by the publication year

- A partition of works – complete description (1) / ISI name only (0)

Similar programs exist also for other bibliographic sources/formats including: Scopus, BibTeX, Zentralblatt Math, Google Scholar, DBLP, and IMDB.

A.1.1.1   Cleaning the data

Given the construction of WoS files, there are errors which make it necessary to clean these data. The directed centrality citation network obtained by using the above procedures is labeled `central.net`. Getting the basic characteristics of the network, after it is read into Pajek, is done with the following commands:

```
Network / Info / General / Input 1 or 2 numbers [0] / OK
```

This network, for the 2010 version, has 548,600 vertices linked by 996,962 arcs. By definition, it cannot have loops (self citations) nor multiple lines (multiple citations). However,

---

[1] For works without complete description only the first author is known.

[2] This is essentially a partition; we use network because in Pajek partitions it is not possible to give names to classes (clusters).

networks obtained from WoS can have both problems, because different articles can get the same WoS name. For example, part of the original WoS data contains:

```
GRANOVET.MS, 1973, AM J SOCIOL, V78, P1360
GRANOVETTER M, 1983, SOCIOLOGICAL THEORY, V1, P203

BORGATTI SP, 2002, UGINET WINDOWS SOFTW
BORGATTI S, 1999, UCINET V USERS GUIDE

CANTANZARO M, 2005, PHYS REV E, V71, UNSP 027103
CANTAZARO M, 2005, PHYS REV E, V71, UNSP 056104
CATANZARO M, 2005, PHYS REV E 2, V71, ARTN 056104
```

In all three groups, the name of the first author is written differently. In the final trio of names, the last pair show how one article can have different ISI names. Also, the use of the same short author name can represent different articles creating loops and multiple arcs in a citation network. These problems can be partially resolved in WoS2Pajek by introducing short names of articles. Details for doing this are provided in the WoS2Pajek manual. In principle, most of these inconsistencies can be detected and repaired, but this is a very time-consuming task, especially for networks of this size. We did some of this for the WoS data. In doing so, it was necessary to make a trade-off between the time taken and obtaining 'clean' data. It was useful to take a shortcut: these inconsistencies were considered as noise by removing the loops and transforming multiple arcs to single arcs. The commands for doing each of these in Pajek are:

```
Network/Create New Network/Transform/Remove/Loops
Network/Create New Network/Transform/Remove/Multiple lines/Single line
```

We also partially identified journal names corresponding to the same journal and shrank them to a single representative. To construct the partition describing the journals names equivalence relation we sorted the names using different keys – for example, the initials of words composing the journal name: QUART J ROY METEOROL SOC → QJRMS . The potentially equivalent journal names form groups in the sorted list. The decisions were made manually.

A.1.2 SNA literature

In a similar way the networks **SN5** for the SNA literature were obtained in 2008 from the Web of Science for a query `"social network*"` and expanded with existing descriptions of the most frequent references and the bibliographies of around 100 social networkers. Using the program **WoS2Pajek** (Batagelj 2007) the corresponding collection of network data was produced. The sizes of the sets are as follows: works $|W| = 193,376$, works with complete description $|C| = 7950$, authors $|A| = 75,930$, journals $|J| = 14,651$, keywords $|K| = 29,267$. The dataset was used for the Viszards session at the SunBelt XXVIII, 22–27 January 2008, St Pete Beach, Florida.

The procedures for constructing the SNA bibliographic network data are basically the same as for the centrality literature.

A.2 Patent data

The patent data come from NBER Patent Data Project[3] (2010) and comprise all US patents granted from 1976 through 2006 together with all citations between them in this period. The full dataset has more than 3.2 million patents with text descriptions[4] and over 23 million citations between them. The description of an earlier dataset from United States Patent and Trademark Office (USPTO) patents (for 1963–1999) is in Hall et al. (2001). The two datasets differ. The previous dataset included all citations from each patent granted from 1963 onwards. We use citations from 1976 to focus on a more recent 20-year period.

The data are very rich. In addition to the sheer quantity of patents and citations, additional information for each patent, including the time of application and when the patent was granted, the technological area in which it belongs, the inventors, and the patents' assignees, exists. This patent information is supplied by its inventors, or assignees, to obtain exclusive rights to inventions. Legally, citations to earlier patents dealing with related topics must be included. Other parties can request citations to 'prior art' before patents are granted (Patent Rules, United States Patent and Trademark Office, USPTO[5]) and to the relevant literature. These citations limit the scope of new inventions (Hall et al. 2001). Further, they enable tracing the development of inventions, examining their spillover effects, facilitate studying the history of technology, and discerning the importance of different technological fields.

The data limitations of the data considered in Chapter 5 are threefold: 1) the time span is short, given that patents have been granted in the USA from before 1800 (Hall et al. 2001); 2) not all inventions are patented (because of having to meet USPTO specifications); and 3) citations of supplementary scientific materials can be absent. An additional complication stems from different policies, and changed USPTO rules, contributing to granting patents and the citing patterns between them.

We used the following parts of this large dataset for our detailed analyses:

- patent numbers

- the year of application for a patent license

- the granting year for each patent. The USPTO review process takes, on average, about 2 years (standard deviation is 1 year). About 95% of patents are granted within three years of the application (Hall et al. 2001): Table 1

- the country of first inventor listed in the patent application

- the assignee type (individual, corporate, government, and other)

- the main US patent class to which patents belong

- the technological category/subcategory: there are 36 subcategories which have been merged into six main categories: chemical (excluding drugs); computers and com-

---

[3] https://sites.google.com/site/patentdataproject/Home/downloads
[4] There are only 1398 patents lacking descriptions.
[5] http://www.uspto.gov/patents/law/index.jsp

munications (C&C); drugs and medical (D&M); electrical and electronics (E&E); mechanical; and other

Additional constructed variables were precomputed using definition from the 1963–1999 dataset of (Hall et al. 2001). They include:

- a measure of *generality* for patents, the formula for which is provided in Section 5.3 – high generality measures result for patents cited by patents from many fields and have greater innovative impact

- a measure of *heterogeneity* for patents for which the formula is also in Section 5.3 – in the original sources this was called originality but heterogeneity is preferable for patents drawing on multiple technological fields

- the mean backward citation lag (the time difference between the granting, or application, year of a patent, and the citations it received) – where not specified specifically, we use the grant year since this is when patents become visible officially

- the percentage of self-citations, the percentage of citations to patents held by the same assignee, plus its upper and lower bounds

- counts of citations made

A large temporal network was constructed from these data. Patents are represented by the vertices of the network with the citations as directed lines between patents. These arcs are directed towards the older (cited) patents. Every vertex and every line in the network has a time stamp denoting its creation date. This never changes. All of the additional data about patents, also time stamped, were saved as variables describing them. This citation network was acyclic as received, for reasons given in Chapter 5.

The titles of patents can also be used to obtain patent descriptions using *keywords*. The titles of all patents were obtained from the USPTO website.[6] These words were delimited by non-alphanumeric characters and were lemmatized using the MontyLingua Python[7] library, with stopwords removed. In this way, 132,829 keywords were assembled for this 1976–2006 patent citation network.

A.3 Supreme Court data

The Supreme Court citation data feature decisions handed down from 1789 through 2002. These data, obtained from James Fowler's website,[8] were the basis for results reported in Fowler and Jeon (2008). The data contain 30,288 cases with 216,758 citations from decisions to earlier decisions. They contain summary information on the decisions themselves including:

- a unique numerical decision ID for each decision which we used[9]

---

[6] http://patft.uspto.gov/netahtml/PTO/srchnum.htm
[7] http://web.media.mit.edu/ hugo/montylingua/
[8] http://jhfowler.ucsd.edu/judicial.htm
[9] There is some duplication of cases, but we worked with the full dataset to maintain comparability rather than prune the data.

- the US Report ID, for example, *Chisholm v. State of Georgia* has 2US419 as its Report ID

- the parties involved in the case (e.g. Chisholm and Georgia)

- a pairing of overruled cases and the cases overruling them – from multiple sources we identified pairings of overruling and overruled cases that were not in the source data set and added them to the data

- two vectors marking important cases according to outside sources – one source was the *Oxford Guide to the Supreme Court* Hall (2005) with the other being the Legal Information Institute's list of important cases.[10] We also included importance markers drawn from Powe (2009); Vile (2010); and Irons (2006) to construct an overall composite vector of 'important decisions'. Finally, we included the listing of 'landmark' Supreme Court decisions from Wikipedia[11]

- outdegrees and indegrees (as of 2002) for all decisions – these distributions are highly skewed: indegrees range from 0 to 266 with a mean of 7.3 (standard deviation, 12.3) and a median of 3; outdegrees range from 0 to 195 with a mean of 7.3 (standard deviation = 11.4) and a median of 3

- other vectors measuring the salience of decisions including hub scores, authority scores and eigenvector centrality, none of which we used.

We created additional information and data:

- a chronology of historical events including both Supreme Court decisions and enacted laws (at both federal and state levels) to couple decisions to the historical contexts in which they were made

- a listing of the substantive domains that were the focus of the cases reaching the court – these were extracted from Irons (2006); Powe (2009); Vile (2010); and a reading of the opinions together with summaries of cases[12]

- a coding of cases based on Vile's (2010) categorization to organize his discussion of 'essential' Supreme Court decisions – it was based on constitutional domains underlying decisions made by this court

- alternative weights for cases based on methods discussed in Chapter 2 where methodological issues are discussed.

---

[10] http://www.law.cornell.edu/

[11] http://en.wikipedia.org/wiki/

[12] The primary sources for the opinions and, in some cases, summaries were: Bulk Resource (https://bulk.resource.org/courts.gov/); The Justia US Supreme Court Center (http://supreme.justia.com/); The Cornell University Legal Information Institute (http://www.law.cornell.edu); The Oyez Project at IIT Chicago-Kent College of Law (http://www.oyez.org/cases/); FindLaw for professionals (http://caselaw.lp.findlaw.com); 4lawnotes (http://www.4lawnotes.com); Casebriefs (http://www.casebriefs.com); OpenJurist (http://openjurist.org); and Wikipedia (http://en.wikipedia.org/wiki/). We emphasize that none of the authors of this book have had legal training but, where necessary, we did consult legal sources.

A.4   Football data

The ideal dataset would consist of an accurate listing of all football players for every country of origin and all of their stays at every club and in all countries. It would cover the period 1888–2011. As such data do not exist, we created a much smaller (but still complex) dataset defined by *all* players who belonged to clubs appearing in the English Premier League (EPL) for its first 15 years. The English league system is divided into the Football League (with four levels) and non-league football (with many levels). There is a promotion and relegation system coupling levels. We kept track of the Football League levels for all clubs in the league system.

The only way for clubs to enter the top level (the EPL for our study period, but currently called the Premiership) is via promotion from the second league level (currently, the Championship). We focus on the stays and moves of players who started their careers in the EPL or moved to the EPL from clubs outside this league. These clubs can be anywhere in the world. Players also reach the EPL by playing for a club promoted to the EPL. Given the player moves between clubs, and countries, we sought data on players, their moves, the clubs to which they belonged, the leagues of these clubs, and the global organization of football. A huge amount of data about these topics is scattered across many data sources. We extracted data from them to study the issues considered in Chapters 7 and 8. The resulting datasets are either part of our *core data* or relevant *ancillary data*. Here, we describe the procedures used to obtain the data on which our analyses are based.

The data sources we used can be categorized as follows: 1) official club websites for clubs in the Football League; 2) very general websites devoted to providing overall information about the global game (including the type we sought) in one place; 3) club-specific unofficial websites devoted to describing former players at these clubs; 4) websites devoted to assembling data on transfers; 5) websites providing information about players from specific countries that were/are playing football outside these countries; 6) official club websites for English clubs in non-league football (if they had one); 7) print news media, both national and local; 8) electronic news media; 9) books about football containing data; and 10) official websites maintained by bodies responsible for the national and international organization of football. We describe the first nine of these data sources in the section for core data. The last category is relevant primarily for the ancillary datasets and is described separately. The number of data sources that we consulted exceeded 750. They cannot all be listed here. In describing our data assembly, we illustrate only these different types of sources and our use of them.

A.4.1   Core data

Our core data were constructed for all players who played in the EPL between 1992/3, its inaugural season, and the end of the 2006/7 season. The constructed files contain data for: 1) *player attributes*; 2) some club attributes when players were there; 3) durations of *player stays at clubs* and, most importantly, 4) player *moves between clubs*. These time-dated player moves define network ties between both clubs and countries. Data collection started in the mid 1990s and continued through to the middle of 2010 with attention restricted to players in the EPL during any of its first 15 seasons. Those players joining EPL clubs after 2006/7 were ignored for two reasons. One was practical: updating player move data is very time-consuming. More importantly, we wanted a mix of players whose careers were finished, players well into their playing careers, and some younger players nearer, but not at, the beginning of their careers in the data.

A.4.1.1   Player attributes

Player attributes include: date of birth (used to construct ages of players); place of birth; nationality; height and weight;[13] and position played. Two of these attributes are ambiguous. The first is nationality. Some players born in the former colonies of former colonial powers can have either the nationality of the place where they were born or the nationality of the former colonial power. This seems particularly relevant for England, France, the Netherlands, Portugal, and Spain. We used FIFA-recognized nationalities for players. To have uniform player career start points, we exclude players younger than 15.

Position played is ambiguous because players, by design or necessity, can and do play in multiple positions for their clubs. When a data source provides a single position descriptor, we used it because there is no ambiguity. When multiple positions are listed for a player, we selected the first position listed as the *primary* position. When different sources provided different positions for a player, we took the most frequently mentioned position across these sources as the primary position of the player. Further ambiguity stems from the use of different terms for the same position. We settled on using four positions: goalkeeper; defender; midfielder; and forward. Positional labels were grouped into these positions. The labels for *goalkeeper* are the simplest: only goal and goalkeeper were used. We categorized the following labels as *defender*: defender; central defender; centre back; full back; right back; left back; defender/centre back; defender/full back; defender/right back; defender/left back; centre/left defender; and centre/right defender. We included also the following four position descriptors as defenders under the 'first position listed' rule: defender/midfielder; central defender/midfielder; left back/midfielder; and defender/forward. For *Midfielder* we included: midfielder; midfielder/defender; left midfielder; right midfielder; and central midfielder. for *forwards* we included: attacking midfielder; winger; left winger; right winger; forward; striker; centre forward; forward/midfielder; and midfielder/forward.[14])

A.4.1.2   Club attributes and stays

For *each* player in our dataset, we recorded the date of his arrival at a club and the date of his departure when they were available. Our goal was to do this for *every* club having a player during his playing career. The club recorded attributes were: country of the club; league level (when a player was at a club); position in the league at the end of each season; a recent promotion; and a recent relegation (where 'recent' was defined as occurring in the prior season).

A.4.1.3   Player moves

We sought complete sequences of moves for players as constructed from their stays. We obtained the *starting* club for each player (age 15, or later depending on the player[15]), the

---

[13] Height is unlikely to change much over a player's career but his weight can change due to aging, being at a stage on the way of recovering fully from an injury, lacking match fitness, and lack of fitness at specific points in time due either to illness or weight gains during the off-season. For players entering the EPL, we take their height and weight as applying to their date of entry.

[14] This is the one exception to the 'first position listed' rule, and is used to capture an emphasis on attacking for the few players described this way.

[15] For many players, we did obtain club data for earlier years but did not use them.

final club for each player whose career was over, and all clubs in between. For those who were still playing, we sought the *current* club (at the end of the 2006/7 season) as the final club in our study period. These data on moves were, by far, the hardest to obtain. *Every potential data source we consulted was incomplete*, many of them seriously so. Many sources also contained inaccuracies. We were surprised by the number of serious problems we encountered. While these problems are easy to list, they were much harder to resolve. Doing so involved many judgment calls. Databases maintained in different countries share a bias of being less attentive to two items: 1) playing careers of players before they reached the country where the database was maintained, and 2) playing careers after players leave that country. One seemingly trivial finding is that many – but far from all – players who have played in the EPL start in their country of origin and return towards the end of their careers. While trivial, every database we consulted, if taken as complete, did not reveal this career feature for many players.

Player moves are part of how football is organized with players moving from club to club. A transfer is a move where the 'ownership' of the player moves from the sending club to the receiving club. How this transfer mechanism operates has changed over time, when the rules regarding transfers were changed. Transfers often include receiving clubs paying 'transfer fees' to sending clubs. There is no source for all transfer fees, as the details are often not revealed. There are also 'free' transfers where no money is exchanged between clubs. Some are called Bosman transfers, following the 1995 Bosman Ruling, while others follow the expiration of a player's contract. We view transfers as creating relationships between the clubs involved, because clubs negotiate the terms of transfers. For us, a transfer is a transfer: we did not distinguish these types.

Loans form another type of move: one club loans a player to another club. The terms of loans are negotiated by clubs, also creating relationships between them. This can be a transitory relationship for just a single loan or part of a formal arrangement between clubs.[16] Alas, sources differ in describing moves. A move can be recorded as a loan in one source but as a transfer in another. Worse, many moves were absent from many databases. Recording data is further complicated by the idea of a loan 'with an option to buy,' where a transfer can follow a loan (if the receiving club exercises that option). As a result, there can be ambiguity in the nature and recording of loans. Some inconsistencies across databases may reflect this ambiguity. Trials, as the name suggests, occur when a player undergoes tests to determine if the testing club sees any value in recruiting him. We started to record trials but concluded that they are seriously under-reported in all data sources. We ignored trials completely also because trials are unlikely to create systematic ties between clubs, especially for the (many) unsuccessful trials.

There are also problems associated with the following phrases: a player being declared as 'surplus to requirements' at a club by its manager; players and clubs parting company 'by mutual agreement'; and players 'being released' by their clubs. All these departures create uncertainty for players, and ambiguities in reports of departures from clubs in databases. Most often, they result in apparent gaps in the playing careers of footballers. When players are, in essence, discarded by their clubs this does not mean the end of their ambitions to continue playing football – even though this does appear to diminish the ambition to fully track such players by those maintaining databases. Some young players vanish without trace and may

---

[16] The link between Manchester United (in England) and Royal Antwerp (in Belgium) had younger players of Manchester United loaned to Royal Antwerp to gain playing experience.

leave the sport having not made the grade at their first club while others cobble together playing careers that remain invisible in many databases. Usually, in England, these players toil outside the Football League. Older players can be split into those hanging up their boots once they leave the top levels and others who continue to play for lower level teams as they age. We wanted to track as many of these younger and older players as possible.[17]

Our data collection started with clubs playing in the EPL in different seasons. For example, for 2005, some of the club-maintained websites[18] we consulted were:

- Arsenal (http://www.arsenal.com);

- Aston Villa (http://www.avfc.premiumtv.co.uk);

- Birmingham City (http://www.blues.premiumtv.co.uk);

- Blackburn Rovers (http://www.rovers.premiumtv.co.uk);

- Bolton Wanderers (http://www.bwfc.premiumtv.co.uk); and

- Charlton Athletic (http://www.cafc.co.uk).

Each such source lists their squad's players, and some provide partial career histories for these players, including position, height, weight, and date and place of birth. Coverage varied greatly over time on club-maintained websites. The overall quality and completeness of information has improved with time. Most EPL – and League Football clubs – have a uniform access format. Even so, some club-maintained sites were more sophisticated and complete than others.

There were three general databases upon which we relied heavily. The first was 11v11, a subscription site maintained by the Association of Football Statisticians (which one of us joined). Their website (http://www.11v11.com/subscribers/index.php/pageID548) (the 'Football Genome Project') was invaluable as a source for corroborating data that we had and for further data collection. Soccerbase (http://www.soccerbase.com/) was useful early in our data collection.[19] The third main source was Wikipedia (http://www.wikipedia.org/). It provides (somewhat mixed) coverage of many football players in a wide variety of countries. It provides also (incomplete) biographies for many, but far from all, players. The latter two databases were accessed most often when searching for players known to have belonged to EPL clubs in our selected period[20] when 11v11's coverage was incomplete.

---

[17] Young players with common names were the hardest to track. Very often this cannot be done. Players with the same name must be carefully distinguished. Examples include an Alan Smith and a Paul Robinson (among others with the same names). Some specialized knowledge is needed to distinguish such cases. In this instance, both players had played for Leeds United, a trivial fact known by one of us, a long-time fan of the club. We encountered confusion in some databases, which combined fragments of one player's career to fragments of another player with the same name into a playing career and recorded this for 'one' player.

[18] We list only the opening page of these websites to save space here. Clicking on internal links in these sites leads to the data we describe. Despite the presence of 'premiumtv' in some of these web addresses, all the material contained in the websites' information was provided by the clubs to inform fans and the public about players, management, games, social events, and club activities.

[19] Currently, it appears to be more associated with bets on the outcomes of real and hypothetical games. Even so, some of the data we obtained from this source were corroborated elsewhere.

[20] Wikipedia entries cite data in Soccerbase while, on occasion, presenting data at variance with this source (and official sources).

A third category of potential data sources are clubs (or their supporters) providing information on players who once played for them. These include:

- Arsenal: (http://www.arseweb.com/history/);

- Everton: (http://www.toffeeweb.com/players/past/);

- Ipswich Town: (http://www.prideofanglia.com/);

- Leeds United: (http://www.ozwhitelufc.net.au/foreword.php;

- Liverpool: (http://www.liverpoolfc.tv/history/past-players);

- Norwich City:[21] http://www.ex-canaries.co.uk/players; and

- Reading (http://www.btinternet.com/rfc1871/watn/senior.htm).

Included in these websites are the names of other teams for their former players, and some also provide dates. More often, these sources are incomplete but they provide both solid data and some useful clues for searching for further information.

Another sources of data is the listings of transfers for clubs.

- For Tottenham Hotspur, there is http://www.mehstg.com/transfers.htm.

- For Coventry City, a club relegated from the EPL in 2001:

  - http://www.footballtransfers.co.uk/club/coventry-city/567/
  - http://www.transfermarkt.co.uk/en/coventry-city/transfers/verein_990.html

Such sites were reached by searching on club names coupled to transfers as search terms. Data sources where English is not the native language were consulted. Sites like http://www.transfermarkt.de/ have value – especially with the use of Google translate: exact translations are not needed for locating dates and club names! We also went to official club websites in other countries, seeking early and later club stays for players having the same nationality as countries where the clubs are located.

Websites devoted to listing current and earlier club locations of players with a specific nationality exist. Some have quasi-official statuses while others appear to be the work of devoted individuals. Some of them persisted while others were taken down or could not be located later. Examples include:

- Australia:

  - http://www.ozsoccerexperience.com/players.php under the heading of 'The Australian Soccer Experience'
  - http://www.australianplayers.webs.com)
  - http://www.ozfootball.net/ark/Abroad/index.html

- the Netherlands: http://www.dutchplayers.nl/

---

[21] Norwich City's nickname is 'the canaries.' Their website is called 'Flown from the Nest.'

- USA:[22]   http://www.ussoccerplayers.com/ussoccerplayers/2011/02/americans-in-europe.html

- Welsh players: http://www.dragonsoccer.co.uk/players/youngunnews0504.htm

- Source regions e.g. http://africannations.sportinglife.com/, a site built for the 2006 African Nations Cup

- Destinations: http://www.beepworld.de/members57/rotermorgenstern/fussball.htm

Another data source consists of official websites for non-league football clubs. A surprising number of them have websites. However, they are less systematic in their organization and vary greatly in the amount and quality of information provided. As with other electronic data sources that we consulted, we tracked these sites for many years. Their primary use was threefold: 1) sources for complete or partial lists of current players; 2) listings of former players (often with a bias towards former notable players – exactly the type of players for whom we sought data); and 3) lists of players by prior seasons, which helped us learn when players we had tracked to these clubs had departed from them. These websites were relevant for tracing players descending from league football to non-league football and down these levels of non-league football. Examples include:

- Fisher Athletic: http://www.fisherathletic.co.uk/news_detail.phpID=84, a semi-professional club in southeast London

- Ramsgate FC, http://www.ramsgate-fc.co.uk/site2/news.phpextend.210), a small club playing in Kent

- Belper Town, http://www.belpertownfc.co.uk/news_0809/080816.htm, a small club in Derbyshire

- Gateshead http://gfcstats.webs.com/az0008.htm), a non-league club in northeast England, which once played in the Football League.

Websites such as these were reached by searching by either club names or player names. For example, the Gateshead website was reached by the latter route and provided exact dates for the player who was the subject of the search. In addition, there are websites devoted to non-league football. These include the following: http://www.pitchero-nonleague.com/; http://www.non-league.org/; and http://www.nonleague.co.uk/.

Surprising data sources were local newspaper stories available over the Internet and, less surprisingly, national newspapers. Many feature two types of stories. The first reported how small team managers 'snapped up' available players or signed them for inclusion in their squads. Players with prior higher level playing experience are sought. Their arrival is used to generate additional local support for the clubs recruiting them. The second were accounts of specific games, where players of interest were mentioned as doing something of note or whose name was contained in listed squads for these games. In the main, these accounts were most useful in tracing player moves through non-league football. Electronic media have

---

[22] This website was created after the period we study.

relevance also, and were followed over time.[23] Prominent examples include:

- The BBC:

  - (http://news.bbc.co.uk/sport2/hi/football/default.stm
  - http://news.bbc.co.uk/sport2/hi/football/gossip_and_transfers/)

- Sky Sports News (http://www.skysports.com/)

- ESPN (http://soccernet.espn.go.com/cc=5901)

- Fox Sports (http://msn.foxsports.com/foxsoccer)

The website, http://www.nigoalkeeping.com/Interviews/GreggShannon.htm, is an example of websites reporting on specific players. This one includes an interview with a particular player, listing some of the clubs for which he played. Some players have their own websites but vary greatly in covering playing histories.[24]

Among the books that we consulted for player and club information were Crouch (2006); Hammond (2005); Rollin and Rollin (2005); Goldblatt (2006); Graham (2005); Harris (2006); Hugman (2005); Oliver (2005); and Vialli and Marcotti (2006).

These nine types of (what we took as legitimate) data sources provide both a trove of wonderful data together with a large amount of speculative entries and misinformation. This is a part of the pressure of reporting life and the different motivations of people constructing potentially useful databases. As a result, no data source was ever treated as being completely accurate and trustworthy. This need not imply that data sources were necessarily untrustworthy. Our point is that all were checked very carefully. The detailed problems we encountered included: 1) ambiguous names that could have been one or more clubs; 2) different clubs with what seemed to be the same name; 3) the same club listed under different names; 4) some sources using the native language of clubs while others use Anglicized versions of these names; 5) player stays present in some data sources but not others; 6) player stays described using different descriptive terms; 7) player stays lacking a start date (usually in the form of 00.00.year); 8) player stays without an end date; 9) stays with neither a start date nor an end date; 11) player stays at different clubs at the same time (excluding loan stays where a player can appear to belong to two clubs at the same time[25]); 12) players leaving a club before they arrived; 13) large gaps in playing careers of some players; and 14) sequences of player moves without dates. Together, these problems created a nightmare with regard to data cleaning and trying to ensure that our data had sufficient quality for the data analytic tools that we used. We describe some of what we did to improve both the completeness and quality of our data.

---

[23] Their focus tends to be more about current news – hence the tracking over time – and these sources cover games played, footballing controversies, and transfer news (plus rumors).

[24] Some websites reached when searching for the history of particular players were never taken seriously. We ignored all blogs and discussion forums set up to discuss players and/or clubs. Some did appear to provide information about player moves but, in the main, were speculative or consisted of 'I do not know' responses to earlier queries about specific players. As forums can be, and often are, replete with nasty rants about players and clubs, they were ignored. All websites purporting to provide football information but offered also access to Russian women, Ukrainian women, Asian women, or any (always young) women were exited quickly.

[25] When players were loaned, the loan stay took precedence: it was where the player was located.

Expecting specific clubs to be in our dataset on multiple occasions, we sought a unique name for every club. There was considerable ambiguity in the listings of club names across different sources (and even within the same data source). Some problems are easier to resolve than others. Club names in different countries can have accents on or under letters, and we removed them. Examples include: FC Bayern München for which we used Bayern Munchen and not Bayern Munich. The club name of Fenerbahçe (a Turkish club) became Fenerbahce in our listing and Gençlerbirliği (another Turkish club) is recorded as Genclerbirligi. Deportivo de La Coruña (a Spanish club) was recorded as Deportivo La Coruna (after removing also the 'de' from the name[26]). Other alphabets have letters not in the English language and we replaced them with the nearest English letter. For example, FC København was recorded as FC Kobenhavn (not FC Copenhagen, a frequently used name for this club). Another Danish team, Næstved IF, was recorded as Nestved. We tried to preserve as much of a club's source language name as possible. Consistent with this, we use Sparta Praha rather than Sparta Prague and Crvena Zvezda for Red Star Belgrade.[27]

More difficult to detect were cases where clubs have frequently used long and short names. From France, we recorded: Bordeaux instead of the full name of FC Girondins de Bordeaux; Rennes for Stade Rennais FC, and FC Nantes for FC Nantes Atlantique. Some clubs having very long names we shortened when the short names were used often in data sources. Examples include: a Greek team, Panthessalonikios Athlitikos Omilos Konstantinoupoliton, which we recorded as PAOK Salonika, as most sources do; another Greek team, Omilos Filathlon Irakleiou, which we recorded as OF Iraklion[28]; Tampereen Pallo-Veikot, playing in Finland, was recorded as TPV Tampere because TPV is frequently used as the club name and it plays in Tampere; Alkmaar Zaanstreek, a club in the Netherlands was recorded as AZ Alkmaar. However, we did not always opt for the shorter of two versions of a club's name. In the last two examples we did not use TPV for TPV Tampere nor did we use AZ for AZ Alkmaar even though these short names are used often. Also, we used Olympique Lyonnais rather than Lyon, a name used frequently for this French club in data sources.

Although we tended to include the name of the place where a club plays in the club name, there are exceptions for some well-known clubs. We followed the convention of using Celtic for Glasgow Celtic, even though there are other clubs (e.g. Stalybridge Celtic) that include the word Celtic in their names, and we used Rangers for Glasgow Rangers. However, Bohemians Dublin was recorded as Bohemians Dublin because it is not so well-known outside Ireland and we wanted to distinguish it from Bohemians Praha, a Czech club.

We never recorded a nickname for a club name in our data: 'the blues' was not used for Birmingham City or for Chelsea, Arsenal was used but not 'the Gunners', Manchester United was not recorded as 'the red devils', Everton was not recorded as 'the toffees', and Tottenham Hotspur was never listed as Spurs in our data. However, nicknames were used in tracing clubs or recognizing them in news accounts.

We tried to incorporate changes in club names in a systematic and consistent fashion. These changes come in various forms. In 2002, Wimbledon was granted permission to move

---

[26] We dropped 'de' from names, e.g. Olympique Marseille replaced Olympique de Marseille.

[27] We were not fully consistent and used Partizan Belgrade for the biggest intra-city rival of Crvena Zvezda. Although we kept Partizan, we used Belgrade rather than Beograd.

[28] This team has been listed also as OFI Creta and OF Irakleiou. We used OF Iraklion.

to a new ground and opted to make the 90 km move to Milton Keynes in 2003. This was deeply unpopular with the club's fans, and in 2004 the club's name was changed to Milton Keynes Dons.[29] The many disgruntled fans supported a lower-level club, AFC Wimbledon. While most data sources now use Milton Keynes Dons as the name for this club even for times before 2004, we do not. For players with the club before 2004, we used Wimbledon as the club and, for players there after 2004, we used Milton Keynes Dons. We regard them as distinct clubs. In 2007, Gravesend and Northfleet, a non-league club, was renamed Ebbsfleet United. We treated this in the same fashion, depending on when players were with the club, and treated name changes in a similar fashion. Some ambiguity remains however. Very late in our data collection, we learned that FC Nantes had been 'referred to' as FC Nantes Atlantique for the period 1992–2007 but decided to retain FC Nantes as the club's name in our database.

Clubs having, for lack of a better term, coupled histories posed an additional problem. Aldershot was a team that went bankrupt in 1992 and ceased to exist. In 1992, a new team in the same place was formed and named Aldershot Town. The club symbol even shows a phoenix rising from the ashes and can viewed as a kind of continuation of the old team. We use Aldershot for players there before 1992 and Aldershot Town for players at the club after that data. Accrington Stanley, a Lancashire club, went bankrupt in 1966 and a new club was formed in 1968. However, we did not use two names for the club before and after 1966 because our focus was on a later era. Football Club United of Manchester (listed also as FC United of Manchester, or FC United, or FCUoM or FCUM) is not to be confused with Manchester United even though it genesis was fueled by fan disapproval of a change in the ownership of Manchester United. We are sure that we have caught these kinds of distinctions for clubs in leagues where English is the language used, for example Dallas Burn in the USA changing its name to Dallas FC in 2005. For similar changes in other leagues we are less sure.

When more than one football club exists in a given city, other problems arise. We use AC Milan for one of the giant teams in Milan and Inter Milano for the other. (We used neither Inter Milan nor a part of its full official name, Internazionale, to avoid ambiguity.) For cities like Manchester (with Manchester City and Manchester United), Sheffield (with Sheffield United and Sheffield Wednesday) and Bristol (with Bristol City and Bristol Rovers) care is needed to make sure that these teams are not confused. Of course, the more well-known are the clubs in a particular city, the lower is the chance of confusing them.

The potential difficulty of a club being listed with different names or having different clubs being listed with the same name is not the same problem as distinguishing long and short variants of a club's name. In general, this problem involves place names. For example, one data source had both Swansea City and Swansea Town as distinct clubs. There was a transition from the former name to the latter name in 1970/1. We use Swansea City as the club name given how far back in time this club's name change was made. On tracing a player to the Australian city of Perth, we learned that the recording of his club as Perth Glory was incorrect and that he played for Perth SC, a different club. Typing errors in sources we visited were another source of seemingly simple errors. For example, Hartlepool United has been recorded as Hartlepool and Hartlepools in different data sources. Through systemic checking

---

[29] Wimbledon had the nickname 'the dons' and 'Dons' was included in the *formal* name of the new club. So we do not treat it as just a nickname.

and maintaining multiple alternative names for clubs, we are confident that we have unique club names in our database.

We sought good data on player stays and moves. For a player's moves, the sequence of clubs through which he passed had particular importance. We also needed dates of arrivals and departures at clubs. When data sources disagreed about the club of a player, the starting and/or ending dates of stays for a player, or the nature of the stay, our operating rule was 'the majority wins', and confirmation from multiple sources was sought. This included reading narratives of player careers as a way of resolving some discrepancies. Over time, we developed a sense that some sources were more reliable than others and, faced with contradictions between two sources, we opted in favor of the information in the more reliable (or less unreliable) source. But, as noted earlier, no one data source is fully trustworthy. We cannot be sure that we always made the correct choice. It was straightforward to identify nonsense dates such as 00.00.year. As such, they are expressions of ignorance in data sources. Lacking other evidence we changed the 00.00 to the starting date of the season of the year, usually to 01.07.

Recording dates as day/month/year or month/day/year varies across countries, and extra care was needed in determining which convention was being used in each data source. Programs were written to identify logical impossibilities in the data. When the starting date of a player came after his departure for a particular club, this set off another search of the records. Most often, errors were typing errors. Absent a loan, players appearing at two clubs simultaneously signaled data problems to be checked.[30] Many rounds of checking data were needed. Another problem was the presence of long gaps in playing careers, often occurring within single sources. One part of this problem took the form of incorrect dates in sources. These were corrected. They were particularly acute for players descending down the levels of clubs in the second 'half' of their careers. For many, dropping into non-league football begins a period of shorter club stays. Combinations of player narratives, news media stories, and non-league club websites provided the names of clubs for a player. Getting the actual dates was another matter. News stories about games or listings of squads for games provided some clues for determining dates for stays. If there was a long gap and we knew the sequence of clubs, we divided the gap into equal periods simply to have start and end dates for stays.

For some players we had no information beyond the presence of gaps.[31] When there are long enough gaps – beyond 3 months – in playing careers that are not over, we created an 'imaginary club' to cover/describe these gaps. In our search through many data sources, we frequently encountered the term 'without a club' for players. Earlier, we described players being released by clubs. Once released, players seeking to continue playing have to search for clubs willing to hire them. Being released often precedes 'having no club', and these are real periods in careers. Whimsy prompted calling such gaps as an imaginary club. However, we are serious and offer the following justifications: 1) when there are real long gaps in careers, it is simpler to record this if no further temporal information exists; 2) for some analyses we needed times for all stages of a career, and the imaginary club provides this coverage; and 3) this is most often present towards the ends of careers. The first rationale is important because

---

[30] We did encounter instances of players being jointly owned by clubs, and players owned jointly by agents and clubs. We focused solely on the series of clubs and the sequence of moves.

[31] Some apparent gaps signal effective ends of playing careers. Some players do retire from professional football and continue to play as amateurs. Other players retire for several years before attempting, most often without success, comebacks. For our purposes, the initial retirement marks the end of a playing career and we ignore these gaps.

it rules out defining moves between clubs when no relations exist between successive clubs. The term 'imaginary' appears in some of our club-to-club network diagrams in Chapter 8. This signals being without a club and highlights the practice of discarding players.

A.4.2   Ancillary data

We obtained league tables for England for the four top levels back to 1958/9 when a simple four-level league system was created.[32] Another ancillary dataset consists of overall rankings of clubs for France, Germany, Italy, and Spain, dating back as far as possible. We restricted this to the top two levels in each country as a practical matter. We obtained rankings for 1946/7 through 2010/11 for France, Italy, and Spain. For Germany, the league tables for the Bundesliga (back to 1963/4) and 2-Bundesliga (back to 1974/5) were created.[33] As noted earlier, we used unique names for all clubs. Difficulties in doing so re-emerged again for the ancillary data: these data sources often used different names for many clubs.

For France, Italy, and Spain, extracting the relevant tables was straightforward, using a combination of Wikipedia and official sites within these countries. Inconsistencies were resolved in favor of the national official sources. For France, we used:

1. http://en.wikipedia.org/wiki/Football_in_France
2. http://www.fff.fr/
3. http://new.lfp.fr/ligue1 (for Ligue 1)
4. http://new.lfp.fr/ligue2 (for Ligue 2)[34]

For Italy, we used:

1. http://en.wikipedia.org/wiki/Serie_A
2. http://www.legaseriea.it/ (for Serie A)
3. http://en.wikipedia.org/wiki/Serie_B
4. http://www.legaserieb.it/it (for Serie B)

For Spain, we used:

1. http://en.wikipedia.org/wiki/La_Liga
2. http://www.soccer-spain.com/ (also for La Liga)
3. http://www.lfp.es/Estadisticas/ClasificacionHistoricos.aspx
4. http://http://en.wikipedia.org/wiki/Segunda_División (for the second level)

---

[32] Prior to that date, there were two top levels and the third level was divided into two regional (North and South) leagues. Historical tables going back to 1888, across several reorganizations, were obtained from an official website no longer accessible. Others tables were obtained through the league table generator available at the 11v11 website.

[33] We used http://www.bundesliga.de/de/statistik/spieltag/index.php. Season 1991/2 did not fit neatly in the time series of tables: there were two regional tables which we merged to get overall rankings in 2-Bundesliga for that season.

[34] We had sought data at the third level, given that we had such data for England. We learned from an email correspondence with the webmaster of http://www.fff.fr/ that electronic versions of these (regional) data do not exist.

Calculating overall league positions was complicated when a second tier was broken into two or more regional leagues, at different times. These had to be combined to get second tier ranks before appending them to the top-tier ranks. League tables were used also to obtain promotions and relegations between the top leagues[35] and to keep track of the top-tier titles won by specific clubs.

Another dataset was constructed for aggregated flows of foreign players to the top leagues of Europe from the end of WWII through 2010/11. Attention was confined to the top league only. These data were obtained from Wikipedia, converted into Excel files, and transformed to time-marked Pajek network files.[36]

The foreign player flows to the EPL from Wikipedia correspond very closely to our data for 1992/93 through 2006/7. This established, we accurately included foreign player presence in the EPL. These aggregate flows are silent regarding the routes that foreign players took in reaching the EPL.

Many players reach the EPL via journeys starting in many clubs on different continents. A large proportion of them move to an EPL club from another European club. We needed rankings of all European clubs to get a better sense of the clubs from which players arrive in the EPL. These rankings do not exist. However, there is a ranking of the league systems of countries (not national teams although national teams are ranked also). Although the separate leagues operate quite independently from each other they are linked through membership in UEFA. UEFA represents most of the European national football associations. It also organizes competitions involving clubs from different national leagues. These clubs qualify for these competitions by being successful (finishing high enough) in the top-tier league, or by success in national knockout cup competitions. There have been a variety of competitions involving clubs from different countries. The European Cup, the most prestigious, started in 1955 as a knockout competition for which only the champions of the national leagues qualified. UEFA sets the regulations for all its competitions and holds the media rights for them and controls the prize money for each competition.

In 1992 the European Cup was reorganized and called the Champions League. There are 32 clubs in the group stage. This is preceded by four qualifying stages, the last of which is the playoff round. The champions of all countries enter the Champions League. Some teams of high-ranked leagues qualify directly for the group stage, while others have to play in the qualifying rounds. High-ranked countries can send up to four teams.[37] Many champions of

---

[35] There was not an automatic procedure. There were instances when clubs remained in a league despite finishing in a relegation spot at the end of a season if clubs above them were relegated for financial problems or rule violations. In Spain, La Liga reserve teams cannot be promoted to La Liga even if they win the Segunda División: the senior team is already in La Liga. Instead, another club is promoted. France's Ligues 1 and 2 have a similar feature. Also, reserve clubs in the second level are automatically relegated to a lower level when the senior team is relegated.

[36] The sources were text files at:

- http://en.wikipedia.org/wiki/List_of_foreign_Ligue_1_players (France)
- http://en.wikipedia.org/wiki/List_of_foreign_Bundesliga_players (Germany)
- http://en.wikipedia.org/wiki/List_of_foreign_Serie_A_players (Italy)
- http://en.wikipedia.org/wiki/List_of_foreign_La_Liga_players (Spain)
- http://en.wikipedia.org/wiki/List_of_foreign_Premier_League_players (England). Here, the data of flows to the top league is for the EPL only and so covers a much shorter period.

[37] In the 2010/11 Champions League, England, Spain, and Germany had four clubs in the competition, one of which went into the preliminary qualifying round.

weaker leagues go into the qualifying phase for the competition proper. Once the 32 clubs have been determined, the first round has eight groups of four teams for which the first stage is a round-robin tournament to determine which clubs move to the next round. Thereafter, it is a home-and-away knockout competition with the final being a single game. This competition can be very lucrative for successful clubs. Many clubs qualifying for the Champions League are not national champions. This cup has been won by such teams.

The second UEFA competition is the Europa League, formed for the 2009/10 season. The prize monies are much lower. It also has forerunners. Previously, it was known as the UEFA Cup after UEFA incorporated it in 1971. From 1955 to 1971, it was the Inter-Cities Fairs Cup.[38] The Europa Cup is now more complex than its predecessors with many more teams qualifying than before. In general, 'weaker clubs' who finish just outside the top ranks in the national leagues or win internal cup competitions enter the Europa Cup. The Champions League and the Europa Cup are linked: some teams in the Champions League not surviving the group stages drop into the Europa Cup. They are further linked by the winners of the two UEFA leagues taking part in the European Super Cup at the beginning of the following season, another money-making venture for clubs and UEFA.

Detailed placements and seedings for these competitions are based on UEFA coefficients. These are used to score and rank many European clubs and most European leagues.[39] The data come from a separate website[40] as were the methods used to establish the rankings based on the performance of European clubs in UEFA competitions. We extracted these data as preliminary European club rankings and rankings of leagues.

A.5   The USA spatial county network

There are two data source domains. One provided statistical data on counties while the other was used for constructing the adjacency relation for counties. The data for the analyses presented in Chapter 9 come from a variety of sources described in the following sections.

A.5.1.1   Statistical data for counties

For the statistical data we selected county-level variables that we thought were congruent with *Nine Nations of North America* after downloading many variables. The original data contain 6849 variables[41] that were downloaded and stored. The package, xlsReadWrite (http://www.swissr.org/download), was downloaded and used for reading and writing Excel files storing the data. From these data, we selected 1125 variables, as a first step, and stored them also. These data contained information for the USA as a whole, states, and counties. Given our focus on counties, data for states and for the country were removed. There was another complication with a duplication: there was a unit for Dade County which was identical to Miami-Dade, both in Florida. This duplication was removed to leave one county unit in the data. Boston, as a city, was in the data: it was removed also.

---

[38] UEFA jealously guards its domains and this incorporation of the Inter-Cities Fairs Cup, while a natural continuation of an extant competition, is regarded by UEFA as a sharp divide. No results from the Inter-Cities Fairs Cup are recognized by UEFA because their brand was absent.

[39] UEFA's website (http://www.uefa.com/) provides these ranks

[40] http://kassiesa.home.xs4all.nl/bert/uefa/data/method4/crank2011.html

[41] They were downloaded from http://www.census.gov/support/USACdataDownloads.html, and additional data were obtained from the National Oceanic and Atmospheric Administration (http://www.noaa.gov).

The resulting data were reordered to make their order consistent with the ordering of the counties in the file with adjacency links between counties. At this stage, the number of variables was reduced to 92, based on the indicators best matching the narrative of Garreau (1981).

It turned out that not all tables have the same set of units which required detailed searches for data on additional units and matching them. The scheme for doing this is:

```
> setwd("D:/Data/counties/USAC")
> library(xlsReadWritePro)
>
> standard <- function(x)
+   {s <- paste("0000",x,sep=""); substr(s,nchar(s)-4,nchar(s))}
>
> mergeVars <- function(v,t){
+   cat("merge:\n"); flush.console()
+   vid <- v$STCOU; tid <- t$STCOU
+   u <- merge(v,t,by.x="STCOU",by.y="STCOU",all=TRUE)
+   vt <- setdiff(vid,tid); tv <- setdiff(tid,vid)
+   if(length(vt)>0) cat(" missing:",vt,"\n")
+   if(length(tv)>0) cat(" new    :",tv,"\n")
+   u$Areaname <- ifelse(is.na(u$Areaname.x),u$Areaname.y,u$Areaname.x)
+   u$Areaname.x <- u$Areaname.y <- NULL
+   return(u)
+ }
>
> readVars <- function(xlsFiles){
+   first <- TRUE
+   for(xlsFile in xlsFiles){
+     cat(xlsFile,":\n",sep=''); flush.console()
+     xl <- xls.open(xlsFile,readonly=TRUE)
+     ix <- xls.info(xl); sn <- ix$sheet.names
+     for(s in sn){
+       cat("  ",s,": ",sep=''); flush.console()
+       t <- read.xls(xl,sheet=s,colNames=TRUE,stringsAsFactors=FALSE)
+       t$STCOU <- standard(t$STCOU)
+       if(first) {v <- t; first <- FALSE; cat("\n")} else v <- mergeVars(v,t)
+     }
+   }
+   return(v)
+ }
> # -----------------------
> afn <- readVars("./afn01.xls")
./afn01.xls:
  Sheet1:
  Sheet2: merge:
  Sheet3: merge:
> age <- readVars(c("./age01.xls","./age02.xls","./age03.xls","./age04.xls"))
./age01.xls:
  Sheet1:
  Sheet2: merge:
  ...
  Sheet10: merge:
./age04.xls:
  Sheet1: merge:
  Sheet2: merge:
> safn <- c("Areaname","STCOU","AFN220197D","AFN220202D","AFN230197D","AFN230202D",
+       "AFN320197D","AFN320202D","AFN330197D","AFN330202D")
```

```
> afnr <- afn[,safn]
> sage <- c("Areaname","STCOU","AGE010180D","AGE010190D","AGE010200D","AGE050180D",
+       "AGE050190D","AGE050200D","AGE110180D","AGE110190D","AGE110200D",
+       "AGE270180D","AGE270190D","AGE270200D","AGE880190D","AGE880200D")
> ager <- age[,sage]
> cou <- mergeVars(afnr,ager)
merge:
  missing: 02105 02195 02198 02230 02275
  new    : 02201 02232 02280
> write.csv(cou,file="counties1.csv",row.names=FALSE)
```

A final set of statistical variables was then selected in stages: the huge set of variables was reduced to 1125, then to 116 (including computed variables) and then to the 42 variables used in the discriminant analyses and clustering with a relational constraint.

A.5.1.2 Constructing the adjacency relation for US counties

The 3111 mainland US counties neighborhood relation was constructed by Luc Anselin and obtained from http://sal.uiuc.edu/weights/index.html (accessed[42] 11 June 2007).

For visualizing the US counties with a map, it was necessary to extract files with the map data. They were obtained from http://gadm.org/country [country: USA, file format: shapefile] and http://gadm.org/data/shp/USA_adm.zip. For visualization, some R library map tools were used.

To produce the clustering (partition) for Nine Nations, as well as to include corrections, we started with a partition in which complete states were included. The final partition was obtained by corrections to this partition according to the assignments specified for specific counties where boundaries between nation did not coincide with state boundaries. The specific code for doing this was as follows:

```
> setwd("D:/Data/counties/9nations")
> library(maptools)
> gpclibPermit()
> USsta <- readShapeSpatial("USA/USA_adm1.shp")  # state borders
> UScou <- readShapeSpatial("USA/USA_adm2.shp")  # county borders
> load('pq.Rdata')
> state <- read.csv("../pajek/states3110.clu",header=FALSE,skip=1)$V1
> col <- c("red","yellow","green","blue","pink","brown","orange","purple","gray","white")
> nine <- rep(NA,3110)
> nine[state %in% c(25,33,50,44,23)] <- 1                        # New England
> nine[state %in% c(36,34,24,42,39,26)] <- 2                     # Foundry
> nine[state %in% c(54,51,21,37,45,47,13,1,28,22,5,12)] <- 3     # Dixie
> nine[state %in% c(48,35,6)] <- 5                               # Mexamerica
> nine[state %in% c(41,53)] <- 6                                 # Ecotopia
> nine[state %in% c(4,56,30,49,16,32,8)] <- 7                    # The Empty Quarter
> nine[state %in% c(17,19,20,27,31,38,46,40,55,29)] <- 8         # The Breadbasket>
> ids <- read.csv("../usc3110lab.csv",header=FALSE,stringsAsFactors=FALSE)
> id <- ids$V2 <- standard(ids$V2)
> S <- read.csv(file="states.csv",stringsAsFactors=FALSE,sep=";")
> states <- levels(UScou$NAME_1); ps <- match(states,S$name)
> names <- paste(UScou$NAME_2,", ",S$code[ps[as.integer(UScou$NAME_1)]],sep="")
> Name <- rep(NA,3110)
> for(v in 1:3110) {i <- q[[p[[v]]]]; if(!is.na(i)) Name[[v]] <- names[[i]]}
```

---

[42] Unfortunately, this website is no longer available, no doubt due to the geographic migrations of its author. We did not try to locate a more recent file with this relation having already located it.

```
> change <- read.csv(file="9nationsClu.csv",stringsAsFactors=FALSE,sep=";",header=TRUE)
> pos <- match(change$County,Name)
> err <- which(is.na(pos))
> cat(change$County[err],'\n')

> ok <- which(!is.na(pos))
> posok <- pos[ok]
> nine[posok] <- change$Cluster[ok]
> clu <- rep(NA,length(UScou$NAME_1))
> for(v in 1:3110) {i <- q[[p[[v]]]]; if(!is.na(i)) clu[[i]] <- nine[[v]]}
> UScou$clu <- clu
> UScou$clu[which(is.na(UScou$clu))] <- 10
```

The partition was checked by drawing it on the map of the USA with:

```
pdf("USc9nations.pdf",width=11.7,height=8.3,paper="a4r")
plot(UScou,xlim=c(-124,-67),ylim=c(23,48),col=col[UScou$clu],bg="skyblue",
   border="black",lwd=0.05)
plot(USsta,xlim=c(-124,-67),ylim=c(23,48),lwd=0.2,border="violet",add=TRUE)
text(coordinates(UScou),labels=as.character(UScou$NAME_2),cex=0.1)
title("Central US"); dev.off()
save(nine,file='nineClu.Rdata')
out <- file("9nations.clu","w"); cat("*vertices 3110",nine,sep="\n",file=out)
close(out)
```

The correct result is shown in Figure 9.2.

References

Agneessens, F. and Everett, M.G. (eds.) (2013) Advances in two-mode social networks (special issue). *Social Networks*, **35**(2), 145–178.

Ahmed, A., Batagelj, V., Fu, X. et al. (2007) Visualisation and analysis of the internet movie database In *Proceedings of the Asia-Pacific Symposium on Visualisation (APVIS2007)* (ed. Hong SH and Ma KL), pp. 17–24. IEEE Computer Society, Sydney.

Alegi, P. (2010) *African Soccerscapes: How a Continent Changed the World's Game*. Ohio University Press, Athens, OH.

Allen, J.F. (1983) Maintaining knowledge about temporal intervals. *Communications of the ACM*, **26**(11), 832–843.

Anthonisse, J.M. (1971) *The Rush in a Graph*. Mathematische Centrum, Amsterdam.

Asimov, I. (1963) *The Genetic Code*. New American Library, New York.

Barabási, A.L. (2003) *Linked: How Everything is Connected to Everything Else and What It Means for Business, Science, and Everyday Life*. Plume, New York.

Barabási, A.L. and Albert, R. (1999) Emergence of scaling in random networks. *Science*, **286**(5439), 509–512.

Batagelj, V. (1991) Some mathematics of network analysis. Network Seminar, Department of Sociology, University of Pittsburgh, 21 January, 1991. http://vlado.fmf.uni-lj.si/pub/networks/doc/mix/report.pdf.

Batagelj, V. (1994) Semirings for social networks analysis. *The Journal of Mathematical Sociology*, **19**(1), 53–68.

Batagelj, V. (2003) Efficient algorithms for citation network analysis http://arxiv.org/abs/cs.DL/0309023.

Batagelj, V. (2007) *Wos2Pajek – networks from web of science [Data converter]*. http://pajek.imfm.si/doku.php?id=wos2pajek.

Batagelj, V. and Cerinšek, M. (2013) On bibliographic networks. *Scientometrics*, **96**(3), 845–864.

Batagelj, V. and Mrvar, A. (1996–2013) `Pajek` – *program for analysis and visualization of large network*. http://pajek.imfm.si/lib/exe/fetch.php?media=dl:pajekman.pdf.

Batagelj, V. and Mrvar, A. (1998) Pajek – program for large network analysis. *Connections*, **21**(2), 47–57.

Batagelj, V. and Mrvar, A. (2008) Analysis of kinship relations with Pajek. *Social Science Computer Review*, **26**(2), 224–246.

Batagelj, V. and Zaveršnik, M. (2007) Short cycle connectivity. *Discrete Mathematics*, **307**(3-5), 310–318.

Batagelj, V. and Zaveršnik, M. (2011) Fast algorithms for determining (generalized) core groups in social networks. *Advances in Data Analysis and Classification*, **5**(2), 129–145.

Batagelj, V., Ferligoj, A. and Mrvar, A. (2008) Hierarchical clustering in large networks. Workshop on Detection, evolution and visualization of communities in complex networks, Louvain-la-Neuve, 13–14 March, 2008.

Batagelj, V., Mrvar, A. and Zaveršnik, M. (1999) Partitioning approach to visualization of large graphs In *Graph Drawing, 7th International Symposium, GD'99, Stirín Castle, Czech Republic, September 1999, Proceedings* (ed. Kratochvíl J), vol. 1731 of *Lecture Notes in Computer Science*, pp. 90–97. Springer.

Bavelas, A. (1948) A mathematical model of group structures. *Human Organization*, **7**(3), 16–30.

Bavelas, A. (1950) Communication patterns in task-oriented groups. *The Journal of the Acoustical Society of America*, **22**(6), 725–730.

Beauchamp, M.A. (1965) An improved index of centrality. *Behavioral Science*, **10**(2), 161–163.

Bell, M.G.H. and Iida, Y. (1997) *Transportation Network Analysis*. Wiley, Chichester.

Ben-Porat, B. and Ben-Porat, A. (2004) (Un)bound soccer: Globalization and localization of the game in Israel. *International Review for the Sociology of Sport*, **39**(4), 421–436.

Benesh, S.C. and Spaeth, H.J. (2003) The Supreme Court Justice-centered Judicial Databases: The Warren, Burger, and Rehnquist courts (1953–2000 terms).

Berk, M.L. and Monheit, A.C. (1992) The concentration of health expenditures: an update. *Health Affairs*, **11**(4), 145–149.

Berk, M.L. and Monheit, A.C. (2001) The concentration of health expenditures, revisited. *Health Affairs*, **20**(2), 9–18.

Bernal, J.D. (1953) *Science and Industry in the Nineteenth Century*. Routledge, London.

Bickel, P.J. and Doksum, K.A. (1977) *Mathematical Statistics: Basic Ideas and Selected Topics*. Prentice Hall, New Jersey.

Billard, L. and Diday, E. (2006) *Symbolic Data Analysis: Conceptual Statistics and Data Mining*. Wiley, New York.

Bird, S., Klein, E. and Loper, E. (2009) *Natural Language Processing with Python*. O'Reilly Media, Inc., Sebastopol, CA.

Bonacich, P. (1987) Power and centrality: A family of measures. *American Journal of Sociology*, **92**(5), 1170–1182.

Bonacich, P. (2004) The invasion of the physicists. *Social Networks*, **26**(3), 285–288.

Bonhoure, F., Dallery, Y. and Stewart, W.J. (1993) *Algorithms for Periodic Markov Chains* vol. 48 of *The IMA Volumes in Mathematics and its Applications*. Springer.

Borgatti, S.P. and Everett, M.G. (2006) A graph-theoretic framework for classifying centrality measures. *Social Networks* **28**(4), 466–484.

Borgatti, S.P., Everett, M.G. and Johnson, J.C. (2013) *Analyzing Social Networks*. Sage, London.

Brandes, U. (2001) A faster algorithm for betweenness centrality. *Journal of Mathematical Sociology*, **25**(2), 163–177.

Broder, A.Z., Kumar, R., Maghoul, F. et al. (2000) Graph structure in the web. *Computer Networks*, **33**(1-6), 309–320.

Bruynooghe, M. (1977) Méthodes nouvelles en classification automatique des données taxinomiques nombreuses. *Statistique et Analyse des Données*, (3), 24–42.

Burt, R.S. (1980) Models of network structure. *Annual Review of Sociology*, **6**(1), 79–141.

Burt, R.S. (1992) *Structural Holes: The Social Structure of Competition*. Harvard University Press, Cambridge, Mass.

Calero-Medina, C. and Noyons, E. (2008) Combining mapping and citation network analysis for a better understanding of the scientific development: The case of the absorptive capacity field. *Journal of Informetrics*, **2**, 272–279.

Carpenter, M.P, Narin, F. and Wolf, P. (1981) Citation rates to technologically important patents. *World Patent Information*, **3**(4), 160–163.

Cartwright, D. and Harary, F. (1956) Structural balance: A generalization of Heider's theory. *Psychological Review*, **63**, 277–293.

Casteigts, A. and Flocchini, P. (2013) Deterministic Algorithms in Dynamic Networks: Formal Models and Metrics. Technical Report DRDC Ottawa CR 2013-020, Defence R&D Canada.

Casteigts, A., Flocchini, P., Quattrociocchi, W. et al. (2012) Time-varying graphs and dynamic networks. *International Journal of Parallel, Emergent and Distributed Systems*, **27**(5), 387–408.

Chen, C. (1998) Generalised similarity analysis and pathfinder network scaling. *Interacting with Computers*, **10**(2), 107–128.

Chinni, D. and Gimpel, J. (2010) *Our Patchwork Nation: The Surprising Truth about the 'Real' America*. Gotham Books, New York.

Coleman, J.S. (1964) *An Introduction to Mathematical Sociology*. The Free Press, New York.

Conwell, L.J. and Cohen, J.W. (2005) Characteristics of people with high medical expenses in the U.S. civilian noninstitutionalized population, 2002, Statistical Brief #73 http://meps.ahrq.gov/mepsweb/data_files/publications/st73/stat73.pdf.

Cook, K.S., Emerson, R.M., Gilmore, M.R. et al. (1983) The distribution of power in exchange networks: Theory and experimental results. *American Journal of Sociology*, **89**(2), 275–305.

Cormen, T.H., Leiserson, C.E. and Rivest, R.L. (2001) *Introduction To Algorithms* 2 edn. MIT Press, Cambridge.

Correa, J.R. and Stier-Moses, N.E. (2011) Wardrop equilibria In *Wiley Encyclopedia of Operations Research and Management Science* (ed. Cochran JJ, Cox LAJ, Keskinocak P, Kharoufeh JP and Smith JC) John Wiley & Sons, Inc.

Cox, A. (1987) *The Court and the Constitution*. Houghton Mifflin, Boston, MA.

Crouch, T. (2006) *The World Cup: The Complete History*. Aurum Press, London.

Davis, A. and Gardner, B.B. (1941) *Deep South*. University of Chicago Press, Chicago.

de Nooy, W., Mrvar, A. and Batagelj, V. (2012) *Exploratory Social Network Analysis with Pajek (Structural Analysis in the Social Sciences)* revised and expanded second edn. Cambridge University Press, Cambridge.

Dechter, R. (2003) *Constraint Processing*. Morgan Kaufmann, San Francisco.

Diday, E. (1979) *Optimisation en classification automatique, Tome 1., 2..* INRIA, Rocquencourt (in French).

Dieudonné, J. (1960) *Foundations of modern analysis*. Academic Press, New York.

Doreian, P. (1979–1980) On the evolution of group and network structure. *Social Networks*, **2**(3), 235–252.

Doreian, P. 2006 Exploratory social network analysis with Pajek. *Social Networks*, **28**(3), 269–274.

Doreian, P. and Mrvar, A. (1996) A partitioning approach to structural balance. *Social Networks*, **18**, 149–168.

Doreian, P., Batagelj, V. and Ferligoj, A. (2005) *Generalized Blockmodeling*. Cambridge University Press, Cambridge.

Doreian, P. and Stokman, F.N. (eds.) (1997) *Evolution of Social Networks*. Gordon and Breach Publishers, Amsterdam.

Duke, V. (1994) The flood from the east? Perestroika and the migration of sports talent from Eastern Europe In *The Global Sports Arena: Athletic Talent Migration in an Interdependent World* (ed. Bale J and Maguire J), pp. 151–167. Frank Cass, London.

Dunbar, R.I.M. (1992) Neocortex size as a constraint on group size in primates. *Journal of Human Evolution*, **22**(6), 469–493.

Erdrich, L. (2013) Rape on the reservation. New York Times, 26 Feb.

Exall, K.P. (2011) *Who Killed English Football? An Analysis of the State of English Football.* AuthorHouse, Milton Keyne.

Ferligoj, A. and Batagelj, V. (1982) Clustering with relational constraints. *Psychometrika*, **47**(4), 413–426.

Ferligoj, A. and Batagelj, V. (1983) Some types of clustering with relational constraint. *Psychometrika*, **48**(4), 541–552.

Flament, C. (1963) *Applications of Graph Theory to Group Structure*. Prentice-Hall, Englewood Cliffs, NJ.

Fletcher, J.G. (1980) A more general algorithm for computing closed semiring costs between vertices of a directed graph. *Communications of the ACM*, **23**(6), 350–351.

Fowler, J.H. and Jeon, S. (2008) The authority of Supreme Court precedent. *Social Networks*, **30**(1), 16–30.

Freeman, L.C. (1977) A set of measures of centrality based on betweenness. *Sociometry*, **40**(1), 35–41.

Freeman, L.C. (1979) Centrality in social networks: Conceptual clarification. *Social Networks*, **1**(3), 215–239.

Freeman, L.C. (2004) *The Development of Social Network Analysis: A Study in the Sociology of Science.* ΣP Empirical Press, Vancouver, BC.

Freeman, L.C, Borgatti, S.P. and White, D.R. (1991) Centrality in valued graphs: A measure of betweenness based on network flow. *Social Networks*, **13**(2), 141–154.

Gan, G., Ma, C. and Wu, J. (2007) *Data Clustering – Theory, Algorithms, and Applications*. SIAM, Philadelphia.

Garey, M.R. and Johnson, D.S. (1979) *Computers and Intractability: A Guide to the Theory of NP-Completeness*. W.H. Freeman & Co., New York.

Garfield, E. (1979) *Citation indexing – its theory and application in science, technology, and humanities.* John Wiley & Sons, Inc.

Garfield, E., Sher, I. and Torpie, R. (1964) The use of citation data in writing the history of science. http://www.garfield.library.upenn.edu/papers/useofcitdatawritinghistofsci.pdf

Garreau, J. (1981) *The Nine Nations of North America*. Avon, New York.

George, B., Kim, S. and Shekhar, S. (2007) Spatio-temporal network databases and routing algorithms: A summary of results In *Advances in Spatial and Temporal Databases* (ed. Papadias D, Zhang D and Kollios G) vol. 4605 of *Lecture Notes in Computer Science* Springer Heidelberg, New York, Dordrecht, London pp. 460–477.

Goldblatt, D. (2006) *The Ball is Round: A Global History of Football*. Penguin-Viking, London.

Graham, A. (2005) *Football in Italy, a statistical record 1898–2005*. Soccer Book Ltd, Cleethorpes, UK.

Granovetter, M. (1973) The strength of weak ties. *American Journal of Sociology*, **78**(6), 1360–1380.

Granovetter, M. (1985) Economic action and social structure: The problem of embeddedness. *American Journal of Sociology*, **91**(3), 481–510.

Greene, D., Doyle, D. and Cunningham, P. (2010) Tracking the evolution of communities in dynamic social networks In *International Conference on Advances in Social Networks Analysis and Mining (ASONAM 2010)* (ed. Memon N and Alhajj R), pp. 176–183. IEEE, Los Alamitos.

Griliches, Z. (ed.) (1984) *R&D, Patents, and Productivity*. University of Chicago Press, Chicago.

Griliches, Z. (1990) Patent statistics as economic indicators: A survey. *Journal of Economic Literature*, **28**(4), 1661–1707.

Guerrero-Bote, V.P, Zapico-Alonso, F., Espinosa-Calvo, M.E. et al. (2006) Binary pathfinder: An improvement to the pathfinder algorithm. *Information Processing and Management*, **42**(6), 1484–1490.

Hall, B.H., Jaffe, A.B. and Trajtenberg, M. (2002) The NBER patent-citations data file: Lessons, insights, and methodological tools In *Patents, Citations, and Innovations: A Window on the Knowledge Economy* (ed. Jaffe AB and Trajtenberg M) The MIT Press Cambridge, MA pp. 403–459.

Hall, K. (1999) *The Oxford Guide to United States Supreme Court Decisions*. Oxford University Press.

Hall, K.L. (ed.) (2005) *The Oxford Guide to the Supreme Court* second edn. Oxford University Press, New York.

Hammond, M. (ed.) (2005) *The European Book of Football, 2005/2006: A Complete Guide to the Continental Game*. m.press Ltd, Romford, UK.

Harary, F. (1969) *Graph Theory*. Addison-Wesley, Reading, MA.

Harary, F., Norman, R.Z. and Cartwright, D. (1965) *Structural models – an introduction to the theory of directed graphs*. Wiley, New York.

Harris, N. (2006) *The Foreign Revolution: How Overseas Footballers Changed the English Game*. Aurum, London.

Hartigan, J.A. (1975) *Clustering algorithms*. Wiley-Interscience, New York.

Heider, F. (1946) Attitudes and cognitive organization. *Journal of Psychology*, **21**, 107–112.

Hidalgo, C.A., Klinger, B., Barabási, A.L. et al. (2007) The product space conditions the development of nations. *Science*, **317**(5837), 482–487.

Hobsbawn, E. (1995) *Age of Extremes: The Short Twentieth Century, 1914–1991*. Abacus, London.

Holben, B.N., Eck, T.F., Slutsker, I. et al. (1998) Aeronct – a federated instrument network and data archive for aerosol characterization. *Remote Sensing of Environment*, **66**(1), 1–16.

Holland, P., Lasky, K. and Leinhardt, S. (1983) Stochastic blockmodels: First steps. *Social Networks*, **5**, 109–137.

Holme, P. and Saramäki, J. (2012) Temporal networks. *Physics Reports*, **519**(3), 97–125.

Holme, P. and Saramäki, J. (eds.) (2013) *Temporal Networks* Understanding Complex Systems. Springer, Heidelberg, New York, Dordrecht, London.

Holme, P., Kim, B.J., Yoon, C.N. et al. (2002) Attack vulnerability of complex networks. *Physical Review E*, **65**(5), 056109.

Hughes, R. (2013) A very un-British fight for the English title. *New York Times* pp. 20 October 2013.

Hugman, B. (2005) *The PFA Footballers' Who's Who, 2005–6*. Queen Anne Press, Harpenden, UK.

Hummon, N.P. and Doreian, P. (1989) Connectivity in a citation network: The development of DNA theory. *Social Networks*, **11**(1), 39–63.

Hummon, N.P. and Doreian, P. (1990) Computational methods for social network analysis. *Social Networks*, **12**, 273–288.

Hummon, N.P., Doreian, P. and Freeman, L.C. (1990) Analyzing the structure of the centrality-productivity literature created between 1948 and 1979. *Knowledge: Creation, Diffusion, Utilization*, **11**(4), 459–480.

Hyman, H.M. (1959) *To try men's souls: Loyalty oaths in American history*. University of California Press, Berkeley.

Interlink (1990) *Interlink – Tools for Pathfinder Network Analysis*. http://www.interlinkinc.net/.

Irons, P. (2006) *A People's History of the Supreme Court*. Penguin, London.

ISSN (2013) International standard serial number. http://www.issn.org/.

JAS (2013) Journal abbreviation sources. http://www.abbreviations.com/jas.php.

Jennings, A. (2006) *FOUL! The Secret World of FIFA: Bribes, Vote Rigging and Ticket Scandals*. HarperSport, London.

Joly, S. and Le Calvé, G. (1986) Etude des puissances d'une distance. *Statistique et Analyse de Données*, **11**(3), 30–50.

Jones, B. (2002) Computational geometry database. ftp://ftp.cs.usask.ca/pub/geometry/.

KDD Cup 2003 The Stanford Linear Accelerator Center SPIRES-HEP database data. http://www.cs.cornell.edu/projects/kddcup/index.html.

Kejžar, N., Korenjak-Černe, S. and Batagelj, V. (2009) *R package clustddist*. http://r-forge.r-project.org/projects/clustddist/.

Kejžar, N. (2005) Analysis of U.S. patents network: Development of patents over time. *Metodološki zvezki*, **2**(2), 195–208.

Kejžar, N., Korenjak-Černe, S. and Batagelj, V. (2011) Clustering of distributions: A case of patent citations. *Journal of Classification*, **28**(2), 156–183.

Kempe, D., Kleinberg, J. and Kumar, A. (2000) Connectivity and inference problems for temporal networks *Proceedings of the Thirty-second Annual ACM Symposium on Theory of Computing*, pp. 504–513. ACM, New York.

Kessler, M.M. (1963) Bibliographic coupling between scientific papers. *American Documentation*, **14**(1), 10–25.

Kleinberg, J. (1998) Authoritative sources in a hyperlinked environment *Proceedings of the Ninth Annual ACM-SIAM Symposium on Discrete Algorithms*, pp. 668–677. Society for Industrial and Applied Mathematics, Philadelphia.

Kolaczyk, E.D. (2009) *Statistical Analysis of Network Data: Methods and Models*. Springer, New York.

Kronmal, R.A. (1993) Spurious correlation and the fallacy of the ratio standard revisited. *Journal of the Royal Statistical Society. Series A (Statistics in Society)*, **156**(3), 379–392.

Kuhn, T. (1970) *The Structure of Scientific Revolutions*. Chicago University Press, Chicago.

Kuper, S. and Szymanski, S. (2009) *Soccernomics: Why England Loses, Why Germany and Brazil Win, and Why the U.S., Japan, Australia, Turkey – and even Iraq – are Destined to Become the Kings of the World's Most Popular Sport*. Nation Books, Philadelphia.

Lanfranchi, P. (1994) The migration of footballers: The case of France 1932-82 In *The Global Sports Arena: Athletic Talent Migration in an Interdependent World* (ed. Bale J and McGuire JA) Frank Cass London pp. 66–77.

Laumann, E.O., Marsden, P.V. and Prensky, D. (1979) *The boundary specification problem in network analysis*. George Mason University Press, Fairfax, VA.

Lenski, G. (1966) *Power and Privilege: A Theory of Social Statification*. McGraw-Hill, New York.

Leydesdorff, L. (2013) Software and data. http://www.leydesdorff.net/software.htm.

Liu, H. (2004) Montylingua. http://web.media.mit.edu/ hugo/montylingua/.

Luce, R.D. (1950) Connectivity and generalized cliques in sociometric group structure. *Psychometrika* (15), 169–190.

Lucio-Arias, D. and Leydesdorff, L. (2008) Main-path analysis and path-dependent transitions in histcite-based historiograms. *Journal of the American Society for Information Science and Technology*, **59**(12), 1948–1962.

Magee, J. and Sugden, J. (2002) The world at their feet: Professional football and international labor migration. *Journal of Sport and Social Issues* **26**(4), 421–437.

Maguire, J. and Stead, D. (1998) Border crossings: Soccer labour migration and the European Union. *International Review for the Sociology of Sport* **33**(1), 59–73.

Marsden, P.V. (1990) Network data and measurement. *Annual Review of Sociology*, **16**(1), 435–463.

Mayer, E., Kosmin, B.A. and Keysar, A. (2001) *American Jewish Identity Survey 2001: Report*. City University of New York.

McGovern, P. (2002) Globalization or internationalization? Foreign footballers in the English league. *Sociology*, **36**(1), 23–42.

Mizruchi, M.S. and Bunting, D. (1981) Influence in corporate networks: An examination of four measures. *Administrative Science Quarterly*, **26**(3), 475–489.

Moder, J.J. and Phillips, C.R. (1970) *Project Management with CPM and PERT* second edn. Van Nostrand Reinhold Company, New York.

Moon, J.W. and Moser, L. (1965) On cliques in graphs. *Israel Journal of Mathematics*, **3**(1), 23–28.

Moxley, R.L. and Moxley, N.F. (1974) Determining point centrality in uncontrived social networks. *Sociometry*, **37**(1), 122–130.

Murray, B. (1996) *The World's Game: A History of Soccer*. Illinois University Press, Urbana-Champaign.

Murtagh, F. (1985) *Multidimensional Clustering Algorithms* vol. 4 of *Compstat lectures*. Physica-Verlag, Vienna.

Newman, M.E.J. (2001) Scientific collaboration networks. ii. shortest paths, weighted networks, and centrality. *Physical Review E*, **64**(1), 016132.

Newman, M.E.J. (2003) The structure and function of complex networks. *SIAM Review*, **45**(2), 167–256.

Newman, M.E.J. (2005) Power laws, Pareto distributions and Zipf's law. *Contemporary Physics*, **46**(5), 323–351.

Newman, M.E.J, Barabási, A.L. and Watts, D.J. (2006) *The Structure and Dynamics of Networks*. Princeton University Press, Princeton.

Nicosia, V., Tang, J., Mascolo, C. et al. (2013) Graph metrics for temporal networks In *Temporal Networks* (ed. Holme P and Saramäki J) Understanding Complex Systems Springer Heidelberg, New York, Dordrecht, London pp. 15–40.

Oliver, G. (2005) *Almanac of World Football, 2006: The Definitive and Essential Guide to the Global Game*. Headline Book Publishing, London.

Palla, G., Barabasi, A.L. and Vicsek, T. 2007 Quantifying social group evolution. *Nature*, **446**(7136), 664–667.

Park, J. and Newman, M.E.J. (2003) Origin of degree correlations in the internet and other networks. *Physical Review E*, **68**(2), 026112.

Perkins, J. (2010) *Python Text Processing with Nltk 2.0 Cookbook – Use Python's Nltk Suite of Libraries to Maximize Your Natural Language Processing Capabilities*. Packt Publishing Ltd, Birmingham.

Poli, R. (2010) Understanding globalization through football: The new international division of labour, migratory channels and transnational trade circuits. *International Review for the Sociology of Sport*, **45**(4), 491–506.

Popping, R. (2000) *Computer-Assisted Text Analysis*. SAGE.

Powe, L.A. (2009) *The Supreme Court and the American Elite*. Harvard University Pres, Cambridge, MA.

Puzyn, T., Leszczynski, J. and Cronin, M.T. (2010) *Recent Advances in QSAR Studies – Methods and Applications* 2010 edn. Springer, Berlin, Heidelberg.

Qi, X., Duval, R.D., Christensen, K. et al. (2013) Terrorist networks, network energy and node removal: A new measure of centrality based on Laplacian energy. *Social Networking* (2), 19–31.

Quirin, A., Cordón, O., Guerrero-Bote, V.P. et al. (2008a) A quick MST-based algorithm to obtain pathfinder networks (∞, $n - 1$). *Journal of the American Society for Information Science and Technology*, **59**(12), 1912–1924.

Quirin, A., Cordón, O., Santamaria, J. et al. (2008b) A new variant of the pathfinder algorithm to generate large visual science maps in cubic time. *Information Processing & Management* **44**(4), 1611–1623.

Ripley, R.M., Snijders, T.A.B., Boda, Z. et al. (2013) *Manual for SIENA version 4.0 (version November 1, 2013)* University of Oxford, Department of Statistics; Nuffield College Oxford.

Robinson, W.C. (1890) *The Law of Patents for Useful Inventions*. Little, Brown.

Rollin, G. and Rollin, J. (eds.) (2005) *Rothmans Football Yearbook, 2002–2003*. Headline Book Publishing, London.

Rosengren, K.E. (1968) *Sociological Aspects of the Literary System*. Natur och Kultur, Stockholm, Sweden.

Sabidussi, G. (1966) The centrality index of a graph. *Psychometrika* **31**(4), 581–603.

Schmookler, J. (1966) *Invention and Economic Growth*. Harvard University Press, Cambridge.

Schumpeter, J. (1942) *Capitalism, Socialism and Democracy*. Harper, New York.

Schvaneveldt, R.W. (ed.) (1990) *Pathfinder Associative Networks: Studies in Knowledge Organization*. Ablex, Norwood, NJ.

Schvaneveldt, R.W., Durso, F.T. and Dearholt, D.W. (1989) Network structures in proximity data In *The psychology of learning and motivation: Advances in research and theory* (ed. Bower G) vol. 24 Academic Press New York pp. 249–284.

Scott, J. (2000) *Social Network Analysis: A Handbook (Second Edition)*. SAGE.

Seidman, S.B. 1983 Network structure and minimum degree. *Social Networks*, **5**(3), 269–287.

Sharma, M., Ibe, H. and Ozeki, T. (1997) WDM ring network using a centralized multiwavelength light source and add-drop multiplexing filters. *IEEE Journal of Lightwave Technology*, **15**(6), 917–929.

Shimbel, A. (1953) Structural parameters of communication networks. *Bulletin of Mathematical Biophysics*, **15**(4), 501–507.

Small, H.G. (1973) Co-citation in the scientific literature: A new measure of the relationship between two documents. *Journal of the American Society for Information Science*, **24**(4), 265–269.

Stephenson, K. and Zelen, M. (1989) Rethinking centrality: Methods and examples. *Social Networks*, **11**(1), 1–37.

Tada, Y., Kobayashi, Y., Yamabayashi, Y. et al. (1996) OA&M framework for multiwavelength photonic transport networks. *IEEE Journal on Selected Areas in Communications*, **14**(5), 914–922.

Taylor, M. (2006) Global players? Football, migration and globalization, c. 1930–2000. *Histrorical Social Research*, **31**(1), 7–30.

TePaske-King, B. and Richert, N. (2001) The identification of authors in the mathematical reviews database; issues in science and technology librarianship. http://www.istl.org/01-summer/databases.html.

Todeschini, R. and Consonni, V. (2009) *Molecular Descriptors for Chemoinformatics* 2. edn. John Wiley & Sons, New York.

Trajtenberg, M. (2002) A penny for your quotes: Patent citations and the value of innovations In *Patents, Citations, and Innovations: A Window on the Knowledge Economy* (ed. Jaffe AB and Trajtenberg M) The MIT Press Cambridge, MA pp. 403–459.

Ulrichsweb (2013) Global serials directory. http://ulrichsweb.serialssolutions.com/.

Valente, T. (1996) Social network thresholds in the diffusion of innovations. *Social Networks*, **18**, 69–89.

Valente, T.W. and Foreman, R.K. (1998) Integration and radiality: Measuring the extent of an individual's connectedness and reachability in a network. *Social Networks*, **20**(1), 89–105.

Valente, T.W., Watkins, S., Jato, M.N. et al. (1997) Social network associations with contraceptive use among Cameroonian women in voluntary associations. *Social Science and Medicine*, **45**(5), 677–687.

Vavpetič, A., Batagelj, V. and Podpečan, V. (2009) An implementation of the pathfinder algorithm for sparse networks and its application on text networks *Proceedings of the 12th International Multiconference Information Society (IS 2009), 12–16 Oct 2009, Ljubljana, Slovenia*, pp. 236–239. IJS, Ljubljana.

Vialli, G. and Marcotti, G. (2006) *The Italian Job: A Journey in the Heart of Two Great Footballing Cultures*. Bantam Press, London.

Vilain, M., Kautz, H. and van Beek, P. (1990) Constraint propagation algorithms for temporal reasoning: A revised report In *Readings in Qualitative Reasoning About Physical Systems* (ed. Weld DS and de Kleer J) Morgan Kaufmann Publishers Inc. San Francisco, CA, USA pp. 373–381.

Vile, J.R. (2010) *Essential Supreme Court Decisions: Summary of Leading Cases in U.S. Constitutional Law*. Rowman and Littlefield, Plymouth.

Ward, J.H. (1963) Hierarchical grouping to optimize an objective function. *Journal of the American Statistical Association*, **58**(301), 236–244.

Wasserman, S. and Faust, K. (1994) *Social network analysis: methods and applications*. Cambridge University Press, Cambridge, UK.

Watts, D.J. and Strogatz, S.H. (1998) Collective dynamics of 'small-world' networks. *Nature*, **393**(6684), 440–442.

Wikipedia (2013) Chinese names. http://wikipedia.org/wiki/List of common_Chinese_surnames.

Woodard, C. (2011) *American Nations: A History of the Eleven Rival Regional Cultures of North America*. Viking, New York.

Wu, T.H. (1994) A passive protected self-healing mesh network architecture and applications. *IEEE/ACM Transactions on Networking*, **2**(1), 40–52.

Xuan, B.B., Ferreira, A. and Jarry, A. (2003) Computing shortest, fastest, and foremost journeys in dynamic networks. *International Journal of Foundations of Computer Science*, **14**(2), 267–285.

Yallop, D. (1999) *How They Stole the Game*. Constable and Robinson, London.

Yu, W.W. and Ezzati-Rice, T.M. (2005) Statistical brief #81: Concentration of health care expenditures in the U.S. civilian noninstitutionalized population. http://meps.ahrq.gov/mepsweb/data_files/publications/st81/stat81.shtml.

Zaveršnik, M. (2004) Razčlembe omrežij (Network Decompositions). PhD thesis. University of Ljubljana.

Person index

Ahmed, A., 94
Albert, R., 163, 166
Alegi, P., 264
Allen, G., 69
Allen, J., 161
Allen, J.F., 108
Anselin, L., 417
Anthonisse, J.M., 63, 117
Asimov, I., 69, 79

Barabási, A.L., 4, 12, 39, 114, 163, 166
Batagelj, V., 10, 12, 18, 28, 31, 53, 54, 58–60,
    75, 83, 85, 87, 96, 107, 112, 163, 353, 360,
    364, 365, 368, 396, 399
Baum, J.A.C., 129
Bavelas, A., 3, 117, 118, 135
Beauchamp, M.A., 135
Bell, M.G.H., 108
Belle, D., 163
Ben-Porat, A., 264, 267
Ben-Porat, B., 264, 267
Benesh, S.C., 218
Berk, M.L., 189
Bernal, J.D., 69
Bernard, H.R., 163
Bickel, P.J., 207
Bienenstock, E.J., 163
Billard, L., 103
Bird, S., 396
Boda, Z., 108
Bonacich, P., 4, 135, 163
Bonhoure, F., 51
Borgatti, S.P., 139, 163, 166
Brandes, U., 63, 135, 137
Breiger, R., 166
Broder, A.Z., 50
Bruynooghe, M., 365

Bunting, D., 135, 137, 139
Burt, R.S., 121, 134, 135, 137, 139, 166

Calero-Medina, C., 69
Cantwell, J., 240
Carpenter, M.P., 175
Cartwright, D., 48
Casteigts, A., 107, 110, 111, 113
Cerinšek, M., 84
Chen, C., 100
Chi, I., 163
Chinni, D., 10, 353, 357, 358, 376, 377, 381
Chou, K., 163
Cohen, J.W., 189
Coleman, J.S., 135
Consonni, V., 61
Conwell, L.J., 189
Cook, K.S., 135, 137, 139
Cormen, T.H., 77
Correa, J.R., 108
Cox, A., 239
Cronin, M.T., 61
Crouch, T., 409
Cunningham, P., 114

Dallery, Y., 51
Davis, A., 28
de Nooy, W., 10, 18, 85
Dearholt, D.W., 96, 100
Dechter, R., 108
Diday, E., 103
Dieudonné, J., 367
Doksum, K.A., 207
Doll, L.S., 161
Doreian, P., 2, 3, 10, 12, 28, 31, 69, 75–77, 80,
    85, 114, 118, 119, 128, 129, 135, 155, 163,
    166, 184, 255, 364

Understanding Large Temporal Networks and Spatial Networks: Exploration, Pattern Searching, Visualization and Network Evolution, First Edition. Vladimir Batagelj, Patrick Doreian, Anuška Ferligoj and Nataša Kejžar.
© 2014 John Wiley & Sons, Ltd. Published 2014 by John Wiley & Sons, Ltd.

Subject index

Understanding Large Temporal Networks and Spatial Networks: Exploration, Pattern Searching, Visualization and Network Evolution, First Edition. Vladimir Batagelj, Patrick Doreian, Anuška Ferligoj and Nataša Kejžar.
© 2014 John Wiley & Sons, Ltd. Published 2014 by John Wiley & Sons, Ltd.